The Mitochondrion in Aging and Disease

The Mitochondrion in Aging and Disease

Edited by John Binnington

hayle medical

New York

Hayle Medical,
750 Third Avenue, 9th Floor,
New York, NY 10017, USA

Visit us on the World Wide Web at:
www.haylemedical.com

ISBN: 978-1-64647-527-8

Cataloging-in-Publication Data

The mitochondrion in aging and disease / edited by John Binnington.
p. cm.
Includes bibliographical references and index.
ISBN 978-1-64647-527-8
1. Mitochondrial pathology. 2. Mitochondria. 3. Aging. 4. Metabolism--Disorders.
I. Binnington, John.
RB147.5 .M58 2023
616.07--dc23

Table of Contents

Preface

The mitochondrion is a double-membrane-bound cell organelle, which is responsible for processing oxygen and converting food substances into energy. It is essential for cellular bioenergetics and consequently plays a significant role in aging related energy deterioration. Aging is a natural biological process, which occurs in all living organisms and is characterized by declining bioenergetics. A loss of mitochondrial activity and quality is also linked with the development of a variety of age-related diseases. Aging has a negative effect on the mitochondrial heath, as it creates imbalances in the main mitochondrial-regulated pathways. Mitochondrial dysfunction occurs when the mitochondrion fails to function properly as a result of another disease or condition. Secondary mitochondrial dysfunction may be caused by a variety of diseases including diabetes, muscular dystrophy, cancer, Alzheimer's disease, and Lou Gehrig's disease. This book provides significant information to help develop a good understanding of the role of mitochondrion in aging and disease. It will serve as a reference to a broad spectrum of readers.

Significant researches are present in this book. Intensive efforts have been employed by authors to make this book an outstanding discourse. This book contains the enlightening chapters which have been written on the basis of significant researches done by the experts.

Finally, I would also like to thank all the members involved in this book for being a team and meeting all the deadlines for the submission of their respective works. I would also like to thank my friends and family for being supportive in my efforts.

Editor

1

Oxidative Stress Induced Mitochondrial Failure and Vascular Hypoperfusion as a Key Initiator for the Development of Alzheimer Disease

Gjumrakch Aliev [1,2,3,*], Hector H. Palacios [4], Eldar Gasimov [5], Mark E. Obrenovich [6], Ludis Morales [2], Jerzy Leszek [7], Valentin Bragin [3], Arturo Solís Herrera [8] and Dmitry Gokhman [9]

1 School of Health Science and Healthcare Administration, University of Atlanta, 6685 Peachtree Industrial Blvd., Atlanta, Georgia, 30360, USA

2 Department of Nutrition and Biochemistry, Faculty of Sciences, Javeriana University, Bogotà D.C., Colombia

3 Stress Relief and Memory Training Center, Brooklyn, New York, NY 11235, USA

4 Department of Biology, College of Sciences, University of Texas at San Antonio, One UTSA Circle, San Antonio, TX 78249–1664, USA

5 Department of Cytology, Histology and Embryology, Azerbaijan Medical University, 25 Street Bakhikhanov, Baku AZ10 25, Azerbaijan

6 Department of Pathology, School of Medicine, Case Western Reserve University, WRB 5301, Cleveland, Ohio, 44106, USA

7 Department of Psychiatry, Wroclaw Medical University, 25 St. Kraszewskiego, Wroclaw, 50–229, Poland

8 Dirección de Investigación y desarrollo, Centro de Estudios de la Fotosíntesis Humana, S.C. López Velarde 108 y 109, Centro, Aguascalientes, Aguascalientes, 20000, México

9 Department of Mathematics, College of Sciences, University of Texas at San Antonio, One UTSA Circle, San Antonio, TX 78249, USA

* Author to whom correspondence should be addressed; E-Mail: aliev03@gmail.com or GAliev@UofA.edu.

Abstract: Mitochondrial dysfunction may be a principal underlying event in aging, including age-associated brain degeneration. Mitochondria provide energy for basic metabolic processes. Their decay with age impairs cellular metabolism and leads to a decline of cellular function. Alzheimer disease (AD) and cerebrovascular accidents (CVAs) are two leading causes of age-related dementia. Increasing evidence strongly supports the theory that oxidative stress, largely due to reactive oxygen species (ROS), induces mitochondrial damage, which arises from chronic hypoperfusion and is primarily responsible for the pathogenesis that underlies both disease processes. Mitochondrial membrane potential, respiratory control ratios and cellular oxygen consumption decline with age and correlate with increased oxidant production. The sustained hypoperfusion and oxidative stress in brain tissues can stimulate the expression of nitric oxide synthases (NOSs) and brain endothelium probably increase the accumulation of oxidative stress products, which therefore contributes to blood brain barrier (BBB) breakdown and brain parenchymal cell damage. Determining the mechanisms behind these imbalances may provide crucial information in the development of new, more effective therapies for stroke and AD patients in the near future.

Keywords: oxidative stress; Alzheimer disease; antioxidants; hypometabolism; mitochondria; metabolism; neurodegeneration

---

1. Introduction

Alzheimer disease (AD) and cerebrovascular accidents (CVAs) are two leading causes of age-related dementia. Increasing evidence supports the notion that chronic hypoperfusion is primarily responsible for the pathogenesis that underlies both disease processes. In this regard, hypoperfusion appears to induce oxidative stress, which is largely due to the formation of reactive oxygen species. Oxidative imbalance is also associated with other age-related degenerative disorders such as atherosclerosis, ischemia/reperfusion, and rheumatic disorders.

We have found that a chronic injury stimulus induces the hypoperfusion seen in the microcirculation of vulnerable brain regions. This leads to energy failure, which is manifested by damaged mitochondrial ultrastructure, the formation of a large number of non-mature or "young" electron dense "hypoxic" mitochondria and by the overproduction of mitochondrial DNA (mtDNA) deletions. Moreover, these mitochondrial abnormalities coexist with increased redox metal activity, lipid peroxidation and RNA oxidation. This oxidative stress occurs within various cellular compartments, in various parenchymal cells in the brain and most notably in the vascular endothelium, and in mitochondria found therein, which is associated with atherosclerotic damage. Further, the associated pathology is accompanied by neuronal and glial damage, known to be a part of the development of AD pathology. In addition, vascular wall cell pathology in the AD brain correlates linearly with the degree of neuronal and glial cell damage. Mitochondrial lesions in all of these cellular compartments show the same pattern, namely DNA deletions, the overexpression of oxidative stress and appears strongly to be the central target for brain damage in AD, due to high energy demand and susceptibility to oxidation. The result is manifested as energy failure and results in cognitive impairment and memory decline. In this review we outline recent evidence, as well as our own experimental data, indicating that chronic injury–stimulus induces hypoperfusion in the microcirculation of vulnerable brain regions, which leads to energy failure.

2. Vascular Changes and Their Influence in the Pathology Seen in AD

Recent findings demonstrate that there is a similarity between the ultrastructural features of both vascular lesions and mitochondria in brain vascular wall cells from human AD brain biopsies, human short postmortem brain tissues, yeast artificial chromosome (YAC R140) and C57B6/SJL Tg+ mice overexpressing Aß precursor protein (PP) [1,2]. Performing in situ hybridization using mtDNA probes for human wild type, 5kb deleted and mouse mtDNA, and immunocytochemistry using antibodies against APP, 8-hydroxyl-2'-guanosine (8OHG) and cytochrome c oxidase subunit 1 (COX) provide congruent ultrastructural localization [1,2]. A higher degree of amyloid deposition in the vascular walls of the human AD, YAC and C57B6/SJL Tg (+) mice exists compared to age-matched controls [1]. Severely damaged vessels exhibit immunopositive staining for APP. More mitochondrial abnormalities are present in human AD, YAC and C57B6/SJL Tg (+) mouse microvessels where lesions occur [1,2]. Undamaged regions of human AD tissues, YAC and C57B6/SJL Tg (+) mouse tissues and in age-matched control subjects lack these features, while damaged vessels manifest cells possessing clusters of wild and deleted mtDNA–containing positive probes [1,2]. Our observations demonstrate that vascular wall cells, especially their mitochondria, appear to be central targets for oxidative stress-induced damage before the development of AD pathology [1,2]. On the other hand, the positive correlation between AD and cholesterol levels suggests that antioxidant therapy and cholesterol–lowering drugs could delay the occurrence of AD [1]. However, despite their frequencies, the pathophysiological and morphological changes in brain microcirculation that accompany AD remain poorly understood, and the specific factors controlling vascular tone in AD remain unknown [2].

3. Features That Influence the Development and Prognosis of AD during the Interactions between Cerebrovascular Diseases and Dementia

The role of tobacco smoking in the pathogenesis of AD is still unclear and controversial. ROS are generated at sites of injury and/or inflammation. The vascular endothelium, which regulates the passage of macromolecules and circulating cells from blood to tissue, is a major target of oxidant stress and plays a critical role in the pathophysiology of several vascular diseases. In addition, the vascular endothelium, neurons, and glia can synthesize, store, and release ROS and vasoactive substances in response to certain stimuli, especially to chronic hypoxia/hypoperfusion. Their contribution to the pathophysiology of stroke, cerebrovascular disease and AD is extremely important. Moreover, the role of hypoperfusion as a key factor for vascular lesions that causes oxidative stress, appears to be widely accepted as an initiator of AD [1,2]. This idea is based on a positive correlation between AD and cardiovascular diseases [1,3–6].

Specifically, accumulated oxidative stress increases vascular endothelial permeability and promotes leukocyte adhesions, which is coupled with alterations in endothelial signal transduction and redox–regulated transcription factors [for the references and review see: [1,3,4]]. We hypothesize that the cellular and molecular mechanisms by which Hypoperfusion-induced ROS accumulation results in the development of AD is through impairing endothelial barrier function, promoting leukocyte adhesion and altering normal vascular function. The sustained hypoperfusion and oxidative stress of brain tissues also could stimulate the secondary overexpression of iNOS and nNOS and endothelin–1 (ET-1) in brain cells [7]. It is likely that the increased accumulation of oxidative stress products probably contributes to damage to brain parenchymal cells and the decompensation of the blood brain barrier (BBB), which normally prevents permeability of large molecules from passing to the cerebrospinal fluid. Therefore, determining the mechanisms behind these disturbances in experimental animals may provide crucial information in the development of new, more effective therapies for the treatment of cerebrovascular and neurodegenerative diseases, including AD.

Many common underlying risk factors play key roles in the development of cardiovascular, cerebrovascular and neurodegenerative diseases [4,8–10]. Cigarette smoking causes chronic hypoxic conditions and the formation of a large amount of free oxygen radicals that appear to be key factors in the development of these diseases. Latest evidence indicates that continuous exposure to free oxygen radicals induce cellular damage and decreases antioxidant defenses [11].

Several recent studies indicate that cigarette smoking is a cofactor in the initiation of AD via its effect on the vasculature, as previously discussed. Nicotine via nicotinic receptor activation may counter these effects in part. Vascular insufficiency/hypoperfusion has been considered as a pathogenetic factor in the development of AD, and the positive relationship between cerebrovascular diseases such as stroke and especially cerebrovascular atherosclerosis indicates the latter may also be linked to the pathogenesis of AD [4].

4. The Influence of Oxidative Stress on the Function of Brain Microvessels in AD

ROS can function as signaling intermediates at low levels and regulate fundamental cell activities including growth and adaptive responses [11]. However, at higher concentrations, ROS can cause cell injury and death. Vascular endothelium modulates the passage of macromolecules and circulating cells from blood to tissue and is a major target of oxidant stress [12]. Specifically, oxidative stress increases vascular endothelial permeability and promotes leukocyte adhesions, which are coupled with alterations in endothelial signal transduction and redox-regulated transcription factors [12]. Based on these recent findings, we hypothesize that impairing endothelial barrier function and promoting leukocyte adhesion also induce alterations in normal vascular endothelial cell (EC) function, resulting in AD progression [12].

Compared to other organs or tissues, the brain is more vulnerable to ROS-induced damage due to its high rate of oxygen consumption, high polyunsaturated lipid content, and relative paucity of classic antioxidant enzymes [13]. The AD brain contains increased regional levels of oxidative stress indicators [14–20]. Studies demonstrate a decline in polyunsaturated fatty acids (PUFA) [21–23], increased levels of lipid peroxidation markers [19,21], as well as protein oxidation [24,25], DNA oxidation [26–28] and RNA oxidation [1,29–31] during AD. Additionally, the presence of advanced glycation end products (AGE), glycoxidative end products such as N–ε–carboxy–methyl–lysine and lipid peroxidation adducts are present in both neurofibrillary tangles (NFT) and senile plaques (SP) in AD [1,15,18,19,24,25,29,30,31–34] as well as in post–ischemic tissues [35–39].

Vascular aging correlates with both structural and functional changes that can take place at the level of the endothelium, vascular smooth muscle cells (vSMC) and the extracellular matrix of blood vessels. In the endothelium, reduced vasodilatation in response to agonists occurs in large conduit arteries, as well as in resistance arteries as a result of aging [40]. Furthermore, enhanced oxidative stress by hypoperfusion contributes significantly to the deleterious effects of aging on the endothelium by means of NO breakdown due to ROS. The relative contribution of the above phenomenon to age-related endothelial dysfunction is highly dependent on the species and the type of vascular bed involved [9,40–42].

Cortical, subcortical, meningeal gray matter and blood vessels (congophilic angiopathy) all contain Aβ deposits and are prominent features of AD [5,6,9,40–42]. *In vitro* experimental evidence shows that these Aβ deposits induce cerebrovascular dysfunction in the rat brain [43], and that the beta amyloid (Aβ) peptide produces endothelial dysfunction in cerebral microvessels via ROS. This occurs when the ROS superoxide–scavenging enzyme, superoxide dismutase, prevents acetylcholine–induced endothelium–dependent vasodilation [43]. In addition, accumulating evidence supports the idea that the Aß peptide is responsible for the cerebrovascular effects of the upstream molecule beta amyloid precursor protein (AßPP) and its overexpression [44,45].

A study by Iadecola and coworkers shows how transgenic (Tg) mice overexpressing AßPP have a profound and selective impairment in endothelium–dependent regulation of neocortical microcirculation [39]. Moreover, peptides derived from AßPP processing may contribute to the alterations in cerebral blood flow (CBF) and neuronal dysfunction during AD [44]. The study confirmed that Aß1-40 did not influence the increasing CBF produced by the endothelium–independent vasodilators [40]. In contrast, Aß1-42 did not reduce resting CBF or the increasing CBF produced by endothelium–dependent vasodilators. The superoxide scavengers, SOD and MnTBAP, reversed the cerebrovascular effects of Aß1-40 [40–41]. These data strongly suggests that soluble amyloid beta protein (Aß1-40), but not amyloid aggregate (Aß1-42), produces the cerebrovascular alterations seen in transgenic AßPP mouse, and thus, Aß1-40 could play a role in the cerebrovascular alterations observed in AD [6,45]. This study supports recent evidence that microvessels isolated from the AD brain kill neurons *in vitro* [46].

The growing body of evidence suggests that AD shares many common underlying etiologies with other neurodegenerative disorders. For example, multiple sclerosis (MS) is a relatively common disease with no cure. It is the leading cause of neurological disability in young adults, affecting over two million people worldwide (reviewed in reference [47]). Traditionally, MS has been considered a chronic, inflammatory disorder of the central white matter in which ensuing demyelination results in physical disability. Recently, MS has become increasingly viewed as a neurodegenerative disorder in which axonal injury, neuronal loss, and atrophy of the central nervous system leads to permanent neurological and clinical disability. The latest developments on MS research, includes etiology, pathology, genetic association, EAE animal models, mechanisms of neuronal injury and axonal transport, and therapeutics [47]. Moreover, the mechanisms of mitochondrial dysfunction that are involved in MS, including mitochondrial DNA defects and mitochondrial structural/functional changes that accompanies this devastative disease have been able to open new and much more effective treatment strategies [47]. However, despite all the research on the effects of unknown etiology as well as Aß, the source of the potential ROS *in vivo* and its link to mitochondrial dependent hypoperfusion is not completely understood.

5. The Role of Mitochondrial Abnormalities during the Development of AD

In aerobic cells 90–95% of the total amount of adenosine triphosphate (ATP) production requires oxygen. The synthesis of ATP via the mitochondrial respiratory chain is the result of electron transport across the electron transport chain coupled to oxidative phosphorylation [48]. The main radical produced by mitochondria is superoxide anion. Intramitochondrial antioxidant systems scavenge this radical to avoid oxidative damage, which can lead to impaired ATP production [49–52]. During aging and some neurodegenerative diseases, including AD, damaged mitochondria are unable to maintain the energy demands of the cell [53,54]. This can lead to an increased production of free radicals, inducing the interruption of oxidative phosphorylation, and resulting in decreased levels of ATP [51]. Both processes, defective ATP production and increased oxygen radicals, may induce mitochondria–dependent cell death [51].

Animal studies using mitochondrial toxins provide the association between neurodegeneration with mitochondrial dysfunction and oxidative damage [2,51]. These consequences are implicated in the pathogenesis of human as well as animal models of neurodegenerative diseases [55–58] and in particular AD [1,49,50,53,59–62]. After long–term ischemia/reperfusion the mitochondria ultra-structure disintegrates *in vivo* and *in vitro* [9,38,39]. Apoptosis of degenerating neurons occurs in association with the accumulation of perikaryal mitochondria and oxidative damage to the nucleus [63]. This same pattern of mitochondrial lesions is observed in human AD brain biopsy samples [53,59]. The reduced expression of both mtDNA and nuclear DNA encoded genes is consistent with a physiological down–regulation of the mitochondria respiratory chain in response to declining neuronal activity [49–58,61,62,64]. However, the role of somatic cells and mtDNA mutations in the pathogenesis of mitochondria failure during AD is still controversial [50, 58, 61, 62].

The deleted mtDNA increases at least 3-fold in AD cases as compared to controls in humans [53]. Moreover, mtDNA isolated from the brains of AD patients includes oxidative modifications containing 8-hydroxy-2'-deoxyguanosine (8OHdG) [26–28]. Studies using in situ markers for 8OHdG and 8OHG showed that RNA oxidation is a prominent feature of damaged neurons in AD [29–31]. Quantitative analysis revealed a strong positive correlation between mtDNA deletions and 8OHG ($r = 0.934$) [53]. However, given that mtDNA (even DNA containing the 5kb deletion) is spared relative to 8OHG, we suspect that mitochondrial abnormalities correlate, but do not directly produce ROS. Therefore, it is important to recognize that 8OHG is formed by the direct attack of OHo. These OHo radicals have only a 2 nm sphere of diffusion and thus are unable to diffuse through the mitochondrial membrane [53].

More recently polarographic studies by Cormier and coworkers describe the effects of nicotine on respiratory chain in rat brain mitochondria [65]. The measurements of oxygen consumption show significant concentration–dependent inhibition by nicotine. Nicotine binds to complex I of the respiratory chain, inhibits NADH-Ubiquinone reductase activity and competes with NADH for complex I [65]. Furthermore, nornicotine, but not cotinine, the main nicotine metabolite, inhibits mitochondrial respiration. Complex I generates superoxide anion, nicotine, and was able to inhibit this ROS generation [65]. This may explain a part of the beneficial and protective effects of nicotine in a few neurodegenerative diseases, as suggested by many epidemiological studies [65]. However, future studies should focus on elucidating the effect of nicotine on the mitochondria functions as well as

DNA overexpression and/or deletion during the development of neurodegenerative disorders including AD. The exact cellular mechanisms behind vascular lesions and their relation to oxidative stress markers identified by RNA oxidation, lipid peroxidation, or mtDNA deletion remain unknown [66]. Future studies comparing the spectrum of oxidative stress-induced damage during reperfusion injury or, more importantly, during hypoxia/hypoperfusion, with AD damage are warranted [67].

6. Co–Factors for Oxidative Stress-Induced Cerebrovascular Lesions

6.1. Hypoperfusion-Induced Oxidative Stress as a Key Factor for the Development of AD

Hypoperfusion-induced oxidative stress in vascular abnormalities coincides with the pathogenesis of AD [62]. Several studies conclude that chronic cerebral hypoperfusion in AD is secondary to oxygen reduction [10,68–71]. However, recent evidence reveals that a greater fraction of oxygen is removed from the vasculature in AD patients as compared to non–AD controls [72]. This suggests that low vascular blood flow is a prominent feature of the brain during chronic hypoxia/hypoperfusion and may be a main initiating factor during the development of AD [4,73,74]. An impairment of energy metabolism characterizes the AD brain [60]. Positron emission tomography (PET) reveals a decline in the cerebral metabolic rate of the parietal and temporal lobes during AD [25,75]. These metabolic defects are present before AD symptoms develop in ApoE ε4 homozygote patients [25]. De la Torre [73,76] proposes that advanced aging with a comorbid condition, such as a vascular risk factor that further decreases cerebral perfusion, promotes a critically attained threshold of cerebral hypoperfusion (CATCH). With time, CATCH induces brain capillary degeneration and suboptimal delivery of energy substrates to neuronal tissue [73,76]. Because glucose is the main fuel of brain cells, its impaired delivery, together with a deficient delivery of oxygen, compromise neuronal stability because the substrates for aerobic glycolysis fail to meet brain tissue demand. The outcome of CATCH is a metabolic cascade that involves, among other things, mitochondrial dysfunction, oxidative stress, decreased ATP production and increased calcium entry, abnormal protein synthesis, cell ionic pump deficiency, signal transduction defects, and neurotransmission failure. These events contribute to the characteristic progressive cognitive decline of patients with AD, as well as regional anatomic pathology, consisting of synaptic loss, SP, NFT, tissue atrophy, and neurodegeneration. CATCH characterizes the clinical heterogenic pattern of AD and provides compelling evidence that a multitude of etiopathophysiologic vascular risk factors, in the presence of advanced aging, can lead to AD [2–4,10,73,76–79]. Therefore, we hypothesize that taken together with vascular EC and SMC atrophy, hypoperfusion is a key factor in the development of AD.

6.2. Cerebrovascular Lesions Observed During Ischemia/Reperfusion Induced Oxidative Stress

The risk for Alzheimer dementia and stroke are known to increase at comparable rates with age. Recent advances suggest that vascular risk factors linked to cerebrovascular disease and stroke in the elderly significantly increase this risk [6]. Although some vascular lesions such as cerebral amyloid angiopathy, endothelial degeneration, and periventricular white matter lesions are evident in most AD cases, one third will exhibit cerebral infarction. Longitudinal clinical studies suggest that stroke and AD occur in tandem more often than randomly [80]. Strokes often occur in patients with AD and have been linked to the pathogenesis of dementia [6]. Nevertheless, the nature of this relationship remains unexplored. Cerebral ischemia is a possible causal factor for AD. Irrespective of the ultimate pathogenic mechanism, these findings suggest that managing vascular disease is important in the treatment and prevention of AD [8,10] or mixed dementia [6].

Chronic hypoxia can alter cerebral microvessels ultrastructure, but this effect is heterogeneous and in some cases capillaries can respond to hypoxia independently of the arteriole [81]. Exposure to three weeks of hypobaric hypoxia results in increased capillary density in rat models [82]. Capillary segment elongation plays a role in this increase in the deeper layers of the cerebral cortex [82]. Therefore, prolonged hypoxia results in structural and functional adaptive responses that improve tissue oxygen delivery [83]. Mitochondria of brain capillary EC maintain normal density in hypoxia, but the number of mitochondria in the surrounding neuropil decreases significantly about 30% [84]. Moreover, exposure to hypobaric hypoxia yields an increase in basic fibroblastic growth factor (bFGF) mRNA in brain tissue [85]. During moderate hypobaric hypoxia, increased brain vasculature is associated with increased density of the brain capillary glucose transporter (Glut-1). However, this change is reversible and dependent on hypoxia exposure time [86]. This same pattern has been observed in the microvascular system of the human AD brain [1,87–89]. Based on these findings, the relationship between oxidative stress markers and extracellular matrix binding ligands in the hypoxic brain during stroke and AD deserves further investigation. In addition, the injury induced by reperfusion after chronic hypoxia is important to note because the oxidative products that accumulate during hypoxia induce more tissue and cellular damage than the hypoxia itself.

Ischemia/reperfusion is a systemic process affecting the whole organ or tissue. Different types of blood cells may contribute to the pathogenesis of ischemia/reperfusion, including platelets, monocytes, neutrophils and others [35]. According to Bednar and coworkers [90] neutrophils might be important contributors to ischemia–induced brain injury whereas the role of platelets is more nebulous. In fact, systemic depletion of neutrophils reduced the volume of cerebral infarct after transient middle cerebral artery occlusion in the rat [91]. EC affected by ischemia during the early stages is completely reversible and dependent upon reperfusion. Eventually, however, injured tissue passes a *"point of no–return"* and the damage becomes irreversible [92]. Initially, cells strive to increase their surface area for gas and nutrient exchange by expressing cytoplasmic microvilli [9,35,36,38,39] or by extending membrane protrusions into the vessel lumen [9,35,36,38,39,93]. The appearance of these micro-vascular changes corresponds to the duration of the ischemia and may be an adaptive EC response to altered hemodynamic conditions [9,93]. The functional significance of microvilli, microblebs and other morphological changes is not clear, but they may have a role in the production of delayed, post–ischemic hypoperfusion by increasing vascular resistance [9,93]. The extent of EC injury depends on the duration of ischemia and on the metabolic needs of the affected vascular system. The duration of experimental ischemia or acute anoxia required to cause damage varies for different organs. It takes approximately 10–15 minutes for irreversible damage to occur in brain [93–95]. After long–term ischemia and the following reperfusion, the decreased number of active capillary vessels is proportional to the ultrastructural lesions in ischemic vessels and underlying tissues and cells [9,35,36,38,39]. Cada and coworkers demonstrate that decreased CBF in aging rats produces deficits in visuospatial behavior after permanent surgical occlusion of both carotid arteries [96]. This deficit is coupled with metabolic abnormalities of the brain as visualized by quantitative COX histochemical mapping [96]. These results suggest that deficits in visuospatial learning are not exclusively the result of hippocampal dysfunction, but may be directly involved with altered oxidative energy metabolism in other integrative visuomotor regions identified in this study. They also suggest that chronic cerebrovascular ischemia in this aged rat model produces neurometabolic and behavioral alterations that may be relevant risk factors for the development of AD [96].

7. The Potential Role of Vasoactive Substances in the Endothelial Content during Ischemia/ Reperfusion

The synthesis and release of vasoactive substances, such as the endothelium–derived vasodilator NO and vasoconstrictor ET-1, regulate EC's role in controlling vascular tone [65,97–99]. Aside from vascular tone NO regulates platelet aggregation, leukocyte adhesion, SMC proliferation, synaptic neurotransmission and cytotoxic/cytostatic actions of macrophages [84,99–106]. This labile molecule may carry out important biological roles both within the cell in which it is synthesized, and by interacting with nearby cells and molecules [107,108]. Three distinct isoforms of NOS derived from different genes generate NO: nNOS, iNOS, and endothelial NOS (eNOS) [97–111]. These isoforms are similar in structure and function [97,108,110,111]. eNOS was first purified and cloned from vascular endothelium, but is found in cardiac myocytes, blood platelets, brain cells [98,99,104,106,110,112] and in cellular compartments such as mitochondria [113,114]. The activity of eNOS is a major determinant of vascular tone and blood pressure. It is altered in diseases such as hypertension, diabetes, atherosclerosis, ischemia/reperfusion [41,87,104,105,115] and AD [73,89].

Excess NO is produced during excitotoxicity, inflammation and ischemia/reperfusion injury [116], and the oxidation products of NO, namely peroxynitrite and peroxynitrate. Also, $ONOO^-$ can generate the highly reactive hydroxyl radical, a more powerful oxidant than NO itself [17,102,116]. The increased nitrotyrosine immunoreactivity in AD is present in the neuronal cytoplasm of the cerebral cortex within regions of neurodegeneration, yet it is undetectable in corresponding control regions [17]. This distribution is essentially identical to that of free carbonyls [15]. The widespread occurrence of nitrotyrosine immunoreactivity in neurons [17] suggests that chronic oxidative damage is not restricted to long–lived polymers such as NFTs, but instead, reflects a generalized oxidative stress contributing to the pathogenesis of AD.

NOS positive neurons are present in subgroups throughout many regions of the brain [102]. Immunostaining for reduced NADPH-diaphorase, as well as nNOS and eNOS, reveals their presence in dendritic and axonal terminals that closely interact with the middle cerebral artery and cerebral microvessels [101–103,115]. The presence of L–arginine in astrocytes *in vivo* suggests that glia may store this chemical for NO production in brain [101,115,117*]*. Moreover, glial cells exhibit an inflammatory response during infection or ischemic disease. They also release pro–inflammatory cytokines and synthesize and release NO [117]. The large amount of NO that is released from glial cells via the expression of iNOS after their stimulation is neurotoxic, because it induces oxidative stress, mitochondrial dysfunction and excitotoxicity [101,106,118]. Hypoxic brain injury (acute or chronic) is associated with the formation of both NO [102,117,119,120–122] and the superoxide anion, which may react to form free radicals [17,106] and cause neurotoxicity [101–103,106, 119,120,123,124]. Further investigations into determining the exact ultrastructural localization of the different NOS isoforms in the brain vascular tree, neurons and glia in post–hypoxic and AD brain are warranted.

7.1. eNOS Involvement in the Cerebrovascular Tone

A dynamic balance of relaxing and constricting factors regulates cerebrovascular tone. Constitutively produced NO normally influences basal cerebral vascular tone, and mediates vascular responses to diverse stimuli [125] and cerebral vasodilation [117]. Vasorelaxation of brain microvessels is a feature of some diseases including chronic hypertension, diabetes, hypercholesterolemia, subarachnoid hemorrhage (SAH), and ischemia [41,117,125]. NO is also involved in regulating the cerebral circulation during hypercapnia [126,127] and focal [115,128-130] or global brain ischemia [127,131-135]. Furthermore, arginine–derived NO mediates the powerful effects of CO_2 on cerebral circulation. NO synthesized by the action of nNOS participates in regulating basal CBF and is the major contributor to the hypercapnic CBF response [136]. Chronic inhibition of constitutive NO production increases EC permeability during various vascular diseases [8,87,95,98,99,110]. Due to its vascular effect, NO might improve tissue perfusion and exert a protective action. Overproduction, either by activation of nNOS by excitatory amino acids [137], or by induction of iNOS in glial, vascular, or blood cells [132–134] during the ischemic episodes, might be deleterious. Mice with eNOS gene knockout exhibit a decrease in vascular relaxation. Thus, NO synthesized by eNOS protects against ischemic damage by increasing CBF, whereas NO produced by nNOS contributes to lesions [138,139]. The inhibition of NO synthesis by EC leads to increased intracellular oxidative stress, which induces neutrophils-EC interactions [9,37] and may promote the development and progression of vascular diseases such as atherosclerosis [41,105] and ischemia/reperfusion injury [6,9,35,117,125,140].

7.2. nNOS Expression and Regulation

Modification of nNOS expression in the entorhinal cortex and hippocampus occurs during AD [141]. Tissues containing the constitutive forms of NOS, like brain, kidney, and endothelium express dimethylargininase [142–144]. It regulates NO production by hydrolyzing free methylated arginine derivatives (effective endogenous inhibitors of NOS) [145]. The expression of dimethylargininase dramatically increases during AD [20]. Dimethylargininase abnormalities in the AD are the result of elevated levels of nitration from effective oxidants peroxynitrite or peroxynitrate [17,119,146]. However, the ultrastructural localization of dimethylargininase immunoreactivity in different cellular compartments of the AD brain or in Tg animal models of AD has yet to be described.

7.3. iNOS as a Mediator of Oxidation During AD

A variety of cells express iNOS in response to lipopolysaccharides, certain cytokines and ROS generators [8,97,99,101–104,112,115,117]. iNOS may be an important mediator of cytotoxicity in the brain because it produces much greater amounts of NO than either eNOS or nNOS [107]. Thorns and collaborators suggest that iNOS plays a role in the formation of NFT [141]. Iadecola and coworkers propose that iNOS contributes to ischemic brain damage [133]. The catalytic activities of iNOS enzymes or mRNA expression are evident in brain tissue after 2 hours of transient focal ischemia or 1–2 days after permanent focal ischemia [132,134].

8. Subcellular Mechanisms Involved in the Development and Maturation of Human AD

Ultrastructural features of the brain biopsy from the age-matched control (Figure 1A) and AD (Figures 1B–D) patients are characterized by heterogeneous morphology. The EC in intact microvessels did not show visible changes. Mitochondria in EC are intact (Figure 1A).

Contrary to this observation, short–post–mortem (<2 h) brain tissues, from AD patients, show microvessels with severe damage, which characterizes the presence of clusters of mitochondria derived lysosomes and necrotic changes in their ultrastructure (Figure 1B). The capillary endothelium shows the presence of a cluster of damaged mitochondria containing positive mitochondrial DNA (mtDNA) signals visualized by using *in situ* hybridization following indirect 17 nm colloidal gold decoration (Figure 1C). In addition, in AD brain microvessels EC occupied only the small part of the vessel wall. Perivascular cells show the presence of large mitochondria derived vacuoles in their matrix. Adhesion of the activated platelets (PLT) to damaged endothelium appeared to be hallmark of these microvessels (Figure 1D). However, undamaged microvessel endothelium did not show any particular changes in their ultrastructure. Mitochondria also were intact (Figure 2A). However, the perivascular spaces contained large vacuolar structures (see Figure 2A). At the same time often the microvessel EC shows the presence of degenerative mitochondria (Figure 2B). The presence of electron-dense hypoxic mitochondria coexists with the formation of mitochondria derived lysosomal structure in the cytoplasmic matrix of EC and perivascular cells (Figure 2C). The mitochondria abnormality appeared to be a permanent feature of vascular endothelium and perivascular cells where damage became visible which characterizes the presence of a hypoxic and completely damaged mitochondria (Figure 2D).

We demonstrate in recent work that cortical neurons from AD brain biopsies have selective localization of mitochondria abnormalities in the cell body [1,3,53,59,87,89]. The majority of the neurons, which closely associate with the lesioned vessels, possess differing degrees of ultrastructural abnormality. The ultrastructural characteristics of neuronal mitochondria damage from AD brain biopsies show the presence of neurons with differing degrees of ultrastructural lesions (Figure 3). In the neuronal cell body partially and completely damaged mitochondria appeared to be hallmark of these neurons (Figures 3A and 3C). The lesioned mitochondria appeared to be a major substrate for the lipofuscin formation. The electron dense hypoxic mitochondria are seen throughout the cell body and characterize the abnormal mitochondrial cristae (in Figures 3B and 3D).

Figure 1. Ultrastructural features of the brain biopsy from the age-matched control (A) and AD (B–D) patients are characterized by heterogeneous morphology: A. An intact microvessel shows the absence of any particular abnormalities in the ultrastructure of endothelial cells (EC) and perivascular cells. Mitochondria in EC are intact. Original magnification: X 13,900. B. Short-post-mortem (<2 h) brain tissue from AD patients shows microvessels with severe damage such as the presence of clusters of mitochondria–derived lysosomes (single arrow) and necrotic changes in the ultrastructure of the EC and perivascular cells. Original magnification: X 8,300. C. Capillary endothelium (from Figure 1–B), under higher magnification, shows the presence of a cluster of damaged mitochondria (single arrows) containing positive mitochondrial DNA (mtDNA) signals visualized by using *in situ* hybridization following indirect colloidal gold decoration (17 nm gold particles). Original magnification: X 46,000. D. AD brain biopsy. EC occupied only the small part of the vessel wall. Perivascular cells show the presence of large mitochondria derived vacuoles in their matrix. Adhesion of the activated platelets (PLT) to damaged endothelium. Original magnification: 8,300. Abbreviations used in figures: EC–Endothelial cells; ER–Erythrocyte; PLT–Platelets; VL–Vessel lumen (reprinted from [172] with permission).

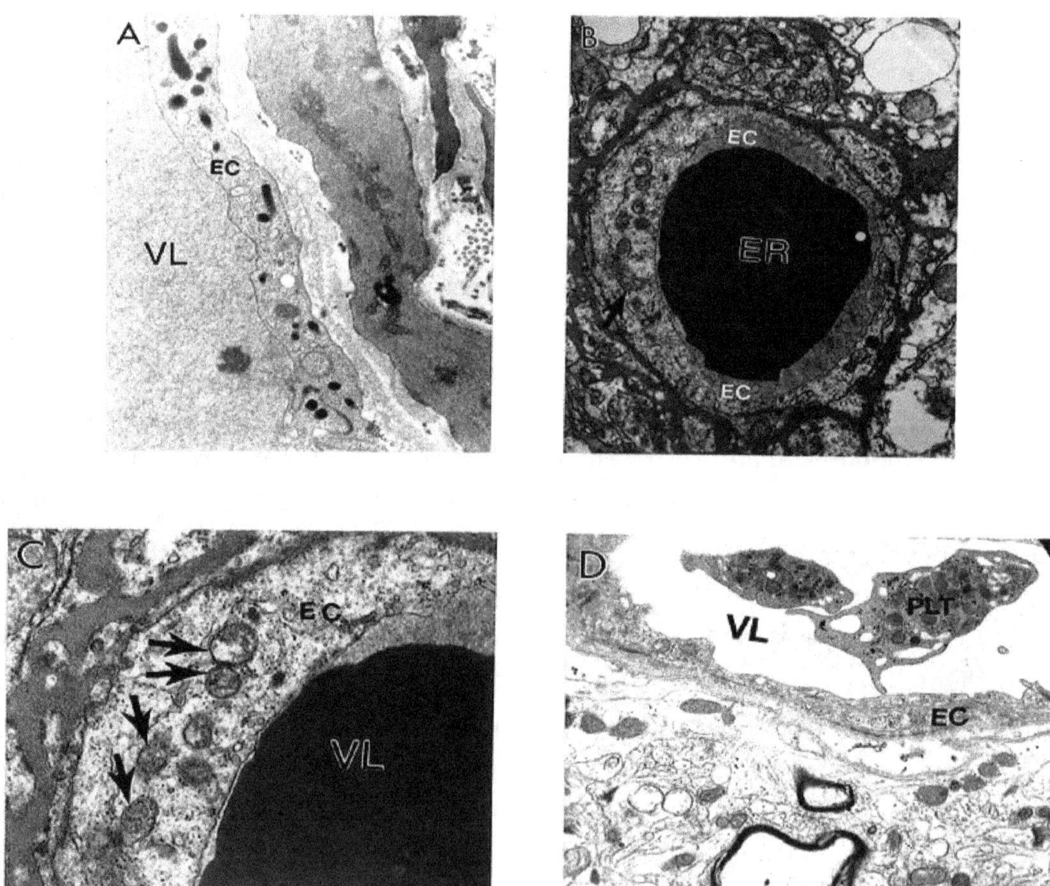

Figure 2. Ultrastructural features of brain microvessels from AD brains characterized by heterogeneous morphology. mitochondria abnormality appeared A. Undamaged microvessel endothelium did not show any particular changes in their ultrastructure. Mitochondria also were intact (single arrows). However, the perivascular spaces contained large vacuolar structures (double arrow). Original magnification X 13,000. B. Vascular EC shows the presence of degenerative mitochondria (double arrow). Original magnification X 6,600. C. The presence of electron-dense hypoxic mitochondria (single arrows) coexists with the formation of mitochondria derived lysosomal structure in the cytoplasmic matrix of EC and perivascular cells (indicated by double arrow). Original magnification X 20,000. D. To be a permanent feature of vascular endothelium and perivascular cells where damage became visible (single and double arrows indicate hypoxic and completely damaged mitochondria, respectively). Original magnification X 20,000. Abbreviations used in figures: BM–basal membrane of endothelium; EC–endothelial cell; ER–erythrocyte; VL–vessel lumen (reprinted from [87] with permission).

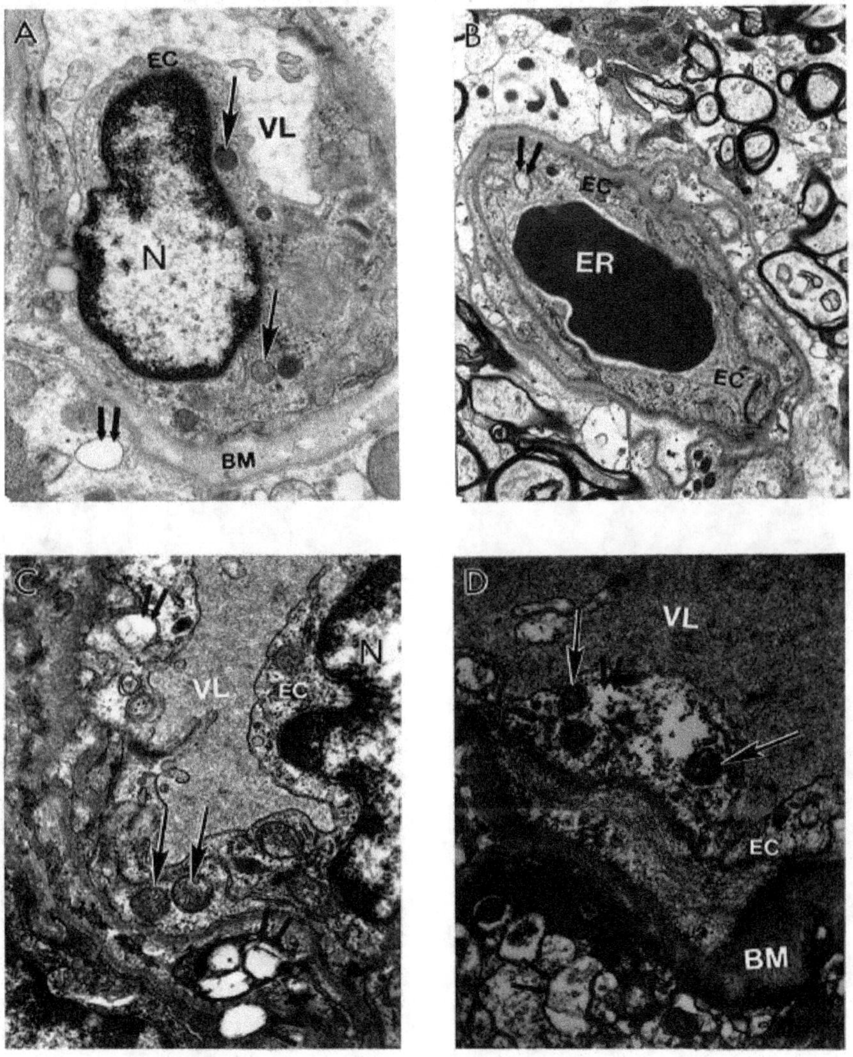

Figure 3. The ultrastructural characteristics of neuronal mitochondria damage from AD brain biopsy. Neurons with different degree of ultrastructural lesions. In the neuronal cell body partially and completely damaged mitochondria (indicated by single arrows and double asterisk respectively in A and C). The lesioned mitochondria appeared to be a major substrate for the lipofuscin formation (double arrow). The electron dense hypoxic mitochondria are seen throughout the cell body and characterize the abnormal mitochondrial cristae. Original magnification: A and B X20,000 respectively. C and D X 16,000 respectively (reprinted from [172] with permission).

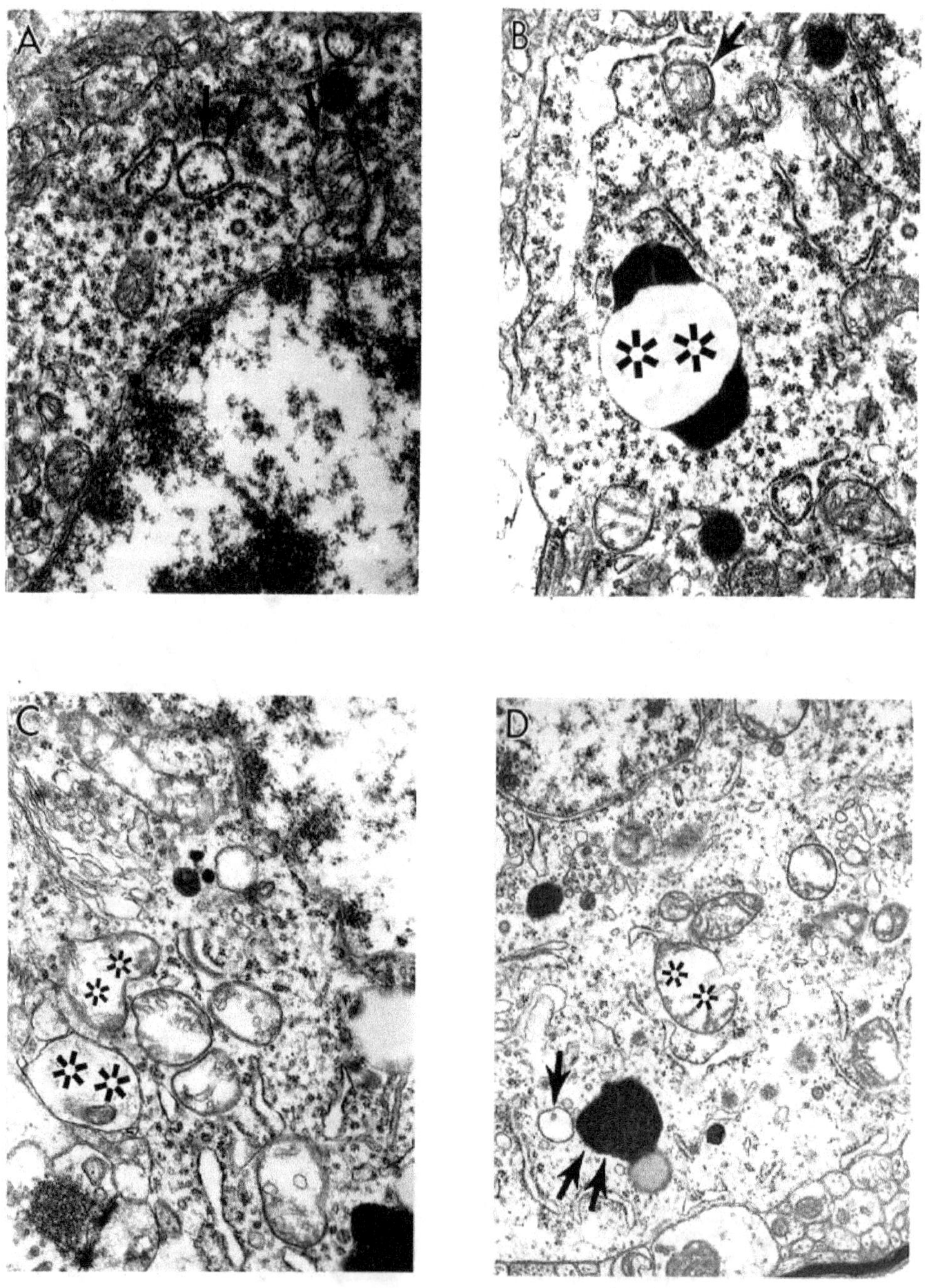

Figure 4. The ultrastructural characteristics of the neuronal mitochondria from AD brain biopsy. A. Neurons with different degrees of ultrastructural lesions. Partially and completely damaged mitochondria are mostly located in the neuronal cell body and coexist with lipofuscin formation. Original magnification X 5,000. B. Large numbers of electron-dense hypoxic mitochondria (indicated by single arrows) were present throughout the cell body and characterized the abnormal mitochondrial cristae. Original magnification: X 20,000. C. Partially (indicated by single arrow) and completely damaged (double arrow) mitochondria. Original magnification X 20,000. D. The neuronal cell body shows the presence of hypoxic mitochondria (indicated by single arrows) close to lipofuscin (double asterisk). Original magnification X 20,000. Abbreviations used in figure: N– neuronal nucleus (reprinted from [87] with permission).

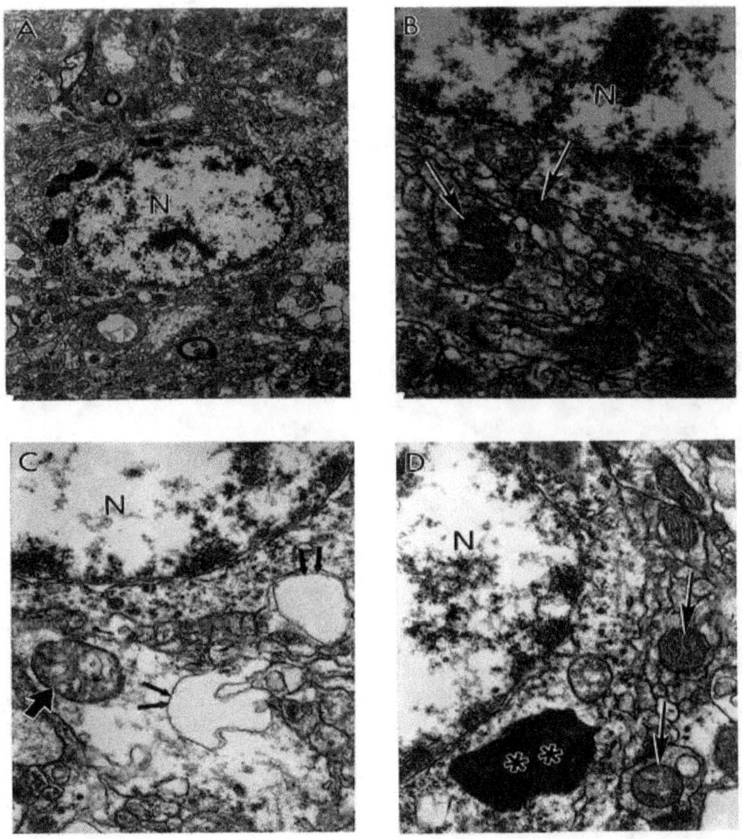

Neurons with different degrees of ultrastructural lesions were seen throughout cortex (Figure 4A). Partially and completely damaged mitochondria are mostly located in the neuronal cell body and coexist with lipofuscin formation (Figure 4A). Moreover, another feature of these neurons appears to be the presence of large numbers of electron-dense hypoxic mitochondria, which were present throughout the cell body and characterized the abnormal mitochondrial cristae (Figure 4B). In addition, these abnormalities coexist with the presence of clusters of the partially and completely damaged mitochondria (Figure 4C). The neuronal cell body always shows the presence of hypoxic mitochondria close to lipofuscin (Figure 4D).

The features of wild type mitochondrial DNA (mtDNA) and 8-OHG staining in the hippocampus of short post–mortem (<2 h) human AD brain shows that wild type mtDNA (17 nm gold) is associated with severely damaged mitochondria and mitochondria derived lysosomes (Figure 5A). However, the area containing lipofuscin did not show any mtDNA containing positive gold particles (Figure 5A). Features of 8-OHG staining in post–mortem AD brain shows that the 8-OHG containing positive signals (17 nm gold particles) was seen throughout neuronal cell body and within in the matrix of damaged mitochondria (Figures 5B–D). However, non–damaged mitochondria (in Figure 5C) and lipofuscin (in Figures 5B and 5D) do not contain 8-OHG positive gold particles.

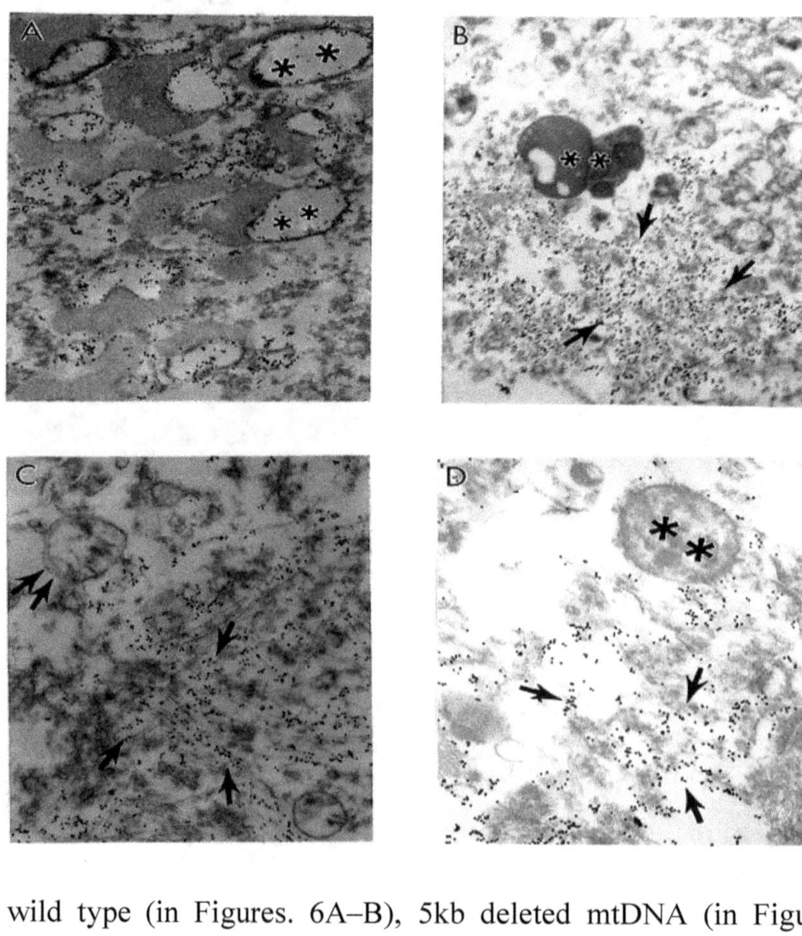

Figure 5. The features of wild type (A) mitochondrial DNA (mtDNA) and 8-OHG (B–D) staining in the hippocampus of short post–mortem (<2 h) human AD brain. Post–mortem AD hippocampus shows that wild type mtDNA (17 nm gold) is associated with severely damaged mitochondria and mitochondria derived lysosomes (double asterisk). Any area containing lipofuscin did not show mtDNA containing positive gold particles. Original magnification X 26,000. B–D: Features of 8-OHG staining in post–mortem AD brain. 8-OHG containing positive signals (17 nm gold particles) was seen throughout neuronal cell body and within in the matrix of damaged mitochondria (single arrow). Non–damaged mitochondria (indicated by double arrows in Figure.5C) and lipofuscin (double asterisk in Figures. B and D) do not contain 8-OHG positive gold particles. Original magnification: X16,000, 26,000 and 33,000 respectively B, C and D (reprinted from [172] with permission).

The features of wild type (in Figures. 6A–B), 5kb deleted mtDNA (in Figure 6C), and COX immunoreactivity (in Figure. 6D) in the hippocampus of a postmortem human AD case shows wild type mtDNA containing positive signals (17 nm colloidal gold) detection were seen in the completely damaged mitochondria or mitochondria derived lysosomes. Any areas containing lipofuscin did not show mtDNA–containing positive signals (in Figures 6A–B). In addition, the 5kb deleted mtDNA containing gold particles (17 nm) were mostly located in mitochondria–derived lysosomes (in Figure 6C). Damaged, abnormal mitochondria shows COX positive containing gold particles in their matrix (in Figure 6D).

Figure 6. The features of wild, 5kb deleted mitochondria DNA (mtDNA), and COX immunoreactivity in the hippocampus of a postmortem human AD case. A and B. Hippocampal neuron shows wild type mtDNA containing positive signals (17 nm colloidal gold) detection were seen in the completely damaged mitochondria or mitochondria derived lysosomes (single arrows). Areas containing lipofuscin (asterisk) did not show any mtDNA containing positive signals. Magnification X 26,000 and X 20,000, respectively A and B. C. 5kb deleted mtDNA containing gold particles (17 nm) were mostly located in mitochondria–derived lysosomes (single arrows). Original magnification X 33,000. D. Damaged, abnormal mitochondria shows COX positive containing gold particles in the matrix (single arrows, colloidal gold 17 nm). Original magnification X 26,000 (reprinted from [172] with permission).

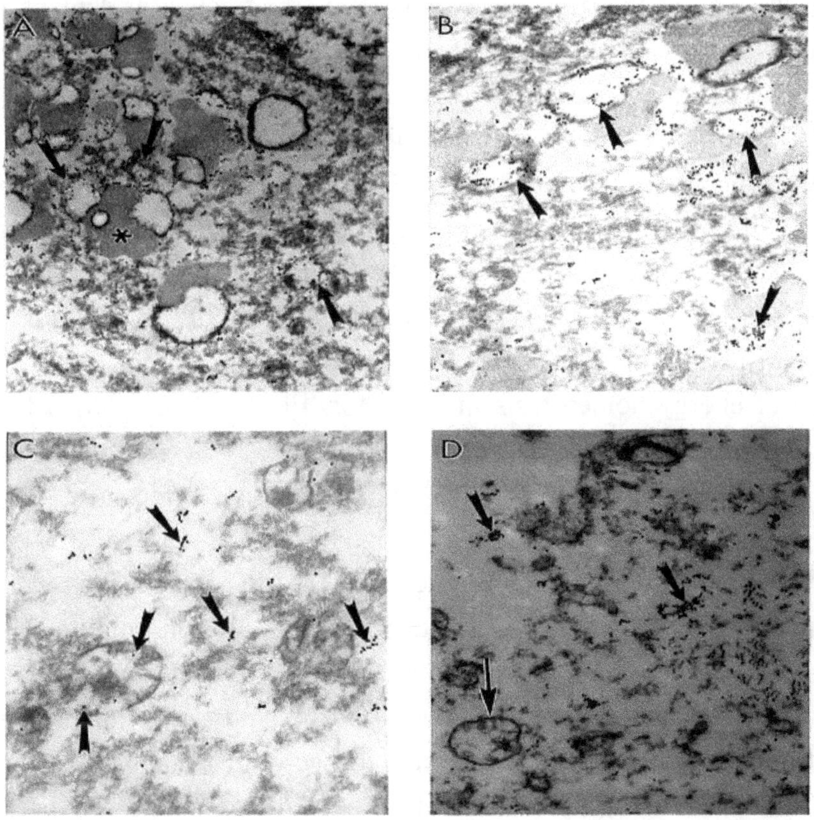

Mitochondrial lesions and lipofuscinogenesis are also present in other cellular compartments of the brain parenchyma. Glial cells at the damaged area, also characterized by the accumulation of lipofuscin and mitochondria–derived lysosomes appear to be a major component and source for these substrates (data not shown). In addition, glial cells also show the intracellular accumulation of different sized amyloid deposits, and they are accompanied by the presence of giant–sized lipid–laden vacuoles and mitochondria–derived lysosomes.

Quantitative morphometric measurements of the percentage of the different types of mitochondria (normal, partially damaged and completely damaged) indicate that age-matched control groups have a significantly higher percentage of normal mitochondria, compared to completely damaged mitochondria from AD cases [2,3,51,56]. No significant differences between partially damaged mitochondria are seen in both groups, indicating that aging induces damage to mitochondria. However, the main differences between the age-matched controls and AD cases appear to be significant differences in the percentage of the normal and completely damaged mitochondria [2,3,51,56].

9. Antioxidant Application for the Treatment of AD

AD treatment has yet to yield a successful therapy that addresses the cause of injury found in AD brain biopsies [11]. Of the various theories proposed for AD etiology, ROS generation is cited as a common factor based on several cellular, molecular, and animal model studies of AD. During aging, ROS may play a large role in cell death, an important factor responsible for disease progression [66]. Efforts to reduce the pathology associated with ROS via antioxidants seem to offer new hope to patients suffering from this devastating disease [11].

Mitochondria has been considered a primary target in the search for age-related cognitively impaired conditions to restore cognitive function including treatments for dementia [11,147–154]. This is because the brain, which is characterized by a high energy metabolism and abundance of oxidizable materials such as polyunsaturated fatty acids and neuropeptides, is exceedingly susceptible to oxidative damage [155], which is known to cause mitochondrial dysfunction [154]. Furthermore, Aβ is reported to promote an excess accumulation of intracellular Ca^{2+} into mitochondria, inducing mitochondrial permeability pores to open, damaging mitochondrial structures. which in return increase the production of defective mitochondria, decrease mitochondrial trafficking, and alter mitochondrial dynamics in neurons affected by AD [156]. The association of mitochondrial activity with the antioxidant capacity of certain micronutrients such as alpha–Lipoic acid (LA), a coenzyme essential for the maintenance of energy homeostasis in mitochondria, has been shown to influence cognitive function in an different animal species [147–154].

Previous studies have demonstrated the potential protective effects of selective mitochondrial antioxidant treatments on brain mitochondria from aged rats [147–150,152,157]. When treating aged rats with selective mitochondrial antioxidants, such as ALCAR and LA, they were able to reduce OS and restore cognitive function and mitochondrial structural abnormalities in all parenchymal cells [147,151,157,158]. It is important to note that the oxidative damage is associated with mitochondria early in AD progression [159]. In addition, to study mitochondrial decay and oxidative damage resulting from aging, an examination into the activities and kinetics of the mitochondrial complexes (a hallmark of the mitochondrial ability to produce energy) was performed, and showed that mitochondrial complexes can be restored by selective mitochondrial antioxidant treatment [158]. This established that in the brain mitochondria of old rats, when compared with that of young rats, there were significantly decreased endogenous antioxidants and less superoxide dismutase activity; more oxidative damage to lipids and proteins; and decreased activities of the mitochondrial complex I, IV and V [158]. In regards to this, mutant proteins associated with AD are reported to block the transport of nuclear–encoded mitochondrial proteins to mitochondria, interact with mitochondrial proteins and disrupt the electron transport chain, induce free radicals, cause mitochondrial dysfunction, and, ultimately, damage neurons [160]. Moreover, the mitochondrial complex I showed a decrease in binding affinity (increase in K(m)) for substrates. Feeding LA/ALCAR to old rats partially restored age-associated mitochondrial dysfunction compared to young rats. These results indicate that oxidative mitochondrial decay plays an important role in brain aging, inducing the generation of free radicals that leads to oxidative damage in postmortem brain neurons from AD patients and in brain neurons from cell models and transgenic mouse models of AD [161], and that a combination of nutrients targeting mitochondria, such as LA/ALCAR, could ameliorate mitochondrial decay through preventing mitochondrial oxidative damage [158]. In a recent study [147] we were able to demonstrate that the integrity of mitochondrial ultrastructure, which is dependent on aging, could be improvement in old rat brain mitochondria when compared to the control group by using antioxidant treatments [147]. In contrast, neurons obtained from aged control groups showed a series of mitochondrial abnormalities, such as the presence of giant mitochondria and mitochondria with partial or complete damaged cristae. Targeting mitochondrial OS improved the overall cognitive ability of aged rats [148–150,152,157,162] and aged dogs [153].

Reid and colleagues [163] noted that there is epidemiological evidence that links vascular diseases, such as hypercholesterolemia, with an increased incidence of AD. While no theory has yielded a satisfactory explanation for the pathological changes that lead to neurodegeneration and cognitive dysfunction [163], vascular risk factors seem to offer the most interesting results [2,164]. The relationship between hypercholesterolemia and AD arose in great extent from ApoE4, a known risk factor for AD and a major carrier of cholesterol in the CNS. The detrimental processes of ApoE4 have been shown to influence AD pathological processes, including lipid homeostasis and NFT formation [165], which suggests that brain vascular alternations play a key role in the progression of AD [163]. ApoE4 mechanisms that contribute to the neurodegeneration of the brain could offer strong insights into AD susceptibility. For example, it was shown in a rat model that ApoE levels would increase as a response to peripheral nerve injury [165], implicating the role of ApoE as a repair mechanism. If the delivery of lipophilic antioxidants is impaired due to ApoE4, this could lead to OS [166]. It has been proven that AL can improve memory deficits in animal models of AD and reduce cognitive deficit in AD patients [155]. The contribution of fatty acids in these cerebrovascular processes, and their effect on AD pathogenesis is still uncertain, but the suggestion that they can elicit neuronal overexcitation and synaptic depression as contributor factors to AD is suggested [167].

Several therapeutic strategies have been developed to treat AD, including anti–inflammatory, anti–oxidant, and herbal treatment approaches. These have been tested in animal and cellular models of AD and in clinical trials with AD subjects. In AD animal models and cell models, herbal extracts appear to have fewer adverse effects than beneficial effects on cognitive functions because of their antioxidant, anti-inflammatory properties [168]. We analyzed the effect of mitochondrial antioxidants ALCAR and LA as a treatment model for AD on ApoE4 transgenic mice [154]. The decrease in cerebrovascular oxygen levels seen in AD patients led to the hypothesis that hypoperfusion in the CBF, which over time causes OS and mitochondrial damage, was the main cause of ApoE–related cognitive deficits was seen in AD patients with ApoE4 overexpression [169–171]. Our study demonstrated for the first time, that ApoE4 caused brain hypoperfusion by gradually reducing CBF when compared to a control group. Structural damage of vascular wall cells, especially in mitochondria, seems to play a key role in the generation of ROS, resulting in oxidative damage to the neuron and inducing pathological factors associated with AD [154]. Therefore, we believe that expanding the focus of study in AD towards mitochondrial pathobiology as a treatment strategy will be able to open new and more successful effective treatment strategies for this devastating disease [62,67,147,154,164,172].

10. Conclusions

In this review we indicate that chronic vascular hypoperfusion is a part of the common underlying mechanisms involved in the initiation and development of neurodegenerative disorders such as stroke and arteriosclerosis. In this regard, it appears that the central initiating factor for vascular abnormality is mitochondrial damage and a sum of elucidators for the imbalance in the activity of NOS isoforms, ET-1, oxidative stress markers, mtDNA and mitochondrial enzymes in the vascular wall and in brain parenchymal cells. This is believed to be due to their predominance in CVA and AD. We hypothesize that an imbalance between the NOS species and the endothelium, along with antioxidant system deficiencies, are predominant brain features of stroke and AD patients. Elevated chronic hypoperfusion and physical distortion of tissue are likely to contribute to the collapse of post–ischemic/hypoxic or AD vessels. We theorize that future eliminating mitochondrial abnormalities can be considered as a new and more effective treatment strategies for this devastative disease in the near future.

References

1. Aliev, G.; Smith, M.A.; Seyidov, D.; Neal, M.L.; Lamb, B.T.; Nunomura, A.; Gasimov, E.K.; Vinters, H.V.; Perry, G.; LaManna, J.C.; Friedland, R.P. The role of oxidative stress in the pathophysiology of cerebrovascular lesions in Alzheimer's disease. *Brain Pathol.* **2002**, *12*, 21–35.
2. Aliev, G.; Gasimov, E.; Obrenovich, M.E.; Fischbach, K.; Shenk, J.C.; Smith, M.A.; Perry, G. Atherosclerotic lesions and mitochondria DNA deletions in brain microvessels: implication in the pathogenesis of Alzheimer's disease. *Vasc. Health Risk Manag.* **2008**, *4*, 721–730.
3. Aliev, G.; Palacios, H.H.; Walrafen, B.; Lipsitt, A.E.; Obrenovich, M.E.; Morales, L. Brain mitochondria as a primary target in the development of treatment strategies for Alzheimer disease. *Int. J. Biochem. Cell Biol.* **2009**, *41*, 1989–2004.
4. De la Torre, J.C. Cerebrovascular and cardiovascular pathology in Alzheimer's disease. *Int. Rev. Neurobiol.* **2009**, *84*, 35–48.
5. Kalaria, R.N. The blood–brain barrier and cerebrovascular pathology in Alzheimer's disease. *Ann. N. Y. Acad. Sci.* **1999**, *893*, 113–125.
6. Kalaria, R.N. The role of cerebral ischemia in Alzheimer's disease. *Neurobiol. Aging* 2000, *21*, 321–330.
7. Aliev, G.; Palacios, H.H.; Lipsitt, A.E.; Fischbach, K.; Lamb, B.T.; Obrenovich, M.E.; Morales, L.; Gasimov, E.; Bragin, V. Nitric oxide as an initiator of brain lesions during the development of Alzheimer disease. *Neurotox. Res.* **2009**, *16*, 293–305.

8. Aliev, G. Is non-genetic Alzheimer's disease a vascular disorder with neurodegenerative consequences? *J. Alzheimer's Dis.* **2002**, *4*, 513–516.
9. Aliev, G.; Cirillo, R.; Salvatico, E.; Paro, M.; Prosdocimi, M. Changes in vessel ultrastructure during ischemia and reperfusion of rabbit hindlimb: implications for therapeutic intervention. *Microvasc. Res.* **1993**, *46*, 65–76.
10. De la Torre, J.C. Alzheimer disease as a vascular disorder: nosological evidence. *Stroke* **2002**, *33*, 1152–1162.
11. Aliev, G.; Obrenovich, M.E.; Reddy, V.P.; Shenk, J.C.; Moreira, P.I.; Nunomura, A.; Zhu, X.; Smith, M.A.; Perry, G. Antioxidant therapy in Alzheimer's disease: theory and practice. *Mini Rev. Med. Chem.* **2008**, *8*, 1395–1406.
12. Lum, H.; Roebuck, K.A. Oxidant stress and endothelial cell dysfunction. *Am. J. Physiol. Cell Physiol.* **2001**, *280*, C719–C741.
13. Coyle, J.T.; Puttfarcken, P. Oxidative stress, glutamate, and neurodegenerative disorders. *Science* **1993**, *262*, 689–695.
14. Perry, G.; Smith, M.A.; McCann, C.E.; Siedlak, S.L.; Jones, P.K.; Friedland, R.P. Cerebrovascular muscle atrophy is a feature of Alzheimer's disease. *Brain Res.* **1998**, *791*, 63–66.
15. Smith, M.A.; Perry, G.; Richey, P.L.; Sayre, L.M.; Anderson, V.E.; Beal, M.F.; Kowall, N. Oxidative damage in Alzheimer's. *Nature* **1996**, *382*, 120–121.
16. Smith, M.A.; Petot, G.J.; Perry, G. Commentary: Diet and Oxidative Stress: A Novel Synthesis of Epidemiological Data on Alzheimer's Disease. *Alzheimer's Dis. Rev.* **1997**, *2*, 58–60.
17. Smith, M.A.; Richey Harris, P.L.; Sayre, L.M.; Beckman, J.S.; Perry, G. Widespread peroxynitrite–mediated damage in Alzheimer's disease. *J. Neurosci.* **1997**, *17*, 2653–2657.
18. Smith, M.A.; Rottkamp, C.A.; Nunomura, A.; Raina, A.K.; Perry, G. Oxidative stress in Alzheimer's disease. *Biochim. Biophys. Acta* **2000**, *1502*, 139–144.
19. Smith, M.A.; Sayre, L.M.; Perry, G. Is Alzheimer's a Disease of Oxidative Stress? *Alzheimer's Dis. Rev.* **1996**, *1*, 63–67.
20. Smith, M.A.; Vasak, M.; Knipp, M.; Castellani, R.J.; Perry, G. Dimethylargininase, a nitric oxide regulatory protein, in Alzheimer disease. *Free Radic. Biol. Med.* **1998**, *25*, 898-902.
21. Lovell, M.A.; Xie, C.; Markesbery, W.R. Decreased glutathione transferase activity in brain and ventricular fluid in Alzheimer's disease. *Neurology* **1998**, *51*, 1562–1566.
22. Prasad, M.R.; Lovell, M.A.; Yatin, M.; Dhillon, H.; Markesbery, W.R. Regional membrane phospholipid alterations in Alzheimer's disease. *Neurochem. Res.* **1998**, *23*, 81–88.
23. Prelli, F.; Castano, E.M.; van Duinen, S.G.; Bots, G.T.; Luyendijk, W.; Frangione, B. Different processing of Alzheimer's beta–protein precursor in the vessel wall of patients with hereditary cerebral hemorrhage with amyloidosis–Dutch type. *Biochem. Biophys. Res. Commun.* **1988**, *151*, 1150–1155.
24. Markesbery, W.R. Oxidative stress hypothesis in Alzheimer's disease. *Free Radic. Biol. Med.* **1997**, *23*, 134–147.
25. Markesbery, W.R.; Carney, J.M. Oxidative alterations in Alzheimer's disease. *Brain Pathol.* **1999**, *9*, 133–146.

26. Mecocci, P.; Beal, M.F.; Cecchetti, R.; Polidori, M.C.; Cherubini, A.; Chionne, F.; Avellini, L.; Romano, G.; Senin, U. Mitochondrial membrane fluidity and oxidative damage to mitochondrial DNA in aged and AD human brain. *Mol. Chem. Neuropathol.* **1997**, *31*, 53–64.
27. Mecocci, P.; MacGarvey, U.; Beal, M.F. Oxidative damage to mitochondrial DNA is increased in Alzheimer's disease. *Ann. Neurol.* **1994**, *36*, 747–751.
28. Mecocci, P.; MacGarvey, U.; Kaufman, A.E.; Koontz, D.; Shoffner, J.M.; Wallace, D.C.; Beal, M.F. Oxidative damage to mitochondrial DNA shows marked age-dependent increases in human brain. *Ann. Neurol.* **1993**, *34*, 609–616.
29. Nunomura, A.; Perry, G.; Aliev, G.; Hirai, K.; Takeda, A.; Balraj, E.K.; Jones, P.K.; Ghanbari, H.; Wataya, T.; Shimohama, S.; Chiba, S.; Atwood, C.S.; Petersen, R.B.; Smith, M.A. Oxidative damage is the earliest event in Alzheimer disease. *J. Neuropathol. Exp. Neurol.* **2001**, *60*, 759–767.
30. Nunomura, A.; Perry, G.; Hirai, K.; Aliev, G.; Takeda, A.; Chiba, S.; Smith, M.A. Neuronal RNA oxidation in Alzheimer's disease and Down's syndrome. *Ann. N. Y. Acad. Sci.* **1999**, *893*, 362–364.
31. Nunomura, A.; Perry, G.; Pappolla, M.A.; Wade, R.; Hirai, K.; Chiba, S.; Smith, M.A. RNA oxidation is a prominent feature of vulnerable neurons in Alzheimer's disease. *J. Neurosci.* **1999**, *19*, 1959–1964.
32. Perry, G.; Nunomura, A.; Hirai, K.; Takeda, A.; Aliev, G.; Smith, M.A. Oxidative damage in Alzheimer's disease: the metabolic dimension. *Int. J. Dev. Neurosci.* **2000**, *18*, 417–421.
33. Sayre, L.M.; Zelasko, D.A.; Harris, P.L.; Perry, G.; Salomon, R.G.; Smith, M.A. 4–Hydroxynonenal–derived advanced lipid peroxidation end products are increased in Alzheimer's disease. *J. Neurochem.* **1997**, *68*, 2092–2097.
34. Smith, M.A.; Taneda, S.; Richey, P.L.; Miyata, S.; Yan, S.D.; Stern, D.; Sayre, L.M.; Monnier, V.M.; Perry, G. Advanced Maillard reaction end products are associated with Alzheimer's disease pathology. *Proc. Natl. Acad. Sci, USA* **1994**, *91*, 5710–5714.
35. Cirillo, R.; Aliev, G.; Hornby, E.J.; Prosdocimi, M. Endothelium as a therapeutical target in peripheral occlusive arterial diseases: consideration for pharmacological interventions. *Pharmacol. Res.* **1994**, *29*, 293–311.
36. Cirillo, R.; Salvatico, E.; Aliev, G.; Prosdocimi, M. Effect of cloricromene during ischemia and reperfusion of rabbit hindlimb: evidence for an involvement of leukocytes in reperfusion–mediated tissue and vascular injury. *J. Cardiovasc. Pharmacol.* **1992**, *20*, 969–975.
37. Granger, D.N.; Benoit, J.N.; Suzuki, M.; Grisham, M.B. Leukocyte adherence to venular endothelium during ischemia–reperfusion. *Am. J. Physiol.* **1989**, *257*, G683–688.
38. Sala, A.; Aliev, G.M.; Rossoni, G.; Berti, F.; Buccellati, C.; Burnstock, G.; Folco, G.; Maclouf, J. Morphological and functional changes of coronary vasculature caused by transcellular biosynthesis of sulfidopeptide leukotrienes in isolated heart of rabbit. *Blood* **1996**, *87*, 1824–1832.
39. Salvatico, E.; Aliev, G.M.; Novello, D.; Prosdocimi, M. Functional depression of isolated perfused rat heart mediated by activated leukocytes: protective effect of cloricromene. *J. Cardiovasc. Pharmacol.* **1994**, *24*, 638–647.

40. Matz, R.L.; Schott, C.; Stoclet, J.C.; Andriantsitohaina, R. Age-related endothelial dysfunction with respect to nitric oxide, endothelium–derived hyperpolarizing factor and cyclooxygenase products. *Physiol. Res.* **2000**, *49*, 11–18.
41. Aliev, G.; Burnstock, G. Watanabe rabbits with heritable hypercholesterolaemia: a model of atherosclerosis. *Histol. Histopathol.* **1998**, *13*, 797–817.
42. Stewart–Lee, A.L.; Ferns, G.A.; Anggard, E.E. Differences in onset of impaired endothelial responses and in effects of vitamin E in the hypercholesterolemic rabbit carotid and renal arteries. *J. Cardiovasc. Pharmacol.* **1995**, *25*, 906–913.
43. Price, J.M.; Sutton, E.T.; Hellermann, A.; Thomas, T. beta–Amyloid induces cerebrovascular endothelial dysfunction in the rat brain. *Neurol. Res.* **1997**, *19*, 534–538.
44. Iadecola, C.; Zhang, F.; Niwa, K.; Eckman, C.; Turner, S.K.; Fischer, E.; Younkin, S.; Borchelt, D.R.; Hsiao, K.K.; Carlson, G.A. SOD1 rescues cerebral endothelial dysfunction in mice overexpressing amyloid precursor protein. *Nat. Neurosci.* **1999**, *2*, 157–161.
45. Niwa, K.; Carlson, G.A.; Iadecola, C. Exogenous a beta1–40 reproduces cerebrovascular alterations resulting from amyloid precursor protein overexpression in mice. *J. Cereb. Blood Flow Metab.* **2000**, *20*, 1659–1668.
46. Grammas, P.; Moore, P.; Weigel, P.H. Microvessels from Alzheimer's disease brains kill neurons *in vitro*. *Am. J. Pathol.* **1999**, *154*, 337–342.
47. Mao, P.; Reddy, P.H. Is multiple sclerosis a mitochondrial disease? *Biochim. Biophys. Acta* **2009**, *1802*, 66–79.
48. Acuna–Castroviejo, D.; Martin, M.; Macias, M.; Escames, G.; Leon, J.; Khaldy, H.; Reiter, R.J. Melatonin, mitochondria, and cellular bioenergetics. *J. Pineal Res.* **2001**, *30*, 65–74.
49. Castellani, R.; Hirai, K.; Aliev, G.; Drew, K.L.; Nunomura, A.; Takeda, A.; Cash, A.D.; Obrenovich, M.E.; Perry, G.; Smith, M.A. Role of mitochondrial dysfunction in Alzheimer's disease. *J. Neurosci. Res.* **2002**, *70*, 357–360.
50. Fiskum, G.; Murphy, A.N.; Beal, M.F. Mitochondria in neurodegeneration: acute ischemia and chronic neurodegenerative diseases. *J. Cereb. Blood Flow Metab.* **1999**, *19*, 351–369.
51. Schulz, J.B.; Matthews, R.T.; Klockgether, T.; Dichgans, J.; Beal, M.F. The role of mitochondrial dysfunction and neuronal nitric oxide in animal models of neurodegenerative diseases. *Mol. Cell Biochem.* **1997**, *174*, 193–197.
52. Vancova, O.; Baciak, L.; Kasparova, S.; Kucharska, J.; Palacios, H.H.; Horecky, J.; Aliev, G. *In vivo* and *in vitro* assessment of brain bioenergetics in aging rats. *JCMM* **2009**, doi:10.1111/j.1582-4934.
53. Hirai, K.; Aliev, G.; Nunomura, A.; Fujioka, H.; Russell, R.L.; Atwood, C.S.; Johnson, A.B.; Kress, Y.; Vinters, H.V.; Tabaton, M.; Shimohama, S.; Cash, A.D.; Siedlak, S.L.; Harris, P.L.; Jones, P.K.; Petersen, R.B.; Perry, G.; Smith, M.A. Mitochondrial abnormalities in Alzheimer's disease. *J. Neurosci.* **2001**, *21*, 3017–3023.
54. Horecky, J.; Baciak, L.; Kasparova, S.; Pacheco, G.; Aliev, G.; Vancova, O. Minimally invasive surgical approach for three–vessel occlusion as a model of vascular dementia in the rat–brain bioenergetics assay. *J. Neurol. Sci.* **2009**, *283*, 178–181.
55. Beckman, K.B.; Ames, B.N. The free radical theory of aging matures. *Physiol. Rev.* **1998**, *78*, 547–581.

56. Beckman, K.B.; Ames, B.N. Mitochondrial aging: open questions. *Ann. N.Y. Acad. Sci.* **1998**, *854*, 118–127.
57. Beckman, K.B.; Ames, B.N. Endogenous oxidative damage of mtDNA. *Mutat. Res.* **1999**, *424*, 51–58.
58. Wallace, D.C. Mitochondrial diseases in man and mouse. *Science* **1999**, *283*, 1482–1488.
59. Aliev, G.; Smith, M.A.; Vinters, H.; Johnson, A.B.; Nunomura, A.; Perry, G. Mitochondria abnormalities mark vulnerable neurons in Alzheimer's disease. *J. Neuropathol. Exper. Neurol.* **1999**, *58*, 511.
60. Beal, M.F. Aging, energy, and oxidative stress in neurodegenerative diseases. *Ann. Neurol.* **1995**, *38*, 357–366.
61. Bonilla, E.; Tanji, K.; Hirano, M.; Vu, T.H.; DiMauro, S.; Schon, E.A. Mitochondrial involvement in Alzheimer's disease. *Biochim. Biophys. Acta* **1999**, *1410*, 171–182.
62. Aliev, G.; Smith, M.A.; Obrenovich, M.E.; de la Torre, J.C.; Perry, G. Role of vascular Hypoperfusion-induced oxidative stress and mitochondria failure in the pathogenesis of Alzheimer's disease. *Neurotox. Res.* **2003**, *5*, 491–504.
63. Al–Abdulla, N.A.; Martin, L.J. Apoptosis of retrogradely degenerating neurons occurs in association with the accumulation of perikaryal mitochondria and oxidative damage to the nucleus. *Am. J. Pathol.* **1998**, *153*, 447–456.
64. Wallace, D.C. Mitochondrial DNA in aging and disease. *Sci. Am.* **1997**, *277*, 40–47.
65. Cormier, A.; Morin, C.; Zini, R.; Tillement, J.P.; Lagrue, G. *In vitro* effects of nicotine on mitochondrial respiration and superoxide anion generation. *Brain Res.* **2001**, *900*, 72–79.
66. Reddy, P.H. Role of mitochondria in neurodegenerative diseases: mitochondria as a therapeutic target in Alzheimer's disease. *C.N.S. Spectr.* **2009**, *14*, 8–13, 16–18.
67. Aliyev, A.; Chen, S.G.; Seyidova, D.; Smith, M.A.; Perry, G.; de la Torre, J.; Aliev, G. Mitochondria DNA deletions in atherosclerotic hypoperfused brain microvessels as a primary target for the development of Alzheimer's disease. *J. Neurol. Sci.* **2005**, *229/230*, 285–292.
68. De Jong, G.I.; De Vos, R.A.; Steur, E.N.; Luiten, P.G. Cerebrovascular hypoperfusion: a risk factor for Alzheimer's disease? Animal model and postmortem human studies. *Ann. N.Y. Acad. Sci.* **1997**, *826*, 56–74.
69. De la Torre, J.C. Hemodynamic consequences of deformed microvessels in the brain in Alzheimer's disease. *Ann. N.Y. Acad. Sci.* **1997**, *826*, 75–91.
70. Friston, K.J.; Frackowiak, R.S. Cerebral function in aging and Alzheimer's disease: the role of PET. *Electroencephalogr. Clin. Neurophysiol. Suppl.* **1991**, *42*, 355–365.
71. Kumar, A.; Schapiro, M.B.; Haxby, J.V.; Grady, C.L.; Friedland, R.P. Cerebral metabolic and cognitive studies in dementia with frontal lobe behavioral features. *J. Psychiatr. Res.* **1990**, *24*, 97–109.
72. Galle, J.; Bengen, J.; Schollmeyer, P.; Wanner, C. Impairment of endothelium–dependent dilation in rabbit renal arteries by oxidized lipoprotein(a). Role of oxygen–derived radicals. *Circulation* **1995**, *92*, 1582–1589.
73. de la Torre, J.C.; Stefano, G.B. Evidence that Alzheimer's disease is a microvascular disorder: the role of constitutive nitric oxide. *Brain Res. Brain Res. Rev.* **2000**, *34*, 119–136.

74. Meguro, K.; Blaizot, X.; Kondoh, Y.; Le Mestric, C.; Baron, J.C.; Chavoix, C. Neocortical and hippocampal glucose hypometabolism following neurotoxic lesions of the entorhinal and perirhinal cortices in the non–human primate as shown by PET. Implications for Alzheimer's disease. *Brain* **1999**, *122* (Pt. 8), 1519–1531.
75. Jagust, W.J.; Friedland, R.P.; Budinger, T.F.; Koss, E.; Ober, B. Longitudinal studies of regional cerebral metabolism in Alzheimer's disease. *Neurology* **1988**, *38*, 909–912.
76. de la Torre, J.C. Pathophysiology of neuronal energy crisis in Alzheimer's disease. *Neurodegener. Dis.* **2008**, *5*, 126–132.
77. de la Torre, J.C. Alzheimer disease as a vascular disorder: nosological evidence. *Stroke* **2002**, *33*, 1152–1162.
78. De la Torre, J.C. Critically attained threshold of cerebral hypoperfusion: can it cause Alzheimer's disease? *Ann. N.Y. Acad. Sci.* **2000**, *903*, 424–436.
79. De la Torre, J.C. Alzheimer's disease prevalence can be lowered with non–invasive testing. *J. Alzheimer's Dis.* **2008**, *14*, 353–359.
80. Hofman, A.; Ott, A.; Breteler, M.M.; Bots, M.L.; Slooter, A.J.; van Harskamp, F.; van Duijn, C.N.; Van Broeckhoven, C.; Grobbee, D.E. Atherosclerosis, apolipoprotein E, and prevalence of dementia and Alzheimer's disease in the Rotterdam Study. *Lancet* **1997**, *349*, 151–154.
81. Weinbrecht, P.; Longmuir, I.; Knopp, J.; Mills, M. Cerebral microcirculatory changes during exposure to hypoxia. *Adv. Exp. Med. Biol.* **1987**, *215*, 259–263.
82. Mironov, V.; Hritz, M.A.; LaManna, J.C.; Hudetz, A.G.; Harik, S.I. Architectural alterations in rat cerebral microvessels after hypobaric hypoxia. *Brain Res.* **1994**, *660*, 73–80.
83. Chavez, J.C.; Agani, F.; Pichiule, P.; LaManna, J.C. Expression of hypoxia–inducible factor–1 alpha in the brain of rats during chronic hypoxia. *J. Appl. Physiol.* **2000**, *89*, 1937–1942.
84. Stewart, P.A.; Isaacs, H.; LaManna, J.C.; Harik, S.I. Ultrastructural concomitants of hypoxia–induced angiogenesis. *Acta Neuropathol. (Berl.)* **1997**, *93*, 579–584.
85. LaManna, J.C.; Boehm, K.D.; Mironov, V.; Hudetz, A.G.; Hritz, M.A.; Yun, J.K.; Harik, S.I. Increased basic fibroblastic growth factor mRNA in the brains of rats exposed to hypobaric hypoxia. *Adv. Exp. Med. Biol.* **1994**, *361*, 497–502.
86. Harik, N.; Harik, S.I.; Kuo, N.T.; Sakai, K.; Przybylski, R.J.; LaManna, J.C. Time–course and reversibility of the hypoxia–induced alterations in cerebral vascularity and cerebral capillary glucose transporter density. *Brain Res.* **1996**, *737*, 335–338.
87. Zhu, X.; Smith, M.A.; Perry, G.; Aliev, G. Mitochondrial failures in Alzheimer's disease. *Am. J. Alzheimer's Dis. Other Demen.* **2004**, *19*, 345-352.
88. Aliev, G.; Smith, M.A.; Turmaine, M.; Neal, M.L.; Zimina, T.V.; Friedland, R.P.; Perry, G.; LaManna, J.C.; Burnstock, G. Atherosclerotic lesions are associated with increased immunoreactivity for inducible nitric oxide synthase and endothelin–1 in thoracic aortic intimal cells of hyperlipidemic Watanabe rabbits. *Exp. Mol. Pathol.* **2001**, *71*, 40–54.
89. Shi, J.; Perry, G.; Berridge, M.S.; Aliev, G.; Siedlak, S.L.; Smith, M.A.; LaManna, J.C.; Friedland, R.P. Labeling of cerebral amyloid beta deposits *in vivo* using intranasal basic fibroblast growth factor and serum amyloid P component in mice. *JNM* **2002**, *43*, 1044–1051.
90. Bednar, M.M.; Raymond, S.; McAuliffe, T.; Lodge, P.A.; Gross, C.E. The role of neutrophils and platelets in a rabbit model of thromboembolic stroke. *Stroke* **1991**, *22*, 44–50.

91. Chen, H.; Chopp, M.; Zhang, Z.G.; Garcia, J.H. The effect of hypothermia on transient middle cerebral artery occlusion in the rat. *J. Cereb. Blood Flow Metab.* **1992**, *12*, 621–628.
92. Olah, L.; Wecker, S.; Hoehn, M. Relation of apparent diffusion coefficient changes and metabolic disturbances after 1 hour of focal cerebral ischemia and at different reperfusion phases in rats. *J. Cereb. Blood Flow Metab.* **2001**, *21*, 430–439.
93. Dietrich, W.D.; Busto, R.; Ginsberg, M.D. Cerebral endothelial microvilli: formation following global forebrain ischemia. *J. Neuropathol. Exp. Neurol.* **1984**, *43*, 72–83.
94. Fischer, E.G.; Ames 3d, A. Studies on mechanisms of impairment of cerebral circulation following ischemia: effect of hemodilution and perfusion pressure. *Stroke* **1972**, *3*, 538–542.
95. Wade, J.G.; Amtorp, O.; Sorensen, S.C. No–flow state following cerebral ischemia. Role of increase in potassium concentration in brain interstitial fluid. *Arch. Neurol.* **1975**, *32*, 381–384.
96. Cada, A.; de la Torre, J.C.; Gonzalez–Lima, F. Chronic cerebrovascular ischemia in aged rats: effects on brain metabolic capacity and behavior. *Neurobiol. Aging* **2000**, *21*, 225–233.
97. Marletta, M.A. Nitric oxide synthase: aspects concerning structure and catalysis. *Cell* **1994**, *78*, 927–930.
98. Moncada, S.; Higgs, E.A. Molecular mechanisms and therapeutic strategies related to nitric oxide. *FASEB J.* **1995**, *9*, 1319–1330.
99. Moncada, S.; Palmer, R.M.; Higgs, E.A. Nitric oxide: physiology, pathophysiology, and pharmacology. *Pharmacol. Rev.* **1991**, *43*, 109–142.
100. Crow, J.P.; Ye, Y.Z.; Strong, M.; Kirk, M.; Barnes, S.; Beckman, J.S. Superoxide dismutase catalyzes nitration of tyrosines by peroxynitrite in the rod and head domains of neurofilament L. *J. Neurochem.* **1997**, *69*, 1945–1953.
101. Dawson, V.L.; Brahmbhatt, H.P.; Mong, J.A.; Dawson, T.M. Expression of inducible nitric oxide synthase causes delayed neurotoxicity in primary mixed neuronal–glial cortical cultures. *Neuropharmacology* **1994**, *33*, 1425–1430.
102. Dawson, V.L.; Dawson, T.M. Nitric oxide in neuronal degeneration. *Proc. Soc. Exp. Biol. Med.* **1996**, *211*, 33–40.
103. Dawson, V.L.; Dawson, T.M.; London, E.D.; Bredt, D.S.; Snyder, S.H. Nitric oxide mediates glutamate neurotoxicity in primary cortical cultures. *Proc. Natl. Acad. Sci. USA* **1991**, *88*, 6368–6371.
104. Michel, T.; Feron, O. Nitric oxide synthases: which, where, how, and why? *J. Clin. Invest.* **1997**, *100*, 2146–2152.
105. Wever, R.M.; Luscher, T.F.; Cosentino, F.; Rabelink, T.J. Atherosclerosis and the two faces of endothelial nitric oxide synthase. *Circulation* **1998**, *97*, 108–112.
106. Xia, Y.; Dawson, V.L.; Dawson, T.M.; Snyder, S.H.; Zweier, J.L. Nitric oxide synthase generates superoxide and nitric oxide in arginine–depleted cells leading to peroxynitrite–mediated cellular injury. *Proc. Natl. Acad. Sci. USA* **1996**, *93*, 6770–6774.
107. Ignarro, L.J. Biosynthesis and metabolism of endothelium–derived nitric oxide. *Annu. Rev . Pharmacol. Toxicol.* **1990**, *30*, 535–560.
108. Stamler, J.S. Redox signaling: nitrosylation and related target interactions of nitric oxide. *Cell* **1994**, *78*, 931–936.

109. Robinson, L.J.; Weremowicz, S.; Morton, C.C.; Michel, T. Isolation and chromosomal localization of the human endothelial nitric oxide synthase (NOS3) gene. *Genomics* **1994**, *19*, 350–357.
110. Sessa, W.C. The nitric oxide synthase family of proteins. *J. Vasc. Res.* **1994**, *31*, 131–143.
111. Stuehr, D.J. Structure–function aspects in the nitric oxide synthases. *Annu. Rev. Pharmacol. Toxicol.* **1997**, *37*, 339–359.
112. Morris, S.M., Jr.; Billiar, T.R. New insights into the regulation of inducible nitric oxide synthesis. *Am. J. Physiol.* **1994**, *266*, E829–E839.
113. Bates, T.E.; Loesch, A.; Burnstock, G.; Clark, J.B. Immunocytochemical evidence for a mitochondrially located nitric oxide synthase in brain and liver. *Biochem. Biophys. Res. Commun.* **1995**, *213*, 896–900.
114. Bates, T.E.; Loesch, A.; Burnstock, G.; Clark, J.B. Mitochondrial nitric oxide synthase: a ubiquitous regulator of oxidative phosphorylation? *Biochem. Biophys. Res. Commun.* **1996**, *218*, 40–44.
115. Dawson, D.A. Nitric oxide and focal cerebral ischemia: multiplicity of actions and diverse outcome. *Cerebrovasc. Brain Metab. Rev.* **1994**, *6*, 299–324.
116. Bredt, D.S.; Snyder, S.H. Nitric oxide: a physiologic messenger molecule. *Annu. Rev. Biochem.* **1994**, *63*, 175–195.
117. Faraci, F.M.; Brian, J.E., Jr. Nitric oxide and the cerebral circulation. *Stroke* **1994**, *25*, 692–703.
118. Almeida, A.; Bolanos, J.P.; Medina, J.M. Nitric oxide mediates glutamate–induced mitochondrial depolarization in rat cortical neurons. *Brain Res.* **1999**, *816*, 580–586.
119. Beckman, J.S. The double–edged role of nitric oxide in brain function and superoxide–mediated injury. *J. Dev. Physiol.* **1991**, *15*, 53–59.
120. Beckman, J.S.; Beckman, T.W.; Chen, J.; Marshall, P.A.; Freeman, B.A. Apparent hydroxyl radical production by peroxynitrite: implications for endothelial injury from nitric oxide and superoxide. *Proc. Natl. Acad. Sci. USA* **1990**, *87*, 1620–1624.
121. Beckman, J.S.; Chen, J.; Ischiropoulos, H.; Crow, J.P. Oxidative chemistry of peroxynitrite. *Methods Enzymol.* **1994**, *233*, 229–240.
122. Cazevieille, C.; Muller, A.; Meynier, F.; Bonne, C. Superoxide and nitric oxide cooperation in hypoxia/reoxygenation–induced neuron injury. *Free Radic. Biol. Med.* **1993**, *14*, 389–395.
123. Radi, R.; Beckman, J.S.; Bush, K.M.; Freeman, B.A. Peroxynitrite oxidation of sulfhydryls. The cytotoxic potential of superoxide and nitric oxide. *J. Biol. Chem.* **1991**, *266*, 4244–4250.
124. Lafon–Cazal, M.; Pietri, S.; Culcasi, M.; Bockaert, J. NMDA–dependent superoxide production and neurotoxicity. *Nature* **1993**, *364*, 535–537.
125. Faraci, F.M. Role of endothelium–derived relaxing factor in cerebral circulation: large arteries vs. microcirculation. *Am. J. Physiol.* **1991**, *261*, H1038-H1042.
126. Fabricius, M.; Lauritzen, M. Examination of the role of nitric oxide for the hypercapnic rise of cerebral blood flow in rats. *Am. J. Physiol.* **1994**, *266*, H1457–H1464.
127. Iadecola, C. Does nitric oxide mediate the increases in cerebral blood flow elicited by hypercapnia? *Proc. Natl. Acad. Sci. USA* **1992**, *89*, 3913–3916.

128. Buisson, A.; Margaill, I.; Callebert, J.; Plotkine, M.; Boulu, R.G. Mechanisms involved in the neuroprotective activity of a nitric oxide synthase inhibitor during focal cerebral ischemia. *J. Neurochem.* **1993**, *61*, 690–696.
129. Buisson, A.; Plotkine, M.; Boulu, R.G. The neuroprotective effect of a nitric oxide inhibitor in a rat model of focal cerebral ischaemia. *Br. J. Pharmacol.* **1992**, *106*, 766–767.
130. Samadani, A.F.; Dawson, T.M.; Dawson, V.L. Nitric oxide synthase in model of focal ischemia. *Stroke* **1997**, *28*, 1283–1288.
131. Hamada, Y.; Hayakawa, T.; Hattori, H.; Mikawa, H. Inhibitor of nitric oxide synthesis reduces hypoxic–ischemic brain damage in the neonatal rat. *Pediatr. Res.* **1994**, *35*, 10–14.
132. Iadecola, C.; Li, J.; Ebner, T.J.; Xu, X. Nitric oxide contributes to functional hyperemia in cerebellar cortex. *Am. J. Physiol.* **1995**, *268*, R1153–R1162.
133. Iadecola, C.; Zhang, F.; Casey, R.; Clark, H.B.; Ross, M.E. Inducible nitric oxide synthase gene expression in vascular cells after transient focal cerebral ischemia. *Stroke* **1996**, *27*, 1373–1380.
134. Iadecola, C.; Zhang, F.; Xu, S.; Casey, R.; Ross, M.E. Inducible nitric oxide synthase gene expression in brain following cerebral ischemia. *J. Cereb. Blood Flow Metab.* **1995**, *15*, 378-384.
135. Nakashima, M.N.; Yamashita, K.; Kataoka, Y.; Yamashita, Y.S.; Niwa, M. Time course of nitric oxide synthase activity in neuronal, glial, and endothelial cells of rat striatum following focal cerebral ischemia. *Cell Mol. Neurobiol.* **1995**, *15*, 341–349.
136. Wang, Q.; Pelligrino, D.A.; Baughman, V.L.; Koenig, H.M.; Albrecht, R.F. The role of neuronal nitric oxide synthase in regulation of cerebral blood flow in normocapnia and hypercapnia in rats. *J Cereb. Blood Flow Metab.* **1995**, *15*, 774–778.
137. Garthwaite, J.; Beaumont, P.S. Excitatory amino acid receptors in the parallel fibre pathway in rat cerebellar slices. *Neurosci. Lett.* **1989**, *107*, 151–156.
138. Hara, H.; Huang, P.L.; Panahian, N.; Fishman, M.C.; Moskowitz, M.A. Reduced brain edema and infarction volume in mice lacking the neuronal isoform of nitric oxide synthase after transient MCA occlusion. *J. Cereb. Blood Flow Metab.* **1996**, *16*, 605–611.
139. Huang, Z.; Huang, P.L.; Panahian, N.; Dalkara, T.; Fishman, M.C.; Moskowitz, M.A. Effects of cerebral ischemia in mice deficient in neuronal nitric oxide synthase. *Science* **1994**, *265*, 1883–1885.
140. Eliasson, M.J.; Sampei, K.; Mandir, A.S.; Hurn, P.D.; Traystman, R.J.; Bao, J.; Pieper, A.; Wang, Z.Q.; Dawson, T.M.; Snyder, S.H.; Dawson, V.L. Poly(ADP–ribose) polymerase gene disruption renders mice resistant to cerebral ischemia. *Nat. Med.* **1997**, *3*, 1089–1095.
141. Thorns, V.; Hansen, L.; Masliah, E. nNOS expressing neurons in the entorhinal cortex and hippocampus are affected in patients with Alzheimer's disease. *Exp. Neurol.* **1998**, *150*, 14–20.
142. Bogumil, R.; Knipp, M.; Fundel, S.M.; Vasak, M. Characterization of dimethylargininase from bovine brain: evidence for a zinc binding site. *Biochemistry* **1998**, *37*, 4791–4798.
143. Kimoto, M.; Tsuji, H.; Ogawa, T.; Sasaoka, K. Detection of NG,NG–dimethylarginine dimethylaminohydrolase in the nitric oxide–generating systems of rats using monoclonal antibody. *Arch. Biochem. Biophys.* **1993**, *300*, 657–662.
144. Kimoto, M.; Whitley, G.S.; Tsuji, H.; Ogawa, T. Detection of NG,NG–dimethylarginine dimethylaminohydrolase in human tissues using a monoclonal antibody. *J. Biochem. (Tokyo)* **1995**, *117*, 237–238.

145. MacAllister, R.J.; Parry, H.; Kimoto, M.; Ogawa, T.; Russell, R.J.; Hodson, H.; Whitley, G.S.; Vallance, P. Regulation of nitric oxide synthesis by dimethylarginine dimethylaminohydrolase. *Br. J. Pharmacol.* **1996**, *119*, 1533–1540.
146. Beckman, J.S.; Carson, M.; Smith, C.D.; Koppenol, W.H. ALS, SOD and peroxynitrite. *Nature* **1993**, *364*, 584.
147. Aliev, G.; Liu, J.; Shenk, J.C.; Fischbach, K.; Pacheco, G.J.; Chen, S.G.; Obrenovich, M.E.; Ward, W.F.; Richardson, A.G.; Smith, M.A.; Gasimov, E.; Perry, G.; Ames, B.N. Neuronal mitochondrial amelioration by feeding acetyl–L–carnitine and lipoic acid to aged rats. *JCMM* **2009**, *2,* 320–333.
148. Ames, B.N. Delaying the mitochondrial decay of aging. *Ann. N.Y. Acad. Sci.* **2004**, *1019*, 406–411.
149. Ames, B.N. Mitochondrial decay, a major cause of aging, can be delayed. *J. Alzheimer's Dis.* **2004**, *6*, 117–121.
150. Ames, B.N.; Liu, J. Delaying the mitochondrial decay of aging with acetyl L-Carnitine. *Ann. N.Y. Acad. Sci.* **2004**, *1033*, 108–116.
151. Liu, J.; Head, E.; Gharib, A.M.; Yuan, W.; Ingersoll, R.T.; Hagen, T.M.; Cotman, C.W.; Ames, B.N. Memory loss in old rats is associated with brain mitochondrial decay and RNA/DNA oxidation: partial reversal by feeding acetyl–L–carnitine and/or R–alpha –lipoic acid. *Proc. Natl. Acad. Sci. USA* **2002**, *99*, 2356–2361.
152. Liu, J.; Head, E.; Kuratsune, H.; Cotman, C.W.; Ames, B.N. Comparison of the effects of L–carnitine and acetyl–L–carnitine on carnitine levels, ambulatory activity, and oxidative stress biomarkers in the brain of old rats. *Ann. N.Y. Acad. Sci.* **2004**, *1033*, 117–131.
153. Milgram, N.W.; Araujo, J.A.; Hagen, T.M.; Treadwell, B.V.; Ames, B.N. Acetyl–L–Carnitine and alpha–Lipoic acid supplementation of aged beagle dogs improves learning in two landmark discrimination tests. *FASEB J.* **2007**, *21*, 3756–3762.
154. Shenk, J.C.; Liu, J.; Fischbach, K; Xu, K.; Puchowicz, M.; Obrenovich, M.E; Gasimov, E.; Alvarez, L,M.; Ames, B.N.; LaManna J.C.; Aliev, G. The effect of acetyl–L–Carnitine and R–alpha–Lipoic acid treatment in ApoE4 mouse as a model of human Alzheimer's disease. *J. Neurol. Sci.* **2009**, *283*, 199–206.
155. Gomez–Pinilla, F. Brain foods: the effects of nutrients on brain function. *Nat. Rev. Neurosci.* **2008**, *9*, 568-578
156. Reddy, P.H. Amyloid beta, mitochondrial structural and functional dynamics in Alzheimer's disease. *Exp. Neurol.* **2009**, *218*, 286–292.
157. Liu, J.; Killilea, D.W.; Ames, B.N. Age-associated mitochondrial oxidative decay: improvement of carnitine acetyltransferase substrate–binding affinity and activity in brain by feeding old rats acetyl–L–Carnitine and/or R–alpha–Lipoic acid. *Proc. Natl. Acad. Sci. USA* **2002**, *99*, 1876–1881.
158. Long, J.; Gao, F.; Tong, L.; Cotman, C.W.; Ames, B.N.; Liu, J. Mitochondrial decay in the brains of old rats: ameliorating effect of alpha–Lipoic acid and acetyl–L–Carnitine. *Neurochem. Res.* **2009**, *34*, 755–63.
159. Reddy, P.H.; Beal, M.F. Amyloid beta, mitochondrial dysfunction and synaptic damage: implications for cognitive decline in aging and Alzheimer's disease. *Trends Mol. Med.* **2008**, *14*, 45–53.

160. Reddy, P.H. Mitochondrial medicine for aging and neurodegenerative diseases. *Neuromolecular Med.* **2008**, *10*, 291–315.

161. Reddy, P.H. Amyloid precursor protein–mediated free radicals and oxidative damage: implications for the development and progression of Alzheimer's disease. *J. Neurochem.* **2006**, *96*, 1–13.

162. Liu, J.; Atamna, H.; Kuratsune, H.; Ames, B.N. Delaying brain mitochondrial decay and aging with mitochondrial antioxidants and metabolites. *Ann. N.Y. Acad. Sci.* **2002**, *959*, 133–166.

163. Reid, P.C.; Urano, Y.; Kodama, T.; Hamakubo, T. Alzheimer's disease: cholesterol, membrane rafts, isoprenoids and statins. *JCMM* **2007**, *11*, 383–392.

164. Aliev, G.; Shenk, J.C.; Fischbach, K.; Perry, G. Stem cell niches as clinical targets: the future of anti–ischemic therapy? *Nat. Clin. Pract. Cardiovasc. Med.* **2008**, *5*, 590–591.

165. Cedazo–Minguez, A. Apolipoprotein E and Alzheimer's disease: molecular mechanisms and therapeutic opportunities. *JCMM* **2007**, *11*, 1227–1238.

166. Mas, E.; Dupuy, A.M.; Artero, S.; Portet, F.; Cristol, J.P.; Ritchie, K.; Touchon, J. Functional Vitamin E deficiency in ApoE4 patients with Alzheimer's disease. *Dement. Geriatr. Cogn. Disord.* **2006**, *21*, 198-204.

167. Sanchez–Mejia, R.O.; Newman, J.W.; Toh, S.; Yu, G.Q.; Zhou, Y.; Halabisky, B.; Cisse, M.; Scearce–Levie, K.; Cheng, I.H.; Gan, L.; Palop, J.J.; Bonventre, J.V.; Mucke, L. Phospholipase A2 reduction ameliorates cognitive deficits in a mouse model of Alzheimer's disease. *Nat. Neurosci.* **2008**, *11*, 1311–1318.

168. Anekonda, T.S.; Reddy, P.H. Can herbs provide a new generation of drugs for treating Alzheimer's disease? *Brain Res. Brain Res. Rev.* **2005**, *50*, 361–376.

169. Aliev, G.; Castellani, R.J.; Petersen, R.B.; Burnstock, G.; Perry, G.; Smith, M.A. Pathobiology of familial hypercholesterolemic atherosclerosis. *J. Submicrosc. Cytol. Pathol.* **2004**, *36*, 225–240.

170. Mahley, R.W.; Weisgraber, K.H.; Huang, Y. Apolipoprotein E4: a causative factor and therapeutic target in neuropathology, including Alzheimer's disease. *Proc. Natl. Acad. Sci. USA* **2006**, *103*, 5644–5651.

171. Reiman, E.M.; Caselli, R.J.; Chen, K.; Alexander, G.E.; Bandy, D.; Frost, J. Declining brain activity in cognitively normal apolipoprotein E epsilon 4 heterozygotes: A foundation for using positron emission tomography to efficiently test treatments to prevent Alzheimer's disease. *Proc. Natl. Acad. Sci. USA* **2001**, *98*, 3334–3339.

172. Aliev, G.; Smith, M.A.; de la Torre, J.C.; Perry, G. Mitochondria as a primary target for vascular hypoperfusion and oxidative stress in Alzheimer's disease. *Mitochondrion* **2004**, *4*, 649–663.

2

Current Experience in Testing Mitochondrial Nutrients in Disorders Featuring Oxidative Stress and Mitochondrial Dysfunction: Rational Design of Chemoprevention Trials

Giovanni Pagano [1,*], Annarita Aiello Talamanca [1], Giuseppe Castello [1], Mario D. Cordero [2], Marco d'Ischia [3], Maria Nicola Gadaleta [4], Federico V. Pallardó [5], Sandra Petrović [6], Luca Tiano [7] and Adriana Zatterale [8]

[1] Istituto Nazionale Tumori Fondazione G. Pascale—Cancer Research Center at Mercogliano (CROM)—IRCCS, Naples I-80131, Italy; E-Mails: a.aiello@istitutotumori.na.it (A.A.T.); beppe.castello@gmail.com (G.C.)

[2] Research Laboratory, Dental School, Universidad de Sevilla, Sevilla 41009, Spain; E-Mail: mdcormor@us.es

[3] Department of Chemical Sciences, University of Naples "Federico II", Naples I-80126, Italy; E-Mail: dischia@unina.it

[4] National Research Council, Institute of Biomembranes and Bioenergetics, Bari I-70126, Italy; E-Mail: marianicola.gadaleta@uniba.it

[5] CIBERER (Centro de Investigación Biomédica en Red de Enfermedades Raras), University of Valencia—INCLIVA, Valencia 46010, Spain; E-Mail: federico.v.pallardo@uv.es

[6] "Vinca" Institute of Nuclear Sciences, University of Belgrade, Belgrade 11001, Serbia; E-Mail: dragita.sandra@gmail.com

[7] Biochemistry Unit, Department of Clinical and Dental Sciences, Polytechnical University of Marche, Ancona I-60131, Italy; E-Mail: l.tiano@univpm.it

[8] Genetics Unit, Azienda Sanitaria Locale (ASL) Napoli 1 Centro, Naples I-80136, Italy; E-Mail: azatt@tin.it

* Author to whom correspondence should be addressed; E-Mail: gbpagano@tin.it

External Editor: Jaime M. Ross

Abstract: An extensive number of pathologies are associated with mitochondrial dysfunction (MDF) and oxidative stress (OS). Thus, mitochondrial cofactors termed "mitochondrial nutrients" (MN), such as α-lipoic acid (ALA), Coenzyme Q10 (CoQ10), and L-carnitine (CARN) (or its derivatives) have been tested in a number of clinical trials, and this review is focused on the use of MN-based clinical trials. The papers reporting on MN-based clinical trials were retrieved in MedLine up to July 2014, and evaluated for the following endpoints: (a) treated diseases; (b) dosages, number of enrolled patients and duration of treatment; (c) trial success for each MN or MN combinations as reported by authors. The reports satisfying the above endpoints included total numbers of trials and frequencies of randomized, controlled studies, *i.e.*, 81 trials testing ALA, 107 reports testing CoQ10, and 74 reports testing CARN, while only 7 reports were retrieved testing double MN associations, while no report was found testing a triple MN combination. A total of 28 reports tested MN associations with "classical" antioxidants, such as antioxidant nutrients or drugs. Combinations of MN showed better outcomes than individual MN, suggesting forthcoming clinical studies. The criteria in study design and monitoring MN-based clinical trials are discussed.

Keywords: mitochondrial nutrients; mitochondrial dysfunction; oxidative stress; oxidative phosphorylation; Krebs cycle

---

1. Introduction

The functions of mitochondria, unconfined to bioenergetic pathways, were linked to OS by studies in early 1990's reporting on reactive oxygen species (ROS) formation as by-products of oxygen metabolism [1,2]. The implications for OS in Parkinson's disease were discovered by Di Monte *et al.* [3] and associated with MDF by a correlation of complex I–III deficiency with lower-than-normal levels of CoQ10, a cofactor in the oxidative phosphorylation (OXPHOS) pathway [4,5].

The association of OS and MDF has been reported in in mitochondrial diseases [6,7], and our recent reviews evaluated the literature showing that OS/MDF is involved in broad-ranging pathologies, including some genetic diseases [8], aging and age-associated disorders, neurologic and psychiatric diseases, malignancies and autoimmune diseases [9]. This state-of-art has prompted a growing body of literature from clinical studies, aimed at compensating OS/MDF by means of MN administration [10–14]. These endogenous cofactors, such as ALA, CoQ10 and CARN (or CARN derivatives) are essential in mitochondrial functions.

ALA exists in two redox states, and the reduced thiol form is a potent mitochondrial antioxidant, a metal chelator and a glutathione (GSH) repletor. It is involved as an essential cofactor in Krebs cycle for mitochondrial α-ketoacid dehydrogenases. CoQ10 is a benzoquinone derivative playing a central role in the mitochondrial respiratory chain through shuttling between three redox forms, the quinone, the semiquinone and the hydroquinone. It acts as a carrier accepting electrons from mitochondrial complex I and complex II and transferring them to complex III. CARN (bioactive L-form) is biosynthesized from lysine and methionine and is concentrated in tissues that use fatty acids as their primary source of energy. It is involved in the transport of long-chain fatty acids from the intermembraneous space to the matrix in the mitochondria for generation of metabolic energy.

A number of studies have demonstrated MN deficiencies in several diseases [9], either related to deficiencies in OXPHOS activities [15–18] or caused by their degradation due to OS-related by-products as, e.g., by ALA oxidative modification induced by 4-hydroxynonenal in Alzheimer disease brain [19]. The present review is to provide an overall survey of the currently published clinical trials having tested each of the above MN, or their combinations, and/or their associations with "typical" antioxidants, including antioxidant nutrients and/or herbal preparations and/or disease-specific drugs. For each disease, or disease group, the total numbers of clinical trials were recorded, along with the relative frequencies of controlled, randomized trials. No further attempt was made to evaluate the clinical and/or laboratory outcomes of the evaluated reports. The major objective of the present survey was providing an overall state-of-art, and suggesting some prospects in study design and in monitoring the outcomes of forthcoming clinical trials. These should be aimed at verifying any specific MN deficiency(ies) associated to a given disorder, and at prompting adequate follow-up in monitoring the effects, if any, of mitochondria-targeted interventions.

2. Methods

A MedLine retrieval up to July 2014 was carried out for each MN. The papers reporting on clinical trials for each MN were evaluated according to: (a) treated diseases (or disease groups); (b) dosages, number of enrolled patients and duration of treatment; (c) numbers of trials and frequencies of randomized, controlled studies (opposed to open-label and pilot studies); and (d) an empyrical "success ratio" (SR) of trials testing each of the individual MN or their combinations, as reported by authors; SR was calculated by dividing the number of successful results by the total of trials for each agent. No clinical trials involving healthy volunteers were included for evaluation. The reports failing to provide clear-cut data for dosages, the numbers of patients and/or treatment duration were not included in survey, nor were included self-repeating reports of previous or contemporary studies.

3. α-Lipoic Acid

As shown in Table 1, ALA testing was reported in 81 evaluated clinical trials on a total of 2980 patients, since the pioneering report by Marshall *et al.* on alcohol-related liver disease [10], and since early studies in 1990's in counteracting diabetes-associated neuropathies [20–22]. Both Type 1 and Type 2 diabetes mellitus (DM) have been the major focus for ALA administration, encompassing 42 trials (of which 30 controlled trials) on a total of 2980 patients and with a SR = 0.93 [20–61]. A major outcome of ALA treatment in diabetic patients consisted of the amelioration of neurologic damage [22–24,26–31,33–35,37,39,42,46,49,51,52,60,61]. Moreover, ALA-treated diabetic patients displayed decreased serum concentrations of thiobarbituric acid reactive substances (TBARS) [20,43], increased insulin sensitivity [21,25,36,58], an improvement in endothelium function [38,43,55], decreased erectile dysfunction [59] and decreased urinary PGF2α-isoprostanes, a marker of oxidative damage [50].

Beyond the literature of ALA-centered clinical trials in DM, a broad clinical use of ALA is recognized as a prescription generic drug for DM in Germany (German Drug Index), thus the use of ALA in diabetic patients may be seen as an established practice in the specialist community.

Table 1. Clinical studies utilizing α-lipoic acid aimed at compensating oxidative stress (OS)/mitochondrial dysfunction (MDF)-related pathogenetic mechanisms.

Diseases/Conditions	No. Studies (Controlled Studies)	No. Treated Patients	Success Ratio	References
Type 1 and Type 2 diabetes	42 (30)	2980	0.93	[20–61]
Neurological diseases	9 (5)	509	0.89	[62–70]
Liver and metabolic diseases	8 (5)	417	0.86	[10,71–77]
Heart and vessel diseases	4 (4)	137	1.00	[78–81]
Kidney diseases	4 (3)	288	0.25	[82–85]
Genetic and mitochondrial diseases	3 (2)	129	0.33	[86–88]
Burning mouth syndrome	5 (5)	293	0.60	[89–93]
Other diseases §	6 (4)	525	1.00	[94–99]
Total	81	5278		

§ Malignancies [94,95]; HIV [96,97]; vitiligo [98]; osteoporosis [99].

Other clinical trials successfully tested ALA in neurological diseases [62–70], in liver and metabolic diseases [10,71–77], and in heart and vessel diseases [78–81] (Table 1).

Lesser, if any, positive effects were reported for three ALA trials in kidney diseases [82–85]. Other clinical trials tested the effects of ALA administration in genetic and mitochondrial diseases [86–88], burning mouth syndrome [89–93], cancer cachexia [94,95], mitochondrial function in HIV-1-related lipoatrophy [96,97], in vitiligo [98], and in osteoporosis [99].

Among these studies, some reports investigated the comparative effects of different MN formulations, and/or provided evidence for combined effects in clinical and laboratry endpoints.

Li *et al.* [79] showed that ALA administration to patients with acute coronary syndrome caused a decrease in an OS marker, 8-*iso*-prostaglandin F2α and an increase in aldehyde dehydrogenase-2, responsible for acetaldehyde oxidation in ethanol metabolism that also provides protection against OS. Protective effects were reported by Martins *et al.* [87] in sickle cell trait subjects and sickle cell patients following ALA administration, with a significant increase in blood catalase (CAT) and a significant decrease in malondialdehyde (MDA) and in carbonyl levels.

A clinical trial reported by Galasko *et al.* [64] tested the effects of combined ALA with Vit E and Vit C (E/C/ALA), or CoQ10, or placebo in three groups of patients with Alzheimer's disease, finding a decrease in CSF F2-isoprostane levels in the E/C/ALA group that suggested a reduction of OS in the brain. However, this treatment raised the caution of faster cognitive decline.

Altogether, the current body of evidence may suggest further interventions with ALA in several disorders, beyond DM and the other disorders where ALA treatment led both to clinical improvements and to compensation of OS-related endpoints.

4. Coenzyme Q10

A recent review has focused on the multiple implications of CoQ10 deficiency and of CoQ10 administration encompassing an extensive number of disorders [100]. Out of 101 reports on clinical trials testing CoQ10, 39 studies were focused on support to patients with heart and vessel disorders or undergoing heart surgery [11,12,81,101–136], including 32 controlled studies, as shown in Table 2. The pioneering studies by the groups of Tanaka [11] and of Langsjoen [12] since 1982 provided the avenue for a number of clinical trials assessing the successful outcomes of CoQ10 administration to patients with coronary artery disease, cardiomyopathy, heart failure, or heart surgery, with an overall SR = 0.89. Among these studies, Dai *et al.* [105] evaluated mitochondrial function in patients with ischemic left ventricular systolic dysfunction receiving CoQ10 *vs.* placebo, in terms of plasma lactate/pyruvate (LP) ratio. After an 8-week treatment, CoQ10-treated patients had significant increases in plasma CoQ10 concentration, brachial flow-mediated dilation, and decreased LP ratio, showing positive effects both on heart function and in balancing mitochondrial activities [105]. Lee *et al.* [106] tested CoQ10 administration in patients with coronary artery disease and found a decrease in the inflammatory markers C-reactive protein (hs-CRP) and interleukin 6 (IL-6), and decreased MDA and superoxide dismutase (SOD) activities, along with increased CoQ10 levels. Direct evidence for a CoQ10-induced compensation of mitochondrial function was reported by Rosenfeldt *et al.* [133] in patients undergoing heart surgery, which resulted in increased CoQ10 levels in serum, atrial trabeculae, and isolated mitochondria compared with patients receiving placebo, with an improvement in mitochondrial respiration, as adenosine diphosphate/oxygen ratio, and a decrease in mitochondrial MDA content.

Table 2. Clinical studies utilizing coenzyme Q10 aimed at compensating OS/MDF-related pathogenetic mechanisms.

Diseases/Conditions	No. Studies (Controlled Studies)	No. Treated Patients	Success Ratio	References
Heart and vessel diseases	39 (32)	3386	0.89	[11,12,81,101–136]
Genetic and mitochondrial diseases	18 (8)	680	0.75	[13,14,88,137–151]
Neurological diseases	16 (10)	1185	0.87	[65,152–166]
Type 1 and Type 2 Diabetes	9 (7)	370	0.89	[41,166–174]
Malignancies	6 (2)	301	0.67	[175–180]
Kidney diseases	4 (2)	171	0.50	[181–184]
Other diseases §	15 (13)	555	0.59	[76,185–198]
Total	107	6648		

§ Psoriasis [76,196]; metabolic syndrome [185–187]; statin-induced myalgias [188–192]; bronchial asthma [193]; pre-eclampsia [194]; mucocutaneous infections [195]; cataract surgery [197]; idiopathic infertility [198].

Eighteen studies (of which 8 controlled studies) have focused on clinical trials testing CoQ10 in a number of genetic and mitochondrial diseases [13,14,88,137–151] on 680 patients (Table 2) with different success across the different diseases investigated (SR = 0.75).

Among these studies, Friedreich ataxia (FRDA) is relevant for a deficiency in the gene encoding frataxin, a mitochondrial protein implicated in iron metabolism and glutathione balance [139], and lower-than-normal CoQ10 levels [139,140]. FRDA was investigated by Cooper's group [140,141] for CoQ10 and Vit E administration finding a significant improvement in cardiac and skeletal muscle bioenergetics and in International Co-operative Ataxia Ratings Scale. Rodriguez *et al.* [88], as mentioned above, tested the effect of a combination therapy with CoQ10 with ALA and creatine monohydrate on several outcome variables in patients with mitochondrial diseases, both achieving amelioration of clinical and of OS/MDF endpoints [88]. Limited evidence for a CoQ10-induced improvement in brain and muscle bioenergetics in patients with mitochondrial diseases was reported by Barbiroli *et al.* [142] and by Glover *et al.* [145].

A set of studies was focused on 16 clinical trials (of which 10 controlled trials) testing CoQ10 supplementation in 1185 patients with eight neurologic disorders [65,152–166].

The above-mentioned report by Galasko *et al.* [65] tested CoQ10 as an alternative treatment to E/C/ALA in patients with Alzheimer's disease, failing to report any effects of CoQ10, unlike E/C/ALA. A study by Shults *et al.* [152] in patients with Parkinson's disease (PD) tested the effects of three CoQ10 dosages (300, 600, or 1200 mg/d) that were evaluated with the Unified Parkinson Disease Rating Scale at the baseline and up to 16-month visits. Less disability developed in subjects assigned to CoQ10 than in those assigned to placebo, with a significant benefit in patients receiving the highest dosage [152]. Moderate beneficial effects of CoQ10 in PD patients were also reported by Müller *et al.* [153], though at the low dosage (360 mg/d) reported by Shults *et al.* [152]. A study by Storch *et al.* [154] found that nanoparticular CoQ10 at a dosage of 300 mg/d was well tolerated, though failing to display symptomatic effects in midstage PD.

Sanoobar *et al.* [155,156] tested CoQ10 in multiple sclerosis (MS) patients, and found reduced OS and increased antioxidant enzyme activity [155], with a significant decrease of tumor necrosis factor alpha (TNF-α) and IL-6 levels [156] in the CoQ10 group compared to placebo group.

A series of studies focused on the role of CoQ10 in fibromyalgia (FM) in long-term treatment of fibromyalgic patients, resulting in significant pain reduction, fatigue, and morning tiredness [157–160]. CoQ10-treated patients underwent significant reduction in the pain visual scale and in tender points, recovery of inflammation, antioxidant enzymes, mitochondrial biogenesis, and AMPK gene expression levels [157,158]. A recent report by Miyamae *et al.* [159] measured plasma levels of ubiquinone-10, ubiquinol-10, free cholesterol, cholesterol esters, and free fatty acids in patients with juvenile FM *vs.* healthy control subjects. Plasma level of ubiquinol-10 was significantly decreased and the ratio of ubiquinone-10 to total CoQ10 was significantly increased in juvenile FM relative to healthy controls, with compensated plasma levels of lipid endpoints [159]. A previous uncontrolled trial by Lister [160] tested the effects of combined CoQ10 and and *Ginkgo biloba* extract reporting improved quality-of-life in FM patients. Altogether, independent studies support the usefulness of CoQ10 in FM pathogenesis and point to CoQ10 utilization in FM both in inducing clinical improvements and in compensating FM-associated OS/MDF.

No evidence of CoQ10-induced beneficial effects were found in a Phase II clinical trial in patients with amyotrophic lateral sclerosis [161].

Migraine has been investigated for deficiency of CoQ10, which showed lower-than-normal levels in patients with pediatric and adolescent migraine [162]. Clinical trials testing CoQ10 in migraine patients provided overall positive, though apparently transient outcomes [162–164].

Forester *et al.* [165] tested high-dose CoQ10 in patients with geriatric bipolar depression suggesting a reduction in depression symptom severity.

Stamelou *et al.* [166] tested CoQ10 in patients with progressive supranuclear palsy (PSP), and found that patients receiving CoQ10 displayed decreased concentration of low-energy phosphates, with increased ratio of high-energy phosphates to low-energy phosphates. Clinically, the PSP rating scale and the Frontal Assessment Battery improved upon CoQ10 treatment compared to placebo [166].

Nine clinical trials (seven controlled trials) were conducted by testing CoQ10 in 370 patients with Type 1 and Type 2 DM, particularly focusing on DM-associated vascular complications and on statin-induced CoQ10 depletion [167–174]. The outcomes of these studies showed beneficial effects of CoQ10, better if combined with fenofibrate [169]. By considering the established advantages of ALA treatment in DM patients [20–61], one may suggest further clinical trials toward the clinical use of ALA and CoQ10 combinations in DM. This trial design was performed by Palacka *et al.* with successful outcomes both in improving heart left ventricular function and in decreasing lactate dehydrogenase activity in DM patients treated with ALA and CoQ10 [41].

Six clinical trials have been reported testing CoQ10 in association with other agents (e.g., riboflavin and niacin) in patients with advanced malignancies under antineoplastic therapy as supportive or palliative treatments [175–180]. Some of these trials reported survival prolongation [177], and an early study by Lockwood *et al.* [176] reported partial and complete regression of breast cancer in patients treated with CoQ10. A clinical trial by Rusciani *et al.* [180] tested the effects of CoQ10 and interferon α-2b in melanoma patients, and found significantly decreased recurrence rates *vs.* patients receiving interferon only.

Four clinical trials tested the CoQ10 in patients with chronic kidney disease or undergoing hemodialysis, or submitted to statin treatment [181–184]. Sakata *et al.* [182] reported CoQ10 administration in hemodialysis patients as partially effective for suppressing OS. Shojaei *et al.* [184] tested CoQ10 and CARN, separately or in association, in hemodialysis patients who were on statin treatment. This study showed that supplementation with CoQ10 and CARN reduced serum levels of lipoprotein(a) in maintenance hemodialysis patients treated with statins. Altogether, the clinical trials testing CoQ10 in patients with kidney diseases failed to report clear-cut beneficial effects (SR = 0.50).

Controversial results were reported from clinical trials aimed at testing CoQ10 in counteracting various dysmetabolic conditions [185–187], statin-induced myalgias [188–192], bronchial asthma [193], pre-eclampsia [194], cutaneous infections [195], psoriasis [76,196], cataract surgery [197], and idiopathic infertility [198].

5. L-Carnitine and Acetyl- or Propionyl-Carnitine

Multi-decade long investigations showed the crucial roles of CARN in mitochondrial physiology, along with CARN deficiency in several disorders. As reviewed by Gilbert since 1985 [199], CARN deficiency results in accumulation of neutral lipid deposits within skeletal muscle, myocardium and liver, with mitochondria aggregates in skeletal muscle and myocardium. The efficacy of CARN *vs.* acetyl-CARN (ALC) was compared both *in vitro* and in aging rat brain since early studies [200,201]. CARN and ALC were similar in elevating carnitine levels in plasma and brain as well as in increasing ambulatory activity of old rats. However, ALC but not CARN was able to decrease the level of oxidative stress biomarkers in the brain of old rats [201]. Subsequent investigations pointed to a role of ALC in the reactivation of mitochondrial biogenesis in aging through the increased expression of PGC-1α signaling pathway [202,203].

As shown in Table 3, out of 74 evaluated studies, 18 clinical trials (of which 16 controlled trials) tested the effects of CARN in kidney diseases, treating a total of 427 patients with end-stage renal disease under hemodialysis, based on the rationale that CARN levels sharply decrease during hemodialysis, thus impairing response to recombinant human erythropoietin (rHuEPO) [181,184,204–220]. One study utilized both CARN and CoQ10—separately or in association—finding positive effects of both supplements, yet without evidence for improvement effects following combined supplementation [184]. Some reports failed to find significant improvements following CARN administration [205–208,213,220] (overall SR = 0.58). One might note that out of four trials adopting *i.v.* CARN (or PLC) administration, three resulted in significant improvements in the tested endpoints, such as Medical Outcomes Short Form-36, rHuEPO requirement, hemodynamic flow, endothelial profile and homocysteine levels, or inflammatory status as significant decrease in C-reactive protein [210,211,217].

Table 3. Clinical studies utilizing L-carnitine (CARN) (or acetyl-CARN (ALC)) aimed at compensating OS/MDF-related pathogenetic mechanisms.

Diseases/Conditions	No. Studies (Controlled Studies)	No. Treated Patients	Success Ratio	References
Kidney diseases	18 (16)	427	0.58	[184,204–220]
Type 1 and Type 2 Diabetes	13 (9)	1894	1.00	[221–233]
Heart and vessel diseases	9 (6)	359	1.00	[78,114,234–240]
Liver diseases	6 (2)	275	1.00	[241–246]
Neurological diseases	9 (9)	384	1.00	[68,247–254]
Malignancies	7 (6)	699	0.70	[255–261]
Genetic diseases	6 (4)	202	1.00	[262–267]
HIV	4 (1)	93	1.00	[97,268–270]
Other diseases §	2 (2)	99	1.00	[271,272]
Total	74	4432		

§ Major surgery [271]; ulcerative colitis [272].

Unlike renal diseases, all of 13 clinical trials (9 controlled trials) supplementing CARN (or ALC, or PLC) to patients with Type 1 and Type 2 DM to a total of 1894 patients succeeded achieving improvements in disease status [221–233]. Long-term infusion of CARN or ALC in DM patients was tested by the group of Giancaterini [221,222], resulting in improved insulin sensitivity, and in decreased lactate levels, suggesting activation of pyruvate dehydrogenase, whose activity is depressed in the insulin resistant status. Three independent studies found a long-term treatment with ALC or with PLC effective and well tolerated in improving neurophysiological parameters and in reducing pain [223,226,229]. Derosa *et al.* reported that CARN significantly lowered the plasma lipoprotein(a) *vs.* placebo in hypercholesterolemic patients with DM [224]. A clinical trial with PLC in patients with DM and peripheral arterial disease was reported by Ragozzino *et al.* [225], who found significant increases in maximal walking distance, initial claudication distance, and a decrease in dosage of oral antihyperglycemic agents. Two studies investigated the effects of CARN or ALC treatment on lipidemic profile of DM patients [223,224], failing to detect a decrease of lipidemic profile of CARN alone to this endpoint, yet Solfrizzi *et al.* [228] found significant effects from combined treatment with CARN and simvastatin in lowering Lp(a) serum levels in patients with DM than with simvastatin alone. A study by Malaguarnera *et al.* [230] testing CARN in DM patients found significant improvements compared with the placebo group in a number of OS endpoints, proinflammatory markers and lipidemic profile.

A clinical trial was conducted by Ruggenenti *et al.* [231] by testing the effects of ALC on glucose disposal rate in patients with metabolic syndrome and hypertension, reporting that ALC ameliorated arterial hypertension, insulin resistance, glucose tolerance, and hypoadiponectinemia. Patients with Type 1 DM were treated by Uzun *et al.* [233] with CARN to evaluate changes in their neuropathy frequency and nerve conduction velocity, and found that CARN treatment in early stages improved neuropathy in Type 1 DM patients.

Altogether, the body of literature of clinical trials testing CARN—or CARN derivatives—in patients with DM provides substantial evidence for the amelioration of multiple DM-related endpoints that encourage the clinical use of CARN—or CARN derivatives—in DM patients. One may note that no report was found of combined treatments of DM patients using CARN with either ALA or CoQ10, despite the body of positive evidence accumulated in clinical trials testing ALA, or CoQ10 in DM patients (as shown Tables 1 and 2).

A series of 9 clinical trials (6 controlled trials) were reported testing CARN (or ALC, or PLC) in 359 patients with heart and vessel diseases [78,114,234–240]. A study by McMackin *et al.* [78] reported on the effects of combined ALC and ALA treatment in patients with coronary artery disease, and found significant improvement as increased brachial artery diameter, a trend to decreased systolic blood pressure for the whole patient group, with a significant effect in the subgroup with blood pressure above the median and in the subgroup with metabolic syndrome. An analogous investigation was reported by Kumar *et al.* [114], who tested the effects of carni Q-gel (CARN and ubiquinol) in patients with heart failure. Serum concentration of IL-6 was significantly decreased in the intervention group without such changes in the control group. TNF-α, which was comparable at baseline, also showed a greater decline in the carni Q-gel group compared to the placebo group. Serum CoQ10 showed a significant increase in the carni Q-gel group as compared to the control group. The symptom scale indicated that the majority of treated patients showed a significant improvement compared to the placebo group [114]. Altogether, both of these clinical trials using associations of ALC with ALA [78] or CARN with CoQ10 (as ubiquinol) [114] showed definite improvements both in heart and vessel performance and in related laboratory endpoints.

Relevant mechanistic data on PLC treatment in patients with peripheral arterial disease (PAD) were obtained by Loffredo *et al.* [234], who measured serum levels of nitrite and nitrate (NO_x) and 8-hydroxy-2'-deoxyguanosine (8-OHdG), and maximal walking distance (MWD) in PAD patients. Serum levels of 8-OHdG were significantly increased in PAD patients, and serum levels of NO_x were significantly decreased. Patients treated with PLC showed a significant increase of MWD and in NO_x, while 8-OHdG levels underwent a significant decrease, unlike the patients given placebo [234].

Improved short-term exercise capacity, diastolic function and symptoms, WHO heart functional class were reported in patients receiving CARN [237,238], with improvements in diastolic function and in myocardial hypoxanthine concentration [239,240].

A total of 275 patients affected by various liver disorders were treated with CARN or ALC in 6 clinical trials (of which 2 controlled studies) [241–246]. Łapiński and Grzeszczuk [241] tested CARN (or L-ornitine L-aspartate) in patients with liver cirrhosis, and evaluated the outcomes as changes in serum concentrations of ammonia, cholesterol and triglycerides. A significant improvement was observed both in the patients treated with CARN and in those treated with L-ornitine L-aspartate.

However, the individuals treated with CARN showed a significant increase of serum cholesterol and triglyceride concentrations.

Patients with non-alcoholic fatty liver disease were treated with CARN by Lim *et al.* [242]; by measuring a liver function test, peripheral blood mitochondrial DNA and 8-oxo-dG analysis. The results showed a decrease in ALT, AST, and total bilirubin in treated patients, not in controls. In the CARN treated patients, mitochondrial DNA copy number was significantly increased, unlike the control group. 8-oxo-dG levels showed a non-significant decrease in CARN group while it tended to increase in the control group. The group of Malaguarnera *et al.* has conducted a series of clinical trials testing the effects of CARN in patients with hepatic encephalopathy or with HCV-induced chronic hepatitis [243–246]. Beneficial effects in CARN-treated patients included significant differences in AST, ALT, viremia, Hb, RBC, WBC and platelets. Also significant improvements were detected in neurologic scores and fatigue symptoms in hepatic encephalopathy following CARN administration [244,245].

Nine controlled clinical trials for CARN—mostly ALC—were conducted in a total of 384 patients with some neurological disorders, including Alzheimer's disease, multiple sclerosis, migraine, sciatica, fibromyalgia, chronic fatigue syndrome, and amyotrophic lateral sclerosis [68,247–254]. A report by Memeo and Loiero [68] compared the effects of treating sciatica patients with either ALA or ALC, and found significant protective effects from both of these agents detected as improvements from baseline in neuropathy on electromyography.

Tomassini *et al.* [247] tested the efficacy of ALC *vs.* amantadine, a widely used drug in treating patients with multiple sclerosis (MS)-related fatigue. The results showed significant effects of ALC compared with amantadine for the Fatigue Severity Scale, suggesting that ALC is better tolerated and more effective than amantadine for the treatment of MS-related fatigue.

Patients with Alzheimer's disease and vascular dementia were treated with ALC by Gavrilova *et al.* [248], and the outcomes were assessed with MMSE and CGI scales, and a battery of neuropsychological tests. The treatment effect of ALC was significantly higher than in the placebo group.

Tarighat Esfanjani *et al.* [249] evaluated the effects of magnesium, CARN, and concurrent magnesium-CARN supplementation in migraine prophylaxis. Migrainous patients were randomly assigned into three intervention groups: magnesium oxide *vs.* CARN *vs.* Mg-CARN, and a control group. The results showed a significant reduction in all migraine indicators in all studied groups, yet without a clear-cut difference in-between the three supplementation regimens.

Four independent clinical trials tested CARN effects [250–253] in patients with chronic fatigue syndrome (CFS), or fibromyalgia (FM), or narcolepsy (NL). Pistone *et al.* [250] treated elderly subjects with onset of fatigue following slight physical activity. Wessely and Powell fatigue scores decreased significantly in subjects taking CARN *vs.* the placebo group, with significant improvements in total fat mass, total muscle mass, total cholesterol, LDL-C, HDL-C, triglycerides, apoA1, and apoB [250]. Miyagawa *et al.* [251] tested CARN administration in NL patients and found a significant reduction of excessive daytime sleepiness in CARN-treated NL patients *vs.* those receiving placebo. A clinical trial testing the effects of ALC in FM patients was reported by Rossini *et al.* [252]. The "total myalgic score" and the number of positive tender points declined significantly in the ALC-treated patients *vs.* the placebo group, and a significant difference was observed for depression and musculo-skeletal pain [252]. The effects of ALC and PLC were tested in patients with CFS by Vermeulen and

Scholte [253]. This clinical trial compared ALC, PLC, and their combination. Clinical global impression of change after treatment showed considerable improvement in 59% of the patients in the ALC group and 63% in the PLC group, but less in the ALC plus PLC group (37%). In the ALC group the changes in plasma CARN levels correlated with clinical improvement [253].

A clinical trial by Beghi *et al.* tested the effects of ALC plus riluzole *vs.* riluzole and placebo on disability and mortality of patients with amyotrophic lateral clerosis [254]. Patients receiving ALC became non-self-sufficient to significantly less extent compared to those receiving placebo, and median survival was significantly extended in ALC-treated patients *vs.* placebo group [254].

Based on CARN depletion in advanced cancer, palliative treatments by CARN supplementation have been proposed by a series of seven clinical trials, of which six controlled trials [255–261]. Unfortunately, the outcomes showed a treatment-related increase of CARN serum levels, yet CARN or ALC supplementation did not improve fatigue in patients, or even increased chemotherapy-induced peripheral neuropathy [261]; otherwise, positive results were associated with treatments with multiple agents [258,259], thus casting doubts as to the specific efficacy of CARN supplementation.

A few genetic disorders were the focus of six clinical trials, of which 4 controlled trials, using CARN or ALC [262–267]. A clinical trial was carried out by Schöls *et al.* [262] by testing the administration of CARN *vs.* creatine in patients with Friedreich's ataxia. After CARN treatment, patients had a significantly improved phosphocreatine recovery compared to baseline, while creatine effects did not reach significance.

Two independent clinical trials tested the effects of CARN in thalassemic patients, concurring to report improved cardiac function [263,264]. Moreover, Karimi *et al.* [264] tested the effect of combination therapy of hydroxyurea with CARN and magnesium chloride on hematologic parameters and cardiac function of patients with β-thalassemia intermedia. Patients were randomly divided into four groups: group A (hydroxyurea alone); group B (hydroxyurea and CARN); group C (hydroxyurea and magnesium chloride); and group D (hydroxyurea, CARN and magnesium chloride). In groups B, C, and D, mean Hb and hematocrit significantly increased during 6-month treatment [264].

Three independent studies tested the effects of CARN or ALC in patients with autistic spectrum disorders (ASD) [265–267]. Ellaway *et al.* [265] tested CARN in a group of Rett syndrome females, and found significantly improved sleep efficiency and expressive speech, compared to control Rett syndrome patients. A multi-center clinical trial testing ALC on the attention deficit hyperactivity disorder in fragile X syndrome boys was reported by Torrioli *et al.* [266]. Patients treated with ALC, compared with the placebo group, showed a stronger reduction of hyperactivity and improvement of social behavior [266]. Geier *et al.* [267] tested CARN in patients with ASD that were randomly assigned to receive a controlled regimen liquid CARN or placebo for 3 months. Treated patients showed significant improvements in Childhood Autism Rating Scale, modified clinical global impression, and Autism Treatment Evaluation Checklist scores.

Four clinical trials (of which one controlled trial) have tested the CARN, or ALC, aimed at improving different complications of HIV infection [97,268–270]. The above-cited report by Milazzo *et al.* [97] tested ALC *vs.* a combination of ALA and *N*-acetylcysteine (NAC) for a number of OS/MDF and other endpoints in HIV-1-infected patients with lipoatrophy, showing that either ALC, or ALA/NAC supplementation exerts a protective role on mitochondrial function in HIV-1-infected patients [97]. Two closely related studies [268,269] tested the effects of ALC in mitigating the neuropathy associated with antiretroviral toxicity in HIV-infected patients. Mean pain intensity score was significantly reduced, while electrophysiological parameters did not significantly change during treatment [269].

A controlled clinical trial was reported by Pignatelli *et al.* [271] on the effect of CARN on OS and platelet activation after major surgery, whose trauma is associated with an increased ROS production. At baseline and after treatment, OS was evaluated by detection of circulating levels of soluble NOX2-derived peptide, and by analyzing platelet ROS formation. The results showed an increase of OS endpoints in the placebo group compared with the baseline after the surgical intervention, while the CARN-treated group did not significantly differ from the baseline.

Patients with ulcerative colitis (UC) were treated by Mikhailova *et al.* [272] by PLC testing a clinical/endoscopic response. Patients with mild-to-moderate UC receiving stable oral aminosalicylate or thiopurine therapy were randomised to receive PLC or placebo. A significant response was found in patients receiving PLC as clinical/endoscopic response *vs.* those receiving placebo [272].

6. Treatments with Mitochondrial Nutrient Combinations

Out of 262 clinical trials reviewed here (Tables 1–3), only a tiny minority of seven reports referred to the combined use of two MN, ALA and/or CoQ10 and/or CARN [41,68,76,78,81,88,184], as shown in Table 4. Only four clinical trials tested the ALA/CoQ10 combination [41,76,81,88], while three clinical trials tested CARN combined with either ALA or CoQ10 [68,78,184]. Surprisingly, no retrieved report mentioned the combination of the three cofactors in any clinical trial; to the best of our knowledge; only Mattiazzi *et al.* [273] reported on the combined *in vitro* use of ALA, CoQ10 and CARN in improving a mitochondrial (mtDNA T8993G (NARP)) mutation on T8993G mutated cells.

Table 4. Clinical studies utilizing multiple mitochondrial nutrients (α-lipoic acid and/or coenzyme Q10 and/or L-carnitine) aimed at compensating OS/MDF-related pathogenetic mechanisms.

Diseases/Conditions	α-Lipoic Acid	Coenzyme Q10 (Daily Dose, mg/d)	L-Carnitine	References
Heart and vessel diseases	100	100	–	[81]
	400	–	1000	[78]
Mitochondrial diseases	600	240	–	[88]
Type 2 diabetes	100	60	–	[41]
Psoriasis	150	50	–	[76]
Sciatalgia	600	–	1180	[68]
Kidney diseases	–	100	500	[184]

The report by Rodriguez *et al.* [88] tested the combined use of ALA, CoQ10 and creatine in a group of patients with different mitochondrial diseases; these may be viewed as the "ideal" target for administration of MN, in view of supporting known MDF [6,7,9,14,274–276] that involve OS, decreased ATP production, and involvement of anaerobic energetic pathways. The combination therapy resulted in lower resting plasma lactate and OS markers (urinary 8-isoprostanes), and improved energetic performance [88].

Beyond mitochondrial diseases, a substantial body of evidence has been accumulated pointing to the involvement of OS/MDF in a number of heterogeneous diseases [9]..

The combined use of ALA and CoQ10 was tested in patients with Type 2 DM [41] and in cardiac surgery patients [81]. One may note that both the report by Palacka *et al.* [41] and that by Leong *et al.* [81] fulfilled the expected clinical outcomes, along with the finding of decreased plasma lipid peroxides [41].

A number of clinical trials have been conducted by testing MN in combinations with classical antioxidants and/or herbal compounds and/or disease-specific drugs.

A total of 27 reports were evaluated concerning the therapies utilizing ALA or CoQ10 combinations with various antioxidants and/or herbal compounds and/or drugs (Tables 5 and 6). As shown in Table 5, a study utilizing ALA and Vit E alone or in combination failed to find changes in the lipid profile or insulin sensitivity of Type 2 DM patients [44]. Supplementation of quercetin plus Vit C or ALA alone did not change the blood biomarkers of inflammation and disease severity of rheumatoid arthritis patients under conventional treatments [64].

A combined administration with ALA, CoQ10, magnesium orotate, ω-3 polyunsaturated fatty acids (ω-3 PUFA) and selenium was associated by Leong *et al.* [81] with improved redox status and reduced myocardial damage in patients undergoing heart surgery.

A clinical trial in Alzheimer's disease patients showed that supplementation with ALA, Vit E and Vit C reduced OS in the brain [65]. The beneficial effects of combination therapy were also observed in Parkinson's disease (PD) patients receiving homocysteine-lowering therapy (folate and Vit B) and ALA supplementation, suggesting that combination therapy may prevent bone loss in PD patients [277]. A combination therapy of ALA and transdermal testosterone provided amelioration of on erectile dysfunction and quality of life in patients with Type 2 DM [59]. Treatment with ALA and γ-linolenic acid was proposed for controlling symptoms and improving the evolution of carpal tunnel syndrome [67]. Antioxidant supplementation with Vit C, Vit E and ALA was reported as a safe, though ineffective, treatment for dementia in individuals with Down syndrome [86].

Combined supplementations were tested in liver disorders, such as chronic hepatitis virus C infection. Melhem *et al.* [72] reported on a combination of ALA with antioxidative preparations (glycyrrhizin, *Schisandra*, silymarin, Vit C, L-glutathione, and Vit E) in patients with chronic hepatitis C, with favorable response rate.

Twelve clinical trials have reported on combination regimens with CoQ10 and various antioxidants (Table 6). Witte *et al.* [112] tested a combination of CoQ10 and high-dose micronutrients that was found to improve left ventricular function and quality of life in elderly patients with chronic heart failure. Shargorodsky *et al.* [128] tested a combination therapy with CoQ10, Vit C, Vit E and selenium in patients with multiple cardiovascular risk factors, finding an improvement in glucose and lipid metabolism, and decreased blood pressure.

Table 5. Clinical studies utilizing mitochondrial nutrients and antioxidants and/or herbal compounds aimed at compensating OS/MDF-related pathogenetic mechanisms.

Diseases/Conditions	α-Lipoic Acid + Other Agent(s) (Daily Dose)	References
Type 1 and Type 2 diabetes	ALA 100 mg, CoQ10 60 mg, Vit E 200 mg	[41]
	ALA 600 mg, Vit E 800 mg	[44]
	ALA 800 mg, pyridoxine 80 mg	[56]
	ALA 600 mg, transdermal testosterone 50 mg	[59]
	ALA 2 × 600 mg, allopurinol 300 mg, nicotinamide 2 × 750 mg	[60]
Heart surgery	ALA 100 mg, CoQ10 100 mg, Mg orotate 400 mg, ω-3 PUFA 300 mg, Se 200 μg	[81]
Metabolic syndrome	ALA 600 mg, Vit E 100 IU	[77]
Cancer-related anorexia/cachexia	ALA 300 mg, polyphenols 400 mg, carbocysteine 2.7 g, Vit E 400 mg, Vit A 30,000 IU, Vit C 500 mg, (*n*-3)-PUFA 2 cans	[94]
Alzheimer disease	ALA 900 mg, CoQ10 400 mg, Vit C 500 mg, Vit E 800 IU	[65]
Parkinson's disease	ALA 1200 mg, folate 5 mg, Vit B12 1500 μg	[277]
Carpal tunnel syndrome	ALA 600 mg, γ-linolenic acid 360 mg, Vit B6 150 mg, Vit B1 100 mg, Vit B12 500 μg	[67]
Down syndrome	ALA 600 mg, Vit E 900 IU, Vit C 200 mg	[86]
HCV infection	ALA 300 mg, glycyrrhizin 1 g, *Schisandra* 1.5 g, silymarin 750 mg, Vit C 6 g, L-glutathione 300 mg, Vit E 800 IU	[72]
Psoriasis	ALA 600 mg, CoQ10 50 mg, resveratrol 20 mg, Vit E 36 mg, Krill oil 300 mg, Vitis vinifera seed oil 30 mg, Se 27 mg	[76]
Osteoporosis	ALA 2 × 300 mg, Vit C 30 mg, Vit E 5 mg, Se 2.75 mg	[99]
Vitiligo	ALA 100 mg, Vit C 100 mg, Vit E 40 mg, PUFA 12%	[98]

Table 6. Clinical studies utilizing mitochondrial nutrients and antioxidants and/or herbal compounds aimed at compensating OS/MDF-related pathogenetic mechanisms.

Diseases/Conditions	Coenzyme Q10 + Other Agent(s) (Daily Dose)	References
Heart surgery	CoQ10 100 mg, ALA 100 mg, Mg orotate 400 mg, ω-3 PUFA 300 mg, Se 200 μg	[81]
Chronic heart failure	CoQ10 150 mg, Ca 250 mg, Mg 150 mg, Zn 15 mg, Cu 1.2 mg, Se 50 μg, Vit A 800 μg, thiamine 200 mg, riboflavin 2 mg, Vit B_6 200 mg, folate 5 mg, Vit B_{12} 200 μg, Vit C 500 mg, Vit E 400 mg, Vit D 10 μg	[112,124]
Cardiovascular diseases	CoQ10 120 mg, Vit C 1 g, Vit E 400 IU, Se 200 μg	[128]
Cardiovascular mortality	CoQ10 200 mg, organic Se 200 μg	[136]
Breast cancer	CoQ10 100 mg, riboflavin 10 mg, niacin 50 mg	[175]
	CoQ10 90 mg, Vit C 2850 mg, Vit E 2500 IU, β-carotene 32.5 IU, Se 387 mg, γ-linolenic acid 1.2 g, *n*-3 fatty acids 3.5 g	[176]
	CoQ10 300 mg, Vit E 300 IU	[177]
Prostate cancer	CoQ10 200 mg, Vit C 750 mg, Vit E 350 mg, Se 200 μg	[179]
Friedreich ataxia	CoQ10 400 mg, Vit E 2100 IU	[140]
	CoQ10 600 mg, Vit E 2100 IU	[141]
Fibromyalgia	CoQ10 200 mg, Gingko biloba extract 200 mg	[160]

A clinical trial utilizing CoQ10 and a mixture of antioxidants was reported by Hertz and Lister [175] to improve survival of patients with end-stage cancer. An early report by Lockwood *et al.* [176] found that supplementation with CoQ10 and antioxidants, γ-linolenic and *n*-3 fatty acids significantly improved clinical conditions in breast cancer patients. A trial by Premkumar *et al.* [177], also conducted on breast cancer patients, reported that a supplementation with CoQ10, riboflavin and niacin decreased the levels of pro-angiogenic factors and increased the levels of anti-angiogenic factors, and reduced tumor burden. On the other hand, a combination therapy with CoQ10 and Vit E did not result in improvement of fatigue symptoms in breast cancer patients, according to Lesser *et al.* [178]. Negative results were obtained by Hoenjet *et al.* [179] by testing a combined supplement containing CoQ10, Vit E, Vit C and selenium in patients with hormonally untreated carcinoma of the prostate, which did not affect serum level of PSA or hormone levels.

Two studies by Hart and co-workers [140,141] reported on a combined treatment with CoQ10 and Vit E in patients with Friedreich ataxia finding sustained improvement in mitochondrial energy production and a significant improvement in cardiac function. A combination therapy with CoQ10 and *Ginkgo biloba* extract significantly improved quality-of-life in patients with clinically diagnosed fibromyalgia [160].

Taken together, the literature on clinical trials testing MN and "classical" antioxidants may provide crucial data on the therapeutic efficacy of compounds aimed at improving mitochondrial health and several OS/MDF-related endpoints, suggesting the design of improved treatments of OS/MDF-related diseases.

7. State-of-Art: Critical Remarks

The present survey of published clinical trials testing MN may reflect the current state-of-art across broad-ranging disorders that are usually focused in separate medical disciplines. On the other hand, the clinical trials testing MN published to date have been conducted on some selected diseases, far below the potential scope of investigation, by considering the extensive range of OS/MDF-related disorders [9]. Trying to interpret this delay, one might find one or more of the following explanations: (a) a disease is commonly recognized to rely on other etiologic grounds than OS/MDF: this is the case, e.g., of Fanconi anemia and of other genetic diseases [8,278]; (b) very rare diseases discourage patient recruitment in view of clinical trials as, e.g., progerias [8,279]; (c) the specialist community may be committed in established therapeutic strategies, thus disregarding potential adjuvant interventions.

Other two remarks may be addressed to the state-of-art in the literature of clinical trials using MN. First, the reviewed studies were mostly oriented to clinical endpoints of the investigated disorders, by underscoring any relevant mechanistic aspects enabling to elucidate trial's results. Second, and most relevant, the vast majority of the reviewed trials tested one MN only, by disregarding: (a) the complementary roles of these cofactors in mitochondrial functions; (b) their different physico-chemical behaviors in water- *vs.* lipid-phase; and (c) the scarce, if any, knowledge as to specific MDF in a given disorder and the appropriate requirements of ALA, or CoQ10, or CARN (or its derivatives) supplementation. Thus, the as yet prevailing design of trials using only one MN may fail to achieve the desired outcomes of OS/MDF compensation and hence of improving clinical endpoints. This limitation, along with some possible shortage in dosages and/or in trial duration might have contributed to some unsuccessful trial outcomes. One may note that supplement combinations of more than one MN, or the use of ALA or CoQ10 with classical antioxidants may have contributed to the overall success of these trials (see Tables 4–6). Thus, one may anticipate the optimization of future clinical trials in OS/MDF-related disorders by means of combined MN treatments, possibly associated with "classical" antioxidants.

The choice of MN formulations may have critical implications on the expected success of MN-based clinical trials. An outstanding aspect in the design of a MN-based clinical trial is represented by the type of formulation chosen. The oxidative state of molecules, chirality, and the vehiculants used in the formulation may have dramatic implications in terms of bioavailability and interaction with the recipient biological system. In the case of CoQ10, bioavailability is known to be influenced by modality of administration, with higher plasma levels reached by administration of the same amount divided in multiple doses, and by vehiculant type with soft gels containing triglyceride dispersing medium showing a better bioavailability in comparison with the crystalline form [280]. Another issue, also linked with bioavailability, is the oxidative state of the active compound in the formulation. Most of the studies cited used ubiquinone, however recently also the reduced form of CoQ10 (ubiquinol) has been used in several clinical trials [113,114,198]. Direct use of ubiquinol displays the advantages of improved absorption and consequently enhanced bioavailability, and direct exposure of the reduced (and active) antioxidant form to digestive epithelial tissue; finally, ubiquinol may influence the ubiquinone/total CoQ10 ratio in conditions where reductive systems are sub-optimal as it has been shown in the aging process [281].

α-Lipoic acid is characterized by an enhanced bioavailability compared to CoQ10, with absorption rates around 30%–40% of an oral dose of ALA [282]. Moreover, while ALA being synthethized by biological systems is exclusively in the *R*-lipoic acid chiral form, formulations are often (*R*/*S*) or (+/−) ALA, composed by a 50/50 mixture of *R*- and *S*-ALA. It has been shown that bioavailability of the two isomers is different, *R*-lipoic acid showing peak plasma concentrations in pharmacokinetic studies 40%–50% higher than *S*-LA, suggesting that *R*-LA is better absorbed than *S*-LA [282].

Regarding CARN and its derivatives, especially ALC, the state-of-art provides extensive information as to the efficacy of CARN, ALC and PLC. The advantages of using ALC *vs.* CARN have been provided [200–203], along with successful use of ALC [68,78,222,223,226,231,246–248,252–254,261,266–268] or PLC [210,225,235,239,253,272] in a number of clinical trials. Acetyl-L-carnitine probably acts as an acetylating agent and its involvement in epigenetic modifications of proteins might explain its success in aging and in OS/MDF-related disorders [283–285]. Future research on CARN/ALC/PLC supplementation should address the correlation of supplement dosage, changes and maintenance of tissue concentrations, and metabolic and functional changes and outcomes [286,287]. An analogous research strategy was highlighted by Camp *et al.* [288] on nutritional interventions for inborn errors of metabolism, suggesting combined strategies in improving the clinical management of an extensive number of disorders.

8. Prospects of Clinical Trials in OS/MDF-Related Disorders

The scheme depicted in Figure 1 points to the prerequisite of identifying the OS and MDF endpoints that are altered in a given disorder, often incompletely elucidated. A choice of OS endpoints should include some established parameters including oxidative damage to DNA, lipids, carbohydrates and proteins, and changes in the glutathione system. A few selected MDF endpoints should be addressed to verify the deficiency, if any, of ALA (to date almost unexplored), CoQ10, and CARN. Other relevant endpoints in assessing an *in vivo* MDF should include mitochondrial DNA content, ATP production, OXPHOS activities, lactate/pyruvate ratio, and expression of mitochondrial antioxidant activities (MnSOD and peroxiredoxin 3). This set of measurements should clarify the quantitative aspects of the disorder to be focused in a clinical trial design. Once the characterization of OS/MDF endpoints is accomplished, the target outcomes of the clinical trials may be defined.

Figure 1. Preliminary database to be achieved in design of mitochondrial nutrients (MN)-based clinical trials.

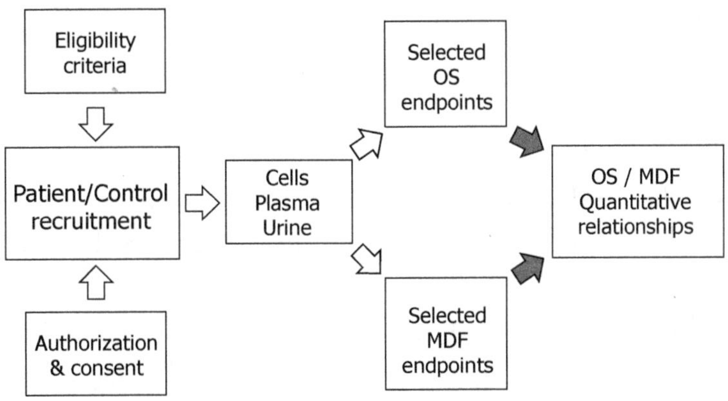

As shown in Figure 2, the OS/MDF endpoints for a given disease, found as abnormal in the preliminary step, shall be the grounds for choosing the best appropriate supplements to be administered to patients. These supplements may include one or—more reasonably—more MN, as discussed above. The follow-up period will be of paramount relevance; indeed, short admnistration regimens may offer hints for efficacy and safety, yet they may conceal long-run effects. Moreover, in diseases with multi-year-long progression, expecting the desired outcomes may be a quite elusive task. Thus, follow-up ought to be designed for a reasonably long duration with, e.g., six-month intervals in results analysis, allowing us to monitor the outcomes of multi-year-long administration. The ultimate goal will be achieved by compensating OS/MDF endpoints, along with the amelioration of disease progression.

Figure 2. Follow-up procedures combining evaluation of MDF and OS endpoints, along with disease specific clinical endpoints.

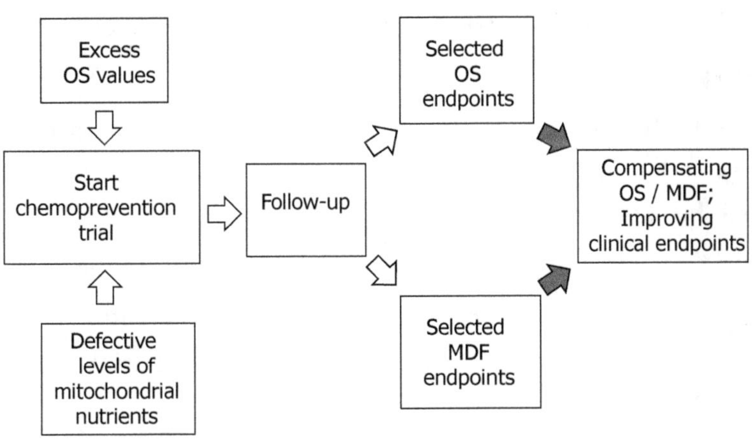

Materializing the above prospect may have dramatic consequences in the success of to-be clinical trials for OS/MDF-related disorders, trespassing from "hopes", or from single-parameter information, toward a rational approach that shall foresee properly targeted interventions, then with rational expectations, rather than attempts.

Acknowledgments

Sandra Petrović was supported by the Ministry of Education, Science and Technological Development of the Republic of Serbia (Grant No. 173046).

References

1. Richter, C.C.; Kass, G.E. Oxidative stress in mitochondria: its relationship to cellular Ca^{2+} homeostasis, cell death, proliferation, and differentiation. *Chem. Biol. Interact.* **1991**, *77*, 1–23.
2. Sohal, R.S.; Brunk, U.T. Mitochondrial production of pro-oxidants and cellular senescence. *Mutat. Res./DNAging* **1992**, *275*, 295–304.
3. Di Monte, D.A.; Chan, P.; Sandy M.S. Glutathione in Parkinson's disease: A link between oxidative stress and mitochondrial damage? *Ann. Neurol.* **1992**, *32*, S111–S115.
4. Beal, M.F. Therapeutic approaches to mitochondrial dysfunction in Parkinson's disease. *Parkinsonism Relat. Disord.* **2009**, *15*, S189–S194.
5. Mischley, L.K.; Allen, J.; Bradley, R. Coenzyme Q10 deficiency in patients with Parkinson's disease. *J. Neurol. Sci.* **2012**, *318*, 72–75.
6. Parikh, S.; Goldstein, A.; Koenig, M.K.; Scaglia, F.; Enns, G.M.; Saneto, R.; Mitochondrial Medicine Society Clinical Directors Working Group; Clinical Director's Work Group. Practice patterns of mitochondrial disease physicians in North America. Part 2: Treatment, care and management. *Mitochondrion* **2013**, *13*, 681–687.
7. Tarnopolsky, M.A. The mitochondrial cocktail: Rationale for combined nutraceutical therapy in mitochondrial cytopathies. *Adv. Drug Deliv. Rev.* **2008**, *60*, 1561–1567.
8. Pallardó, F.V.; Lloret, A.; Lebel, M.; d'Ischia, M.; Cogger, V.C.; le Couteur D.G.; Gadaleta, M.N.; Castello, G.; Pagano, G. Mitochondrial dysfunction in some oxidative stress-related genetic diseases: Ataxia-Telangiectasia, Down Syndrome, Fanconi Anaemia and Werner Syndrome. *Biogerontology* **2010**, *11*, 401–419.
9. Pagano, G.; Aiello Talamanca, A.; Castello, G.; Cordero, M.D.; d'Ischia, M.; Gadaleta, M.N.; Pallardó, F.V.; Petrović, S.; Tiano, L.; Zatterale, A. Oxidative stress and mitochondrial dysfunction across broad-ranging pathologies: Toward a rational design of chemoprevention strategies by means of mitochondrial nutrients. *Oxid. Med. Cell. Longev.* **2014**, *2014*, 541230.
10. Marshall, A.W.; Graul, R.S.; Morgan, M.Y.; Sherlock, S. Treatment of alcohol-related liver disease with thioctic acid: A six month randomised double-blind trial. *Gut* **1982**, *23*, 1088–1093.

11. Tanaka, J.; Tominaga, R.; Yoshitoshi, M.; Matsui, K.; Komori, M.; Sese, A.; Yasui, H.; Tokunaga, K. Coenzyme Q10: The prophylactic effect on low cardiac output following cardiac valve replacement. *Ann. Thorac. Surg.* **1982**, *33*, 145–151.
12. Langsjoen, P.H.; Vadhanavikit, S.; Folkers, K. Response of patients in classes III and IV of cardiomyopathy to therapy in a blind and crossover trial with coenzyme Q10. *Proc. Natl. Acad. Sci. USA* **1985**, *82*, 4240–4244.
13. Ogasahara, S.; Nishikawa, Y.; Yorifuji, S.; Soga, F.; Nakamura, Y.; Takahashi, M.; Hashimoto, S.; Kono, N.; Tarui, S. Treatment of Kearns–Sayre syndrome with coenzyme Q10. *Neurology* **1986**, *36*, 45–53.
14. Scarlato, G.N.; Bresolin, N.I.; Moroni, I.; Doriguzzi, C.; Castelli, E.; Comi, G.; Angelini, C.; Carenzi, A. Multicenter trial with ubidecarenone: Treatment of 44 patients with mitochondrial myopathies. *Rev. Neurol.* **1991**, *147*, 542–548.
15. Mayr, J.A.; Zimmermann, F.A.; Fauth, C.; Bergheim, C.; Meierhofer, D.; Radmayr, D.; Zschocke, J.; Koch, J.; Sperl, W. Lipoic acid synthetase deficiency causes neonatal-onset epilepsy, defective mitochondrial energy metabolism, and glycine elevation. *Am. J. Hum. Genet.* **2011**, *89*, 792–797.
16. Liu, J. The effects and mechanisms of mitochondrial nutrient alpha-lipoic acid on improving age-associated mitochondrial and cognitive dysfunction: An overview. *Neurochem. Res.* **2008**, *33*, 194–203.
17. Maes, M.; Mihaylova, I.; Kubera, M.; Uytterhoeven, M.; Vrydags, N.; Bosmans, E. Coenzyme Q10 deficiency in myalgic encephalomyelitis/chronic fatigue syndrome (ME/CFS) is related to fatigue, autonomic and neurocognitive symptoms and is another risk factor explaining the early mortality in ME/CFS due to cardiovascular disorder. *Neuro Endocrinol. Lett.* **2009**, *30*, 470–476.
18. Tiano, L.; Busciglio, J. Mitochondrial dysfunction and Down's syndrome: Is there a role for coenzyme Q10? *Biofactors* **2011**, *37*, 386–392.
19. Hardas, S.S.; Sultana, R.; Clark, A.M.; Beckett, T.L.; Szweda, L.I.; Murphy, M.P.; Butterfield, D.A. Oxidative modification of lipoic acid by HNE in Alzheimer disease brain. *Redox Biol.* **2013**, *1*, 80–85.
20. Kähler, W.; Kuklinski, B.; Rühlmann, C.; Plötz, C. Diabetes mellitus—A free radical-associated disease. Results of adjuvant antioxidant supplementation. *Z. Gesamte Inn. Med.* **1993**, *48*, 223–232.
21. Jacob, S.; Henriksen, E.J.; Schiemann, A.L.; Simon, I.; Clancy, D.E.; Tritschler, H.J.; Jung, W.I.; Augustin, H.J.; Dietze, G.J. Enhancement of glucose disposal in patients with type 2 diabetes by alpha-lipoic acid. *Arzneimittelforschung* **1995**, *45*, 872–874.
22. Ziegler, D.; Hanefeld, M.; Ruhnau, K.J.; Meissner, H.P.; Lobisch, M.; Schütte, K.; Gries, F.A. Treatment of symptomatic diabetic peripheral neuropathy with the anti-oxidant alpha-lipoic acid. A 3-week multicentre randomized controlled trial (ALADIN Study). *Diabetologia* **1995**, *38*, 1425–1433.
23. Ziegler, D.; Schatz, H.; Conrad, F.; Gries, F.A.; Ulrich, H.; Reichel, G. Effects of treatment with the antioxidant alpha-lipoic acid on cardiac autonomic neuropathy in NIDDM patients. A 4-month randomized controlled multicenter trial (DEKAN Study). Deutsche Kardiale Autonome Neuropathie. *Diabetes Care* **1997**, *20*, 369–373.

24. Strokov, I.A.; Kozlova, N.A.; Mozolevskiĭ, IuV; Miasoedov, S.P.; Iakhno, N.N. The efficacy of the intravenous administration of the trometamol salt of thioctic (alpha-lipoic) acid in diabetic neuropathy. *Zh. Nevrol. Psikhiatr. Im. S. S. Korsakova* **1999**, *99*, 18–22.
25. Jacob, S.P.; Ruus, P.R.; Hermann, R.; Tritschler, H.J.; Maerker, E.; Renn, W.; Augustin, H.J.; Dietze, G.J.; Rett, K. Oral administration of RAC-alpha-lipoic acid modulates insulin sensitivity in patients with type-2 diabetes mellitus: A placebo-controlled pilot trial. *Free Radic. Biol. Med.* **1999**, *27*, 309–314.
26. Ziegler, D.; Hanefeld, M.; Ruhnau, K.J.; Hasche, H.; Lobisch, M.; Schütte, K.; Kerum, G.; Malessa, R. Treatment of symptomatic diabetic polyneuropathy with the antioxidant alpha-lipoic acid: A 7-month multicenter randomized controlled trial (ALADIN III study). ALADIN III Study Group. Alpha-lipoic acid in diabetic neuropathy. *Diabetes Care* **1999**, *22*, 1296–1301.
27. Reljanovic, M.G.; Reichel, G.K.; Rett, K.; Lobisch, M.; Schuette, K.; Möller,W.; Tritschler, H.J.; Mehnert, H. Treatment of diabetic polyneuropathy with the antioxidant thioctic acid (alpha-lipoic acid): A two year multicenter randomized double-blind placebo-controlled trial (ALADIN II). Alpha lipoic acid in diabetic neuropathy. *Free Radic. Res.* **1999**, *31*, 171–179.
28. Ruhnau, K.J.; Meissner, H.P.; Finn, J.R.; Reljanovic, M.; Lobisch, M.; Schütte, K.; Nehrdich, D.; Tritschler, H.J.; Mehnert, H.; Ziegler, D. Effects of 3-week oral treatment with the antioxidant thioctic acid (alpha-lipoic acid) in symptomatic diabetic polyneuropathy. *Diabet. Med.* **1999**, *16*, 1040–1043.
29. Negrişanu, G.; Roşu, M.; Bolte, B.; Lefter, D.; Dabelea, D. Effects of 3-month treatment with the antioxidant alpha-lipoic acid in diabetic peripheral neuropathy. *Rom. J. Intern. Med.* **1999**, *37*, 297–306.
30. Androne, L.; Gavan, N.A.; Veresiu, I.A.; Orasan, R. *In vivo* effect of lipoic acid on lipid peroxidation in patients with diabetic neuropathy. *In Vivo* **2000**, *14*, 327–330.
31. Haak, E., Usadel, K.H.; Kusterer, K.; Amini, P.; Frommeyer, R.; Tritschler, H.J.; Haak, T. Effects of alpha-lipoic acid on microcirculation in patients with peripheral diabetic neuropathy. *Exp. Clin. Endocrinol. Diabetes* **2000**, *108*, 168–174.
32. Evans, J.L.; Heymann, C.J.; Goldfine, I.D.; Gavin, L.A. Pharmacokinetics, tolerability, and fructosamine-lowering effect of a novel, controlled-release formulation of alpha-lipoic acid. *Endocr. Pract.* **2002**, *8*, 29–35.
33. Ametov, A.S.; Barinov, A.; Dyck, P.J.; Hermann, R.; Kozlova, N.; Litchy, W.J.; Low, P.A.; Nehrdich, D.; Novosadova, M.; O'Brien, P.C.; *et al.* The sensory symptoms of diabetic polyneuropathy are improved with alpha-lipoic acid: The SYDNEY trial. *Diabetes Care* **2003**, *26*, 770–776.
34. Hahm, J.R.; Kim, B.J.; Kim, K.W. Clinical experience with thioctacid (thioctic acid) in the treatment of distal symmetric polyneuropathy in Korean diabetic patients. *J. Diabetes Complicat.* **2004**, *18*, 79–85.
35. Ziegler, D.; Ametov, A.; Barinov, A.; Dyck, P.J.; Gurieva, I.; Low, P.A.; Munzel, U.; Yakhno, N.; Raz, I.; Novosadova, M.; *et al.* Oral treatment with alpha-lipoic acid improves symptomatic diabetic polyneuropathy: The SYDNEY 2 trial. *Diabetes Care* **2006**, *29*, 2365–2370.
36. Kamenova P. Improvement of insulin sensitivity in patients with type 2 diabetes mellitus after oral administration of alpha-lipoic acid. *Hormones* **2006**, *5*, 251–258.

37. Liu, F.; Zhang, Y.; Yang, M.; Liu, B.; Shen, Y.D.; Jia, W.P.; Xiang, K.S. Curative effect of alpha-lipoic acid on peripheral neuropathy in type 2 diabetes: A clinical study. *Zhonghua Yi Xue Za Zhi* **2007**, *87*, 2706–2709.
38. Xiang, G.D.; Sun, H.L.; Zhao, L.S.; Hou, J.; Yue, L.; Xu, L. The antioxidant alpha-lipoic acid improves endothelial dysfunction induced by acute hyperglycaemia during OGTT in impaired glucose tolerance. *Clin. Endocrinol.* **2008**, *68*, 716–723.
39. Bureković, A.; Terzić, M.; Alajbegović, S.; Vukojević, Z.; Hadzić, N. The role of alpha-lipoic acid in diabetic polyneuropathy treatment. *Bosn. J. Basic Med. Sci.* **2008**, *8*, 341–245.
40. Heinisch, B.B.; Francesconi, M.; Mittermayer, F.; Schaller, G.; Gouya, G.; Wolzt, M.; Pleiner, J. Alpha-lipoic acid improves vascular endothelial function in patients with type 2 diabetes: A placebo-controlled randomized trial. *Eur. J. Clin. Investig.* **2010**, *40*, 148–154.
41. Palacka, P.; Kucharska, J.; Murin, J.; Dostalova, K.; Okkelova, A.; Cizova, M.; Waczulikova, I.; Moricova, S.; Gvozdjakova, A. Complementary therapy in diabetic patients with chronic complications: A pilot study. *Bratisl. Lek. Listy* **2010**, *111*, 205–211.
42. Gu, X.M.; Zhang, S.S.; Wu, J.C.; Tang, Z.Y.; Lu, Z.Q.; Li, H.; Liu, C.; Chen, L.; Ning, G. Efficacy and safety of high-dose α-lipoic acid in the treatment of diabetic polyneuropathy. *Zhonghua Yi Xue Za Zhi* **2010**, *90*, 2473–2476.
43. Xiang, G.; Pu, J.; Yue, L.; Hou, J.; Sun, H. α-Lipoic acid can improve endothelial dysfunction in subjects with impaired fasting glucose. *Metabolism* **2011**, *60*, 480–485.
44. De Oliveira, A.M.; Rondó, P.H.; Luzia, L.A.; D'Abronzo, F.H.; Illison, V.K. The effects of lipoic acid and α-tocopherol supplementation on the lipid profile and insulin sensitivity of patients with type 2 diabetes mellitus: A randomized, double-blind, placebo-controlled trial. *Diabetes Res. Clin. Pract.* **2011**, *92*, 253–260.
45. Ansar, H.; Mazloom, Z.; Kazemi, F.; Hejazi, N. Effect of alpha-lipoic acid on blood glucose, insulin resistance and glutathione peroxidase of type 2 diabetic patients. *Saudi Med. J.* **2011**, *32*, 584–588.
46. Ziegler, D.; Low, P.A.; Litchy, W.J.; Boulton A.J.; Vinik, A.I.; Freeman, R.; Samigull, R.; Tritschler, H.; Munzel, U.; Maus, J.; *et al.* Efficacy and safety of antioxidant treatment with α-lipoic acid over 4 years in diabetic polyneuropathy: The NATHAN 1 trial. *Diabetes Care* **2011**, *34*, 2054–2060.
47. Haritoglou, C.; Gerss, J.; Hammes, H.P.; Kampik, A.; Ulbig, M.W.; RETIPON Study Group. Alpha-lipoic acid for the prevention of diabetic macular edema. *Ophthalmologica* **2011**, *226*, 127–137.
48. Rahman, S.T.; Merchant, N.; Haque, T.; Wahi, J.; Bhaheetharan, S.; Ferdinand, K.C.; Khan, B.V. The impact of lipoic acid on endothelial function and proteinuria in quinapril-treated diabetic patients with stage I hypertension: Results from the QUALITY study. *J. Cardiovasc. Pharmacol. Ther.* **2012**, *17*, 139–145.
49. Bertolotto, F.; Massone, A. Combination of alpha lipoic acid and superoxide dismutase leads to physiological and symptomatic improvements in diabetic neuropathy. *Drugs R&D* **2012**, *12*, 29–34.

50. Porasuphatana, S.; Suddee, S.; Nartnampong, A.; Konsil, J.; Harnwong, B.; Santaweesuk, A. Glycemic and oxidative status of patients with type 2 diabetes mellitus following oral administration of alpha-lipoic acid: A randomized double-blinded placebo-controlled study. *Asia Pac. J. Clin. Nutr.* **2012**, *21*, 12–21.
51. Ibrahimpasic, K. Alpha lipoic acid and glycaemic control in diabetic neuropathies at type 2 diabetes treatment. *Med. Arch.* **2013**, *67*, 7–9.
52. Tankova, T.; Koev, D.; Dakovska, L. Alpha-lipoic acid in the treatment of autonomic diabetic neuropathy (controlled, randomized, open-label study). *Rom. J. Intern. Med.* **2004**, *42*, 457–464.
53. Huang, E.A.; Gitelman, S.E. The effect of oral alpha-lipoic acid on oxidative stress in adolescents with type 1 diabetes mellitus. *Pediat. Diabetes* **2008**, *9*, 69–73.
54. Mollo, R.; Zaccardi, F.; Scalone, G.; Scavone, G.; Rizzo, P.; Navarese, E.P.; Manto, A.; Pitocco, D.; Lanza, G.A.; Ghirlanda, G.; *et al.* Effect of α-lipoic acid on platelet reactivity in type 1 diabetic patients. *Diabetes Care* **2012**, *35*, 196–197.
55. Morcos, M.; Borcea, V.; Isermann, B.; Gehrke, S.; Ehret, T.; Henkels, M.; Schiekofer, S.; Hofmann, M.; Amiral, J.; Tritschler, H.; *et al.* Effect of alpha-lipoic acid on the progression of endothelial cell damage and albuminuria in patients with diabetes mellitus: An exploratory study. *Diabetes Res. Clin. Pract.* **2001**, *5*, 175–183.
56. Noori, N.; Tabibi, H.; Hosseinpanah, F.; Hedayati, M.; Nafar, M. Effects of combined lipoic acid and pyridoxine on albuminuria, advanced glycation end-products, and blood pressure in diabetic nephropathy. *Int. J. Vitam. Nutr. Res.* **2013**, *83*, 77–85.
57. Hegazy, S.K.; Tolba, O.A.; Mostafa, T.M.; Eid, M.A.; el-Afify, D.R. Alpha-lipoic acid improves subclinical left ventricular dysfunction in asymptomatic patients with type 1 diabetes. *Rev. Diabet. Stud.* **2013**, *10*, 58–67.
58. Huang, Z.; Wan, X.; Liu, J.; Deng, W.; Chen, A.; Liu, L.; Liu, J.; Wei, G.; Li, H.; Fang, D.; *et al.* Short-term continuous subcutaneous insulin infusion combined with insulin sensitizers rosiglitazone, metformin, or antioxidant α-lipoic acid in patients with newly diagnosed type 2 diabetes mellitus. *Diabetes Technol. Ther.* **2013**, *15*, 859–869.
59. Mitkov, M.D.; Aleksandrova, I.Y.; Orbetzova, M.M. Effect of transdermal testosterone or alpha-lipoic acid on erectile dysfunction and quality of life in patients with type 2 diabetes mellitus. *Folia Med.* **2013**, *55*, 55–63.
60. Pop-Busui, R.; Stevens, M.J.; Raffel, D.M.; White, E.A.; Mehta, M.; Plunkett, C.D.; Brown, M.B.; Feldman, E.L. Effects of triple antioxidant therapy on measures of cardiovascular autonomic neuropathy and on myocardial blood flow in type 1 diabetes: A randomized controlled trial. *Diabetologia* **2013**, *56*, 1835–1844.
61. Zhang, X.; Zhang, Y.; Gao, X.; Wu, J.; Jiao, X.; Zhao, J.; Lv, X. Investigating the role of backward walking therapy in alleviating plantar pressure of patients with diabetic peripheral neuropathy. *Arch. Phys. Med. Rehabil.* **2014**, *95*, 832–839.
62. Yadav, V.; Marracci, G.; Lovera, J.; Woodward, W.; Bogardus, K.; Marquardt, W.; Shinto, L.; Morris, C.; Bourdette, D. Lipoic acid in multiple sclerosis: A pilot study. *Mult. Sclerosis* **2005**. *11*, 159–165.

63. Khalili, M.; Azimi, A.; Izadi, V.; Eghtesadi, S.; Mirshafiey, A.; Sahraian, M.A.; Motevalian, A.; Norouzi, A.; Sanoobar, M.; Eskandari, G.; *et al.* Does lipoic acid consumption affect the cytokine profile in multiple sclerosis patients: A double-blind, placebo-controlled, randomized clinical trial. *Neuroimmunomodulation* **2014**, *21*, 291–296.

64. Bae, S.C.; Jung, W.J.; Lee, E.J.; Yu, R.; Sung, M.K. Effects of antioxidant supplements intervention on the level of plasma inflammatory molecules and disease severity of rheumatoid arthritis patients. *J. Am. Coll. Nutr.* **2009**, *28*, 56–62.

65. Galasko, D.R.; Peskind, E.; Clark, C.M.; Quinn, J.F.; Ringman, J.M.; Jicha, G.A.; Cotman, C.; Cottrell, B.; Montine, T.J.; Thomas, R.G.; *et al.* Alzheimer's disease cooperative study. Antioxidants for Alzheimer disease: A randomized clinical trial with cerebrospinal fluid biomarker measures. *Arch. Neurol.* **2012**, *69*, 836–841.

66. Magis, D.; Ambrosini, A.; Sándor, P.; Jacquy, J.; Laloux, P.; Schoenen, J. A randomized double-blind placebo-controlled trial of thioctic acid in migraine prophylaxis. *Headache* **2007**, *47*, 52–57.

67. Di Geronimo, G.; Caccese, A.F.; Caruso, L.; Soldati, A.; Passaretti, U. Treatment of carpal tunnel syndrome with alpha-lipoic acid. *Eur. Rev. Med. Pharmacol. Sci.* **2009**, *13*, 133–139.

68. Memeo, A.; Loiero, M. Thioctic acid and acetyl-L-carnitine in the treatment of sciatic pain caused by a herniated disc: A randomized, double-blind, comparative study. *Clin. Drug Investig.* **2008**, *28*, 495–500.

69. Ranieri, M.; Sciuscio, M.; Cortese, A.M.; Santamato, A.; di Teo, L.; Ianieri, G.; Bellomo, R.G., Stasi, M.; Megna, M. The use of alpha-lipoic acid (ALA), gamma linolenic acid (GLA) and rehabilitation in the treatment of back pain: Effect on health-related quality of life. *Int. J. Immunopathol. Pharmacol.* **2009**, *22*, 45–50.

70. Shinto, L.; Quinn, J.; Montine, T.; Dodge, H.H.; Woodward, W.; Baldauf-Wagner, S.; Waichunas, D.; Bumgarner, L.; Bourdette, D.; Silbert, L.; *et al.* A randomized placebo-controlled pilot trial of omega-3 fatty acids and alpha lipoic acid in Alzheimer's disease. *J. Alzheimers Dis.* **2014**, *38*, 111–120.

71. Dünschede, F.; Erbes, K.; Kircher, A.; Westermann, S.; Seifert. J.; Schad, A.; Oliver. K.; Kiemer, A.K.; Theodor, J. Reduction of ischemia reperfusion injury after liver resection and hepatic inflow occlusion by alpha-lipoic acid in humans. *World J. Gastroenterol.* **2006**, *12*, 6812–6817.

72. Melhem, A.; Stern, M.; Shibolet, O.; Israeli, E.; Ackerman, Z.; Pappo, O.; Hemed, N.; Rowe, M.; Ohana, H.; Zabrecky, G.; *et al.* Treatment of chronic hepatitis C virus infection via antioxidants: Results of a phase I clinical trial. *J. Clin. Gastroenterol.* **2005**, *39*, 737–742.

73. Koh, E.H.; Lee, W.J.; Lee, S.A.; Kim, E.H.; Cho, E.H.; Jeong, E.; Kim, D.W.; Kim, M.S.; Park, J.Y.; Park, K.G.; *et al.* Effects of alpha-lipoic acid on body weight in obese subjects. *Am. J. Med.* **2011**, *124*, 85.e1-8.

74. McNeilly, A.M.; Davison, G.W.; Murphy, M.H.; Nadeem, N.; Trinick, T.; Duly, E.; Novials, A.; McEneny, J. Effect of α-lipoic acid and exercise training on cardiovascular disease risk in obesity with impaired glucose tolerance. *Lipids Health Dis.* **2011**, *10*, 217.

75. Zhang, Y.; Han, P.; Wu, N.; He, B.; Lu, Y.; Li, S.; Liu, Y.; Zhao, S.; Liu, L.; Li, Y. Amelioration of lipid abnormalities by α-lipoic acid through antioxidative and anti-inflammatory effects. *Obesity* **2011**, *19*, 1647–1653.
76. Skroza, N.; Proietti, I.; Bernardini, N.; la Viola, G.; Nicolucci, F.; Pampena, R.; Tolino, E.; Zuber, S.; Mancini, M.T.; Soccodato, V.; *et al.* Efficacy of food supplement to improve metabolic syndrome parameters in patients affected by moderate to severe psoriasis during anti-TNFα treatment. *G. Ital. Dermatol. Venereol.* **2013**, *148*, 661–665.
77. Manning, P.J.; Sutherland, W.H.; Williams, S.M.; Walker, R.J.; Berry, E.A.; de Jong, S.A.; Ryalls, A.R. The effect of lipoic acid and vitamin E therapies in individuals with the metabolic syndrome. *Nutr. Metab. Cardiovasc. Dis.* **2013**, *23*, 543–549.
78. McMackin, C.J.; Widlansky, M.E.; Hamburg, N.M. Effect of combined treatment with alpha-lipoic acid and acetyl-L-carnitine on vascular function and blood pressure in patients with coronary artery disease. *J. Clin. Hypertens.* **2007**, *9*, 249–255.
79. Li, R.J.; Ji, W.Q.; Pang, J.J.; Wang, J.L.; Chen, Y.G.; Zhang, Y. Alpha-lipoic acid ameliorates oxidative stress by increasing aldehyde dehydrogenase-2 activity in patients with acute coronary syndrome. *Tohoku J. Exp. Med.* **2013**, *229*, 45–51.
80. Vincent, H.K.; Bourguignon, C.M.; Vincent, K.R.; Taylor, A.G. Effects of alpha-lipoic acid supplementation in peripheral arterial disease: A pilot study. *J. Altern. Complement. Med.* **2007**, *13*, 577–584.
81. Leong, J.Y.; van der Merwe, J.; Pepe, S.; Bailey, M.; Perkins, A.; Lymbury, R.; Esmore, D.; Marasco, S.; Rosenfeldt, F. Perioperative metabolic therapy improves redox status and outcomes in cardiac surgery patients: A randomised trial. *Heart Lung Circ.* **2010**, *19*, 584–591.
82. Chang, J.W.; Lee, E.K.; Kim, T.H.; Min, W.K.; Chun, S.; Lee, K.U.; Kim, S.B.; Park, J.S. Effects of alpha-lipoic acid on the plasma levels of asymmetric dimethylarginine in diabetic end-stage renal disease patients on hemodialysis: A pilot study. *Am. J. Nephrol.* **2007**, *27*, 70–74.
83. Ramos, L.F.; Kane, J.; McMonagle, E.; Le, P.; Wu, P.; Shintani, A.; Ikizler, T.A.; Himmelfarb, J. Effects of combination tocopherols and alpha lipoic acid therapy on oxidative stress and inflammatory biomarkers in chronic kidney disease. *J. Ren. Nutr.* **2011**, *21*, 211–218.
84. Khabbazi, T.; Mahdavi, R.; Safa, J.; Pour-Abdollahi, P. Effects of alpha-lipoic acid supplementation on inflammation, oxidative stress, and serum lipid profile levels in patients with end-stage renal disease on hemodialysis. *J. Ren. Nutr.* **2012**, *22*, 244–250.
85. Himmelfarb, J.; Ikizler, T.A.; Ellis, C.; Wu, P.; Shintani, A.; Dalal, S.; Kaplan, M.; Chonchol, M.; Hakim, R.M. Provision of antioxidant therapy in hemodialysis (PATH): A randomized clinical trial. *J. Am. Soc. Nephrol.* **2014**, *25*, 623–633.
86. Lott, I.T.; Doran, E.; Nguyen, V.Q.; Tournay, A.; Head, E.; Gillen, D.L. Down syndrome and dementia: A randomized, controlled trial of antioxidant supplementation. *Am. J. Med. Genet. A* **2011**, *155A*, 1939–1948.
87. Martins, V.D.; Manfredini, V.; Peralba, M.C.; Benfato, M.S. Alpha-lipoic acid modifies oxidative stress parameters in sickle cell trait subjects and sickle cell patients. *Clin. Nutr.* **2009**, *28*, 192–197.

88. Rodriguez, M.C.; MacDonald, J.R.; Mahoney, D.J.; Parise, G.; Beal, M.F.; Tarnopolsky, M.A. Beneficial effects of creatine, CoQ10, and lipoic acid in mitochondrial disorders. *Muscle Nerve* **2007**, *35*, 235–242.

89. Femiano, F.; Gombos, F., Scully, C.; Busciolano, M.; de Luca, P. Burning mouth syndrome (BMS): Controlled open trial of the efficacy of alpha-lipoic acid (thioctic acid) on symptomatology. *Oral Dis.* **2000**, *6*, 274–277.

90. Carbone, M.; Pentenero, M.; Carrozzo, M.; Ippolito, A.; Gandolfo, S. Lack of efficacy of alpha-lipoic acid in burning mouth syndrome: A double-blind, randomized, placebo-controlled study. *Eur. J. Pain* **2009**, *13*, 492–496.

91. López-Jornet, P.; Camacho-Alonso, F.; Leon-Espinosa, S. Efficacy of alpha lipoic acid in burning mouth syndrome: A randomized, placebo-treatment study. *J. Oral. Rehabil.* **2009**, *36*, 52–57.

92. Marino, R.; Torretta, S.; Capaccio, P.; Pignataro, L.; Spadari, F. Different therapeutic strategies for burning mouth syndrome: Preliminary data. *J. Oral Pathol. Med.* **2010**, *39*, 611–616.

93. López-D'Alessandro, E.; Escovich, L. Combination of alpha lipoic acid and gabapentin, its efficacy in the treatment of Burning Mouth Syndrome: A randomized, double-blind, placebo controlled trial. *Med. Oral Patol. Oral Cir. Bucal.* **2011**, *16*, e635–e640.

94. Mantovani, G.; Macciò, A.; Madeddu, C.; Gramignano, G.; Lusso, M.R.; Serpe, R.; Massa, E.; Astara, G.; Deiana, L. A phase II study with antioxidants, both in the diet and supplemented, pharmaconutritional support, progestagen, and anti-cyclooxygenase-2 showing efficacy and safety in patients with cancer-related anorexia/cachexia and oxidative stress. *Cancer Epidemiol. Biomark. Prev.* **2006**, *15*, 1030–1034.

95. Mantovani, G.; Macciò, A.; Madeddu, C.; Gramignano, G.; Serpe, R.; Massa, E.; Dessì, M.; Tanca, F.M.; Sanna, E.; Deiana, L.; *et al.* Randomized phase III clinical trial of five different arms of treatment for patients with cancer cachexia: Interim results. *Nutrition* **2008**, *24*, 305–313.

96. Jariwalla, R.J.; Lalezari, J.; Cenko, D.; Mansour, S.E.; Kumar, A.; Gangapurkar, B.; Nakamura, D. Restoration of blood total glutathione status and lymphocyte function following alpha-lipoic acid supplementation in patients with HIV infection. *J. Altern. Complement. Med.* **2008**, *14*, 139–146.

97. Milazzo, L.; Menzaghi, B.; Caramma, I.; Nasi, M.; Sangaletti, O.; Cesari, M.; Zanone Poma, B.; Cossarizza, A.; Antinori, S.; Galli, M. Effect of antioxidants on mitochondrial function in HIV-1-related lipoatrophy: A pilot study. *AIDS Res. Hum. Retroviruses* **2010**, *26*, 1207–1214.

98. Dell'Anna, M.L.; Mastrofrancesco, A.; Sala, R.; Venturini, M.; Ottaviani, M.; Vidolin, A.P.; Leone, G.; Calzavara, P.G.; Westerhof, W.; Picardo, M. Antioxidants and narrow band-UVB in the treatment of vitiligo: A double-blind placebo controlled trial. *Clin. Exp. Dermatol.* **2007**, *32*, 631–636.

99. Mainini, G.; Rotondi, M.; di Nola, K.; Pezzella, M.T.; Iervolino, S.A.; Seguino, E.; D'Eufemia, D.; Iannicelli, I.; Torella, M. Oral supplementation with antioxidant agents containing alpha lipoic acid: Effects on postmenopausal bone mass. *Clin. Exp. Obstet. Gynecol.* **2012**, *39*, 489–493.

100. Garrido-Maraver, J.; Cordero, M.D.; Oropesa-Avila, M.; Oropesa-Avila, M.; Vega, A.F.; de la Mata, M.; Pavon, A.D.; Alcocer-Gomez, E.; Calero, C.P.; Paz, M.V.; *et al.* Clinical applications of coenzyme Q10. *Front. Biosci.* **2014**, *19*, 619–633.
101. Singh, R.B.; Niaz, M.A. Serum concentration of lipoprotein(a) decreases on treatment with hydrosoluble coenzyme Q10 in patients with coronary artery disease: Discovery of a new role. *Int. J. Cardiol.* **1999**, *68*, 23–29.
102. Singh, R.B.; Niaz, M.A.; Rastogi, S.S.; Shukla, P.K.; Thakur, A.S. Effect of hydrosoluble coenzyme Q10 on blood pressures and insulin resistance in hypertensive patients with coronary artery disease. *J. Hum. Hypertens.* **1999**, *13*, 203–208.
103. Tiano, L.; Belardinelli, R.; Carnevali, P.; Principi, F.; Seddaiu, G.; Littarru, G.P. Effect of coenzyme Q10 administration on endothelial function and extracellular superoxide dismutase in patients with ischaemic heart disease: A double-blind, randomized controlled study. *Eur. Heart J.* **2007**, *28*, 2249–2255.
104. Belardinelli, R.; Muçaj, A.; Lacalaprice, F.; Solenghi, M.; Principi, F.; Tiano, L.; Littarru, G.P. Coenzyme Q10 improves contractility of dysfunctional myocardium in chronic heart failure. *Biofactors* **2005**, *25*, 137–145.
105. Dai, Y.L.; Luk, T.H.; Yiu, K.H.; Wang, M.; Yip, P.M.; Lee, S.W.; Li, S.W.; Tam, S.; Fong, B.; Lau, C.P.; *et al.* Reversal of mitochondrial dysfunction by coenzyme Q10 supplement improves endothelial function in patients with ischaemic left ventricular systolic dysfunction: A randomized controlled trial. *Atherosclerosis* **2011**, *216*, 395–401.
106. Lee, B.J.; Huang, Y.C.; Chen, S.J.; Lin, P.T. Coenzyme Q10 supplementation reduces oxidative stress and increases antioxidant enzyme activity in patients with coronary artery disease. *Nutrition* **2012**, *28*, 250–255.
107. Hofman-Bang, C.; Rehnqvist, N.; Swedberg, K.; Wiklund, I.; Aström, H. Coenzyme Q10 as an adjunctive in the treatment of chronic congestive heart failure. The Q10 Study Group. *J. Card. Fail.* **1995**, *1*, 101–107.
108. Munkholm, H.; Hansen, H.H.; Rasmussen, K. Coenzyme Q10 treatment in serious heart failure. *Biofactors* **1999**, *9*, 285–289.
109. Khatta, M.; Alexander, B.S.; Krichten, C.M.; Fisher, M.L.; Freudenberger, R.; Robinson, S.W.; Gottlieb, S.S. The effect of coenzyme Q10 in patients with congestive heart failure. *Ann. Intern. Med.* **2000**, *132*, 636–640.
110. Berman, M.; Erman, A.; Ben-Gal, T.; Dvir, D.; Georghiou, G.P.; Stamler, A.; Vered, Y.; Vidne, B.A.; Aravot, D. Coenzyme Q10 in patients with end-stage heart failure awaiting cardiac transplantation: A randomized, placebo-controlled study. *Clin. Cardiol.* **2004**, *27*, 295–299.
111. Damian, M.S.; Ellenberg, D.; Gildemeister, R.; Lauermann, J.; Simonis, G.; Sauter, W.; Georgi, C. Coenzyme Q10 combined with mild hypothermia after cardiac arrest: A preliminary study. *Circulation* **2004**, *110*, 3011–3016.
112. Witte, K.K.; Nikitin, N.P.; Parker, A.C.; von Haehling, S.; Volk, H.D.; Anker, S.D.; Clark, A.L.; Cleland, J.G. The effect of micronutrient supplementation on quality-of-life and left ventricular function in elderly patients with chronic heart failure. *Eur. Heart J.* **2005**, *26*, 2238–2244.
113. Langsjoen P.H.; Langsjoen, A.M. Supplemental ubiquinol in patients with advanced congestive heart failure. *Biofactors* **2008**, *32*, 119–128.

114. Kumar, A.; Singh, R.B.; Saxena, M.; Niaz, M.A.; Josh, S.R.; Chattopadhyay, P.; Mechirova, V.; Pella, D.; Fedacko, J. Effect of carni Q-gel (ubiquinol and carnitine) on cytokines in patients with heart failure in the Tishcon study. *Acta Cardiol.* **2007**, *62*, 349–354.
115. Belcaro, G.; Cesarone, M.R.; Dugall, M.; Hosoi, M.; Ippolito, E.; Bavera, P.; Grossi, M.G. Investigation of pycnogenol in combination with coenzyme Q10 in heart failure patients (NYHA II/III). *Panminerva Med.* **2010**, *52*, 21–25.
116. Fumagalli, S.; Fattirolli, F.; Guarducci, L.; Cellai, T.; Baldasseroni, S.; Tarantini, F.; di Bari, M.; Masotti, G.; Marchionni, N. Coenzyme Q10 terclatrate and creatine in chronic heart failure: A randomized, placebo-controlled, double-blind study. *Clin. Cardiol.* **2011**, *34*, 211–217.
117. Kuklinski, B.; Weissenbacher, E.; Fähnrich, A. Coenzyme Q10 and antioxidants in acute myocardial infarction. *Mol. Aspects Med.* **1994**, *15*, s143–s147.
118. Singh, R.B.; Wander, G.S.; Rastogi, A.; Shukla, P.K.; Mittal, A.; Sharma, J.P.; Mehrotra, S.K.; Kapoor, R.; Chopra, R.K. Randomized, double-blind placebo-controlled trial of coenzyme Q10 in patients with acute myocardial infarction. *Cardiovasc. Drugs Ther.* **1998**, *12*, 347–353.
119. Singh, R.B.; Neki, N.S.; Kartikey, K.; Pella, D.; Kumar, A.; Niaz, M.A.; Thakur, A.S. Effect of coenzyme Q10 on risk of atherosclerosis in patients with recent myocardial infarction. *Mol. Cell. Biochem.* **2003**, *246*, 75–82.
120. Permanetter, B.; Rössy, W.; Klein, G.; Weingartner, F.; Seidl, K.F.; Blömer, H. Ubiquinone (coenzyme Q10) in the long-term treatment of idiopathic dilated cardiomyopathy. *Eur. Heart J.* **1992**, *13*, 1528–1533.
121. Langsjoen, P.H.; Folkers, K.; Lyson, K.; Muratsu, K.; Lyson, T.; Langsjoen, P. Pronounced increase of survival of patients with cardiomyopathy when treated with coenzyme Q10 and conventional therapy. *Int. J. Tissue React.* **1990**, *12*, 163–168.
122. Morisco, C.; Trimarco, B.; Condorelli, M. Effect of coenzyme Q10 therapy in patients with congestive heart failure: A long-term multicenter randomized study. *Clin. Investig.* **1993**, *71*, S134–S136.
123. Ma, A.W.; Zhang, W.; Liu, Z. Effect of protection and repair of injury of mitochondrial membrane-phospholipid on prognosis in patients with dilated cardiomyopathy. *Blood Press. Suppl.* **1996**, *3*, 53–55.
124. Sacher, H.L.; Sacher, M.L.; Landau, S.W.; Kersten, R.; Dooley, F.; Sacher, A.; Sacher, M.; Dietrick, K.; Ichkhan, K. The clinical and hemodynamic effects of coenzyme Q10 in congestive cardiomyopathy. *Am. J. Ther.* **1997**, *4*, 66–72.
125. Soongswang, J.; Sangtawesin, C.; Durongpisitkul, K.; Laohaprasitiporn, D.; Nana, A., Punlee, K.; Kangkagate, C. The effect of coenzyme Q10 on idiopathic chronic dilated cardiomyopathy in children. *Pediatr. Cardiol.* **2005**, *26*, 361–366.
126. Caso, G.; Kelly, P.; McNurlan, M.A.; Lawson, W.E. Effect of coenzyme Q10 on myopathic symptoms in patients treated with statins. *Am. J. Cardiol.* **2007**, *99*, 1409–1412.
127. Kocharian, A.; Shabanian, R.; Rafiei-Khorgami, M.; Kiani, A.; Heidari-Bateni, G. Coenzyme Q10 improves diastolic function in children with idiopathic dilated cardiomyopathy. *Cardiol. Young* **2009**, *19*, 501–506.

128. Shargorodsky, M.; Debby, O.; Matas, Z.; Zimlichman, R. Effect of long-term treatment with antioxidants (vitamin C, vitamin E, coenzyme Q10 and selenium) on arterial compliance, humoral factors and inflammatory markers in patients with multiple cardiovascular risk factors. *Nutr. Metab.* **2010**, *7*, 55.
129. Judy, W.V.; Stogsdill, W.W.; Folkers, K. Myocardial preservation by therapy with coenzyme Q10 during heart surgery. *Clin. Investig.* **1993**, *71*, S155–S161.
130. Taggart, D.P.; Jenkins, M.; Hooper, J.; Hadjinikolas, L.; Kemp, M.; Hue, D.; Bennett, G. Effects of short-term supplementation with coenzyme Q10 on myocardial protection during cardiac operations. *Ann. Thorac. Surg.* **1996**, *61*, 829–833.
131. Chello, M.; Mastroroberto, P.; Romano, R.; Castaldo, P.; Bevacqua, E.; Marchese, A.R. Protection by coenzyme Q10 of tissue reperfusion injury during abdominal aortic cross-clamping. *J. Cardiovasc. Surg.* **1996**, *37*, 229–235.
132. Zhou, M.; Zhi, Q.; Tang, Y.; Yu, D.; Han, J. Effects of coenzyme Q10 on myocardial protection during cardiac valve replacement and scavenging free radical activity *in vitro*. *J. Cardiovasc. Surg.* **1999**, *40*, 355–361.
133. Rosenfeldt, F.; Marasco, S.; Lyon, W.; Wowk, M.; Sheeran, F.; Bailey, M.; Esmore, D.; Davis, B.; Pick, A.; Rabinov, M.; *et al.* Coenzyme Q10 therapy before cardiac surgery improves mitochondrial function and *in vitro* contractility of myocardial tissue. *J. Thorac. Cardiovasc. Surg.* **2005**, *129*, 25–32.
134. Keith, M.; Mazer, C.D.; Mikhail, P.; Jeejeebhoy, F.; Briet, F.; Errett, L. Coenzyme Q10 in patients undergoing (coronary artery bypass graft) CABG: Effect of statins and nutritional supplementation. *Nutr. Metab. Cardiovasc. Dis.* **2008**, *18*, 105–111.
135. Makhija, N.; Sendasgupta, C.; Kiran, U.; Lakshmy, R.; Hote, M.P.; Choudhary, S.K.; Airan, B.; Abraham, R. The role of oral coenzyme Q10 in patients undergoing coronary artery bypass graft surgery. *J. Cardiothorac. Vasc. Anesth.* **2008**, *22*, 832–839.
136. Alehagen, U.; Johansson, P.; Björnstedt, M.; Rosén, A.; Dahlström, U. Cardiovascular mortality and *N*-terminal-proBNP reduced after combined selenium and coenzyme Q10 supplementation: A 5-year prospective randomized double-blind placebo-controlled trial among elderly Swedish citizens. *Int. J. Cardiol.* **2013**, *167*, 1860–1866.
137. Miles, M,V.; Patterson, B.J.; Chalfonte-Evans, M.L.; Horn, P.S.; Hickey, F.J.; Schapiro, M.B.; Steele, P.E.; Tang, P.H.; Hotze, S.L. Coenzyme Q10 (ubiquinol-10) supplementation improves oxidative imbalance in children with trisomy 21. *Pediatr. Neurol.* **2007**, *37*, 398–403.
138. Tiano, L.; Padella, L.; Santoro, L.; Carnevali, P.; Principi, F.; Brugè, F.; Gabrielli, O.; Littarru, G.P. Prolonged coenzyme Q10 treatment in Down syndrome patients: Effect on DNA oxidation. *Neurobiol. Aging* **2012**, *33*, 626.e1-8.
139. Sparaco, M.; Gaeta, L.M.; Santorelli, F.M.; Passarelli, C.; Tozzi, G.; Bertini, E.; Simonati, A.; Scaravilli, F.; Taroni, F.; Duyckaerts, C.; *et al.* Friedreich's ataxia: Oxidative stress and cytoskeletal abnormalities. *J. Neurol. Sci.* **2009**, *287*, 111–118.
140. Cooper, J.M.; Korlipara, L.V.; Hart, P.E.; Bradley, J.L.; Schapira, A.H. Coenzyme Q10 and vitamin E deficiency in Friedreich's ataxia: Predictor of efficacy of vitamin E and coenzyme Q10 therapy. *Eur. J. Neurol.* **2008**, *15*, 1371–1379.

141. Hart, P.E.; Lodi, R.; Rajagopalan, B.; Bradley, J.L.; Crilley, J.G.; Turner, C.; Blamire, A.M.; Manners, D.; Styles, P.; Schapira, A.H.; *et al.* Antioxidant treatment of patients with Friedreich ataxia: Four-year follow-up. *Arch. Neurol.* **2005**, *62*, 621–626.
142. Barbiroli, B.; Frassineti, C.; Martinelli, P.; Iotti, S.; Lodi, R.; Cortelli, P.; Montagna, P. Coenzyme Q10 improves mitochondrial respiration in patients with mitochondrial cytopathies. An *in vivo* study on brain and skeletal muscle by phosphorous magnetic resonance spectroscopy. *Cell. Mol. Biol.* **1997**, *43*, 741–749.
143. Abe, K.; Matsuo, Y.; Kadekawa, J.; Inoue, S.; Yanagihara, T. Effect of coenzyme Q10 in patients with mitochondrial myopathy, encephalopathy, lactic acidosis, and stroke-like episodes (MELAS): Evaluation by noninvasive tissue oximetry. *J. Neurol. Sci.* **1999**, *162*, 65–68.
144. Hanisch, F.; Zierz, S. Only transient increase of serum CoQ subset 10 during long-term CoQ10 therapy in mitochondrial ophthalmoplegia. *Eur. J. Med. Res.* **2003**, *8*, 485–491.
145. Glover, E.I.; Martin, J.; Maher, A.; Thornhill, R.E.; Moran, G.R.; Tarnopolsky, M.A. A randomized trial of coenzyme Q10 in mitochondrial disorders. *Muscle Nerve* **2010**, *42*, 739–748.
146. Huntington Study Group. A randomized, placebo-controlled trial of coenzyme Q10 and remacemide in Huntington's disease. *Neurology* **2001**, *57*, 397–404.
147. Huntington Study Group Pre2CARE Investigators; Hyson, H.C.; Kieburtz, K.; Shoulson, I.; McDermott, M.; Ravina, B.; de Blieck, E.A.; Cudkowicz, M.E.; Ferrante, R.J.; Como, P.; *et al.* Safety and tolerability of high-dosage coenzyme Q10 in Huntington's disease and healthy subjects. *Mov. Disord.* **2010**, *25*, 1924–1928.
148. Spurney, C.F.; Rocha, C.T.; Henricson, E.; Florence, J.; Mayhew, J.; Gorni, K.; Pasquali, L.; Pestronk, A.; Martin, G.R.; Hu, F.; *et al.* CINRG pilot trial of coenzyme Q10 in steroid-treated Duchenne muscular dystrophy. *Muscle Nerve* **2011**, *44*, 174–178.
149. Folkers, K.; Wolaniuk, J., Simonsen, R.; Morishita, M.; Vadhanavikit, S. Biochemical rationale and the cardiac response of patients with muscle disease to therapy with coenzyme Q10. *Proc. Natl. Acad. Sci. USA* **1985**, *82*, 4513–4516.
150. Folkers, K.; Simonsen, R. Two successful double-blind trials with coenzyme Q10 (vitamin Q10) on muscular dystrophies and neurogenic atrophies. *Biochim. Biophys. Acta* **1995**, *1271*, 281–286.
151. Linnane, A.W.; Kopsidas, G.; Zhang, C.; Yarovaya, N.; Kovalenko, S.; Papakostopoulos, P.; Eastwood, H.; Graves, S.; Richardson, M. Cellular redox activity of coenzyme Q10: Effect of CoQ10 supplementation on human skeletal muscle. *Free Radic. Res.* **2002**, *36*, 445–453.
152. Shults, C.W.; Oakes, D.; Kieburtz, K.; Beal, M.F.; Haas, R.; Plumb. S.; Juncos, J.L.; Nutt, J.; Shoulson, I.; Carter, J.; *et al.* Effects of coenzyme Q10 in early Parkinson disease: Evidence of slowing of the functional decline. *Arch. Neurol.* **2002**, *59*, 1541–1550.
153. Müller, T.; Büttner, T.; Gholipour, A.F.; Kuhn, W. Coenzyme Q10 supplementation provides mild symptomatic benefit in patients with Parkinson's disease. *Neurosci. Lett.* **2003**, *341*, 201–204.
154. Storch, A.; Jost, W.H.; Vieregge, P.; Spiegel, J.; Greulich, W.; Durner, J.; Müller, T.; Kupsch, A.; Henningsen, H.; Oertel, W.H.; *et al.* Randomized, double-blind, placebo-controlled trial on symptomatic effects of coenzyme Q10 in Parkinson disease. *Arch. Neurol.* **2007**, *64*, 938–944.

155. Sanoobar, M.; Eghtesadi, S.; Azimi, A.; Khalili, M.; Jazayeri, S.; Reza Gohari, M. Coenzyme Q10 supplementation reduces oxidative stress and increases antioxidant enzyme activity in patients with relapsing-remitting multiple sclerosis. *Int. J. Neurosci.* **2013**, *123*, 776–782.
156. Sanoobar, M.; Eghtesadi, S.; Azimi, A.; Khalili, M.; Khodadadi, B.; Jazayeri, S.; Gohari, M.R.; Aryaeian, N. Coenzyme Q10 supplementation ameliorates inflammatory markers in patients with multiple sclerosis: A double blind, placebo, controlled randomized clinical trial. *Nutr. Neurosci.* **2014**, doi:10.1179/1476830513Y.0000000106.
157. Cordero, M.D.; Cano-García, F.J.; Alcocer-Gómez, E.; de Miguel, M.; Sánchez-Alcázar, J.A. Oxidative stress correlates with headache symptoms in fibromyalgia: Coenzyme Q10 effect on clinical improvement. *PLoS One* **2012**, *7*, e35677.
158. Cordero, M.D.; Alcocer-Gómez, E.; de Miguel, M.; Culic, O.; Carrión, A.M.; Alvarez-Suarez, J.M.; Bullón, P.; Battino, M.; Fernández-Rodríguez, A.; Sánchez-Alcazar, J.A. Can coenzyme q10 improve clinical and molecular parameters in fibromyalgia? *Antioxid. Redox Signal.* **2013**, *19*, 1356–1361.
159. Miyamae, T.; Seki, M.; Naga, T.; Uchino, S.; Asazuma, H.; Yoshida, T.; Iizuka, Y.; Kikuchi, M.; Imagawa, T.; Natsumeda, Y.; *et al.* Increased oxidative stress and coenzyme Q10 deficiency in juvenile fibromyalgia: Amelioration of hypercholesterolemia and fatigue by ubiquinol-10 supplementation. *Redox Rep.* **2013**, *18*, 12–19.
160. Lister, R.E. An open, pilot study to evaluate the potential benefits of coenzyme Q10 combined with *Ginkgo biloba* extract in fibromyalgia syndrome. *J. Int. Med. Res.* **2002**, *30*, 195–199.
161. Kaufmann, P.; Thompson, J.L.; Levy, G.; Buchsbaum, R.; Shefner, J.; Krivickas, L.S.; Katz, J.; Rollins, Y.; Barohn, R.J.; Jackson, C.E.; *et al.* Phase II trial of CoQ10 for ALS finds insufficient evidence to justify phase III. *Ann. Neurol.* **2009**, *66*, 235–244.
162. Rozen, T.D.; Oshinsky, M.L.; Gebeline, C.A.; Bradley, K.C.; Young, W.B.; Shechter, A.L.; Silberstein, S.D. Open label trial of coenzyme Q10 as a migraine preventive. *Cephalalgia* **2002**, *22*, 137–141.
163. Hershey, A.D.; Powers, S.W.; Vockell, A.L.; Lecates, S.L.; Ellinor, P.L.; Segers, A.; Burdine, D.; Manning, P.; Kabbouche, M.A. Coenzyme Q10 deficiency and response to supplementation in pediatric and adolescent migraine. *Headache* **2007**, *47*, 73–80.
164. Slater, S.K.; Nelson, T.D.; Kabbouche, M.A.; LeCates, S.L.; Horn, P.; Segers, A.; Manning, P.; Powers, S.W.; Hershey, A.D. A randomized, double-blinded, placebo-controlled, crossover, add-on study of CoEnzyme Q10 in the prevention of pediatric and adolescent migraine. *Cephalalgia* **2011**, *31*, 897–905.
165. Forester, B.P.; Zuo, C.S.; Ravichandran, C.; Harper, D.G.; Du, F.; Kim, S.; Cohen, B.M.; Renshaw, P.F. Coenzyme Q10 effects on creatine kinase activity and mood in geriatric bipolar depression. *J. Geriatr. Psychiatry Neurol.* **2012**, *25*, 43–50.
166. Stamelou, M.; Reuss, A.; Pilatus, U.; Magerkurth, J.; Niklowitz, P.; Eggert, K.M.; Krisp, A.; Menke, T.; Schade-Brittinger, C.; Oertel, W.H.; *et al.* Short-term effects of coenzyme Q10 in progressive supranuclear palsy: A randomized, placebo-controlled trial. *Mov. Disord.* **2008**, *23*, 942–949.

167. Watts, G.F.; Playford, D.A.; Croft, K.D., Ward, N.C.; Mori, T.A.; Burke, V. Coenzyme Q10 improves endothelial dysfunction of the brachial artery in type II diabetes mellitus. *Diabetologia* **2002**, *45*, 420–426.

168. Hodgson, J.M.; Watts, G.F.; Playford, D.A.; Burke, V.; Croft, K.D. Coenzyme Q10 improves blood pressure and glycaemic control: A controlled trial in subjects with type 2 diabetes. *Eur. J. Clin. Nutr.* **2002**, *56*, 1137–1142.

169. Playford, D.A.; Watts, G.F.; Croft, K.D.; Burke, V. Combined effect of coenzyme Q10 and fenofibrate on forearm microcirculatory function in type 2 diabetes. *Atherosclerosis* **2003**, *168*, 169–179.

170. Chew, G.T.; Watts, G.F.; Davis, T.M.; Stuckey, B.G.; Beilin, L.J.; Thompson, P.L.; Burke, V.; Currie, P.J. Hemodynamic effects of fenofibrate and coenzyme Q10 in type 2 diabetic subjects with left ventricular diastolic dysfunction. *Diabetes Care* **2008**, *31*, 1502–1509.

171. Hamilton, S.J.; Chew, G.T.; Watts, G.F. Coenzyme Q10 improves endothelial dysfunction in statin-treated type 2 diabetic patients. *Diabetes Care* **2009**, *32*, 810–812.

172. Kolahdouz Mohammadi, R.; Hosseinzadeh-Attar, M.J.; Eshraghian, M.R.; Nakhjavani, M.; Khorami, E.; Esteghamati, A. The effect of coenzyme Q10 supplementation on metabolic status of type 2 diabetic patients. *Minerva Gastroenterol. Dietol.* **2013**, *59*, 231–236.

173. Suzuki, S.; Hinokio, Y.; Ohtomo, M.; Hirai, M.; Hirai, A.; Chiba, M.; Kasuga, S.; Satoh, Y.; Akai, H.; Toyota, T. The effects of coenzyme Q10 treatment on maternally inherited diabetes mellitus and deafness, and mitochondrial DNA 3243 (A to G) mutation. *Diabetologia* **1998**, *41*, 584–588.

174. Henriksen, J.E.; Andersen, C.B.; Hother-Nielsen, O.; Vaag, A.; Mortensen, S.A.; Beck-Nielsen, H. Impact of ubiquinone (coenzyme Q10) treatment on glycaemic control, insulin requirement and well-being in patients with type 1 diabetes mellitus. *Diabet. Med.* **1999**, *16*, 312–318.

175. Hertz, N.; Lister, R.E. Improved survival in patients with end-stage cancer treated with coenzyme Q10 and other antioxidants: A pilot study. *J. Int. Med. Res.* **2009**, *37*, 1961–1971.

176. Lockwood, K.; Moesgaard, S.; Folkers, K. Partial and complete regression of breast cancer in patients in relation to dosage of coenzyme Q10. *Biochem. Biophys. Res. Commun.* **1994**, *199*, 1504–1508.

177. Premkumar, V.G.; Yuvaraj, S.; Sathish, S.; Shanthi, P.; Sachdanandam, P. Anti-angiogenic potential of Coenzyme Q10, riboflavin and niacin in breast cancer patients undergoing tamoxifen therapy. *Vascul. Pharmacol.* **2008**, *48*, 191–201.

178. Lesser, G.J.; Case, D.; Stark, N.; Williford, S.; Giguere, J.; Garino, L.A.; Naughton, M.J.; Vitolins, M.Z.; Lively, M.O.; Shaw, E.G.; *et al.* A randomized, double-blind, placebo-controlled study of oral coenzyme Q10 to relieve self-reported treatment-related fatigue in newly diagnosed patients with breast cancer. *J. Support Oncol.* **2013**, *11*, 31–42.

179. Hoenjet, K.M.; Dagnelie, P.C.; Delaere, K.P.; Wijckmans, N.E.; Zambon, J.V.; Oosterhof, G.O. Effect of a nutritional supplement containing vitamin E, selenium, vitamin c and coenzyme Q10 on serum PSA in patients with hormonally untreated carcinoma of the prostate: A randomised placebo-controlled study. *Eur. Urol.* **2005**, *47*, 433–439.

180. Rusciani, L.; Proietti, I.; Paradisi, A.; Rusciani, A.; Guerriero, G.; Mammone, A.; de Gaetano, A.; Lippa, S. Recombinant interferon alpha-2b and coenzyme Q10 as a postsurgical adjuvant therapy for melanoma: A 3-year trial with recombinant interferon-alpha and 5-year follow-up. *Melanoma Res.* **2007**, *17*, 177–183.
181. Gazdíková, K.; Gvozdjáková, A.; Kucharská, J.; Spustová, V.; Braunová, Z.; Dzúrik, R. Effect of coenzyme Q10 in patients with kidney diseases. *Cas. Lek. Cesk.* **2001**, *140*, 307–310.
182. Sakata, T.; Furuya, R.; Shimazu, T.; Odamaki, M.; Ohkawa, S.; Kumagai, H. Coenzyme Q10 administration suppresses both oxidative and antioxidative markers in hemodialysis patients. *Blood Purif.* **2008**, *26*, 371–378.
183. Mori, T.A.; Burke, V.; Puddey, I.; Irish, A.; Cowpland, C.A.; Beilin, L.; Dogra, G.; Watts, G.F. The effects of omega-3 fatty acids and coenzyme Q10 on blood pressure and heart rate in chronic kidney disease: A randomized controlled trial. *J. Hypertens.* **2009**, *27*, 1863–1872.
184. Shojaei, M.; Djalali, M.; Khatami, M.; Siassi, F.; Eshraghian, M. Effects of carnitine and coenzyme Q10 on lipid profile and serum levels of lipoprotein(a) in maintenance hemodialysis patients on statin therapy. *Iran J. Kidney Dis.* **2011**, *5*, 114–118.
185. Kaikkonen, J.; Nyyssönen, K.; Tomasi, A.; Iannone, A.; Tuomainen, T.P.; Porkkala-Sarataho, E.; Salonen, J.T. Antioxidative efficacy of parallel and combined supplementation with coenzyme Q10 and d-alpha-tocopherol in mildly hypercholesterolemic subjects: A randomized placebo-controlled clinical study. *Free Radic. Res.* **2000**, *33*, 329–340.
186. Lee, Y.J.; Cho, W.J.; Kim, J.K.; Lee, D.C. Effects of coenzyme Q10 on arterial stiffness, metabolic parameters, and fatigue in obese subjects: A double-blind randomized controlled study. *J. Med. Food* **2011**, *14*, 386–390.
187. Young, J.M.; Florkowski, C.M.; Molyneux, S.L.; McEwan, R.G.; Frampton, C.M.; Nicholls, M.G.; Scott, R.S.; George, P.M. A randomized, double-blind, placebo-controlled crossover study of coenzyme Q10 therapy in hypertensive patients with the metabolic syndrome. *Am. J. Hypertens.* **2012**, *25*, 261–270.
188. Mabuchi, H.; Nohara, A.; Kobayashi, J.; Kawashiri, M.A.; Katsuda, S.; Inazu, A.; Koizumi, J.; Hokuriku Lipid Research Group. Effects of CoQ10 supplementation on plasma lipoprotein lipid, CoQ10 and liver and muscle enzyme levels in hypercholesterolemic patients treated with atorvastatin: A randomized double-blind study. *Atherosclerosis* **2007**, *195*, e182–e189.
189. Young, J.M.; Florkowski, C.M.; Molyneux, S.L.; McEwan, R.G.; Frampton, C.M.; George, P.M.; Scott, R.S. Effect of coenzyme Q10 supplementation on simvastatin-induced myalgia. *Am. J. Cardiol.* **2007**, *100*, 1400–1403.
190. Bookstaver, D.A.; Burkhalter, N.A.; Hatzigeorgiou, C. Effect of coenzyme Q10 supplementation on statin-induced myalgias. *Am. J. Cardiol.* **2012**, *110*, 526–529.
191. Zlatohlavek, L.; Vrablik, M.; Grauova, B.; Motykova, E.; Ceska, R. The effect of coenzyme Q10 in statin myopathy. *Neuro Endocrinol. Lett.* **2012**, *33*, 98–101.
192. Fedacko, J.; Pella, D.; Fedackova, P.; Hänninen, O.; Tuomainen, P.; Jarcuska, P.; Lopuchovsky, T.; Jedlickova, L.; Merkovska, L.; Littarru, G.P. Coenzyme Q10 and selenium in statin-associated myopathy treatment. *Can. J. Physiol. Pharmacol.* **2013**, *91*, 165–170.

193. Gvozdjáková, A.; Kucharská, J.; Bartkovjaková, M.; Gazdíková, K.; Gazdík, F.E. Coenzyme Q10 supplementation reduces corticosteroids dosage in patients with bronchial asthma. *Biofactors* **2005**, *25*, 235–240.
194. Teran, E.; Hernandez, I., Nieto, B.; Tavara, R.; Ocampo, J.E.; Calle, A. Coenzyme Q10 supplementation during pregnancy reduces the risk of pre-eclampsia. *Int. J. Gynaecol. Obstet.* **2009**, *105*, 43–45.
195. De Luca, C.; Kharaeva, Z.; Raskovic, D.; Pastore, P.; Luci, A.; Korkina, L. Coenzyme Q10, vitamin E, selenium, and methionine in the treatment of chronic recurrent viral mucocutaneous infections. *Nutrition* **2012**, *28*, 509–514.
196. Kharaeva, Z.; Gostova, E.; de Luca, C.; Raskovic, D.; Korkina, L. Clinical and biochemical effects of coenzyme Q10, vitamin E, and selenium supplementation to psoriasis patients. *Nutrition* **2009**, *25*, 295–302.
197. Fogagnolo, P.; Sacchi, M.; Ceresara, G.; Paderni, R.; Lapadula, P.; Orzalesi, N.; Rossetti, L. The effects of topical coenzyme Q10 and vitamin E D-α-tocopheryl polyethylene glycol 1000 succinate after cataract surgery: A clinical and *in vivo* confocal study. *Ophthalmologica* **2013**, *229*, 26–31.
198. Safarinejad, M.R.; Safarinejad, S.; Shafiei, N.; Safarinejad, S. Effects of the reduced form of coenzyme Q10 (ubiquinol) on semen parameters in men with idiopathic infertility: A double-blind, placebo controlled, randomized study. *J. Urol.* **2012**, *188*, 526–531.
199. Gilbert, E.F. Carnitine deficiency. *Pathology* **1985**, *17*, 161–171.
200. Siliprandi, N.; Siliprandi, D.; Ciman, M. Stimulation of oxidation of mitochondrial fatty acids and of acetate by acetylcarnitine. *Biochem. J.* **1965**, *96*, 777–780.
201. Liu, J.; Head, E.; Kuratsune, H.; Cotman, C.W.; Ames, B.N. Comparison of the effects of L-carnitine and acetyl-L-carnitine on carnitine levels, ambulatory activity, and oxidative stress biomarkers in the brain of old rats. *Ann. N. Y. Acad. Sci.* **2004**, *1033*, 117–131.
202. Musicco, C.; Capelli, V.; Pesce, V.; Timperio, A.M.; Calvani, M.; Mosconi, L.; Cantatore, P.; Gadaleta, M.N. Rat liver mitochondrial proteome: Changes associated with aging and acetyl-L-carnitine treatment. *J. Proteomics* **2011**, *74*, 2536–2547.
203. Pesce, V.; Nicassio, L.; Fracasso, F.; Musicco, C.; Cantatore, P.; Gadaleta, M.N. Acetyl-L-carnitine activates the peroxisome proliferator-activated receptor-γ coactivators PGC-1α/PGC-1β-dependent signaling cascade of mitochondrial biogenesis and decreases the oxidized peroxiredoxins content in old rat liver. *Rejuvenation Res.* **2012**, *15*, 136–139.
204. Bertoli, M.; Battistella, P.A.; Vergani, L.; Naso, A.; Gasparotto, M.L.; Romagnoli, G.F.; Angelini, C. Carnitine deficiency induced during hemodialysis and hyperlipidemia: Effect of replacement therapy. *Am. J. Clin. Nutr.* **1981**, *34*, 1496–1500.
205. Kletzmayr, J.; Mayer, G.; Legenstein, E.; Heinz-Peer, G.; Leitha, T.; Hörl, W.H.; Kovarik, J. Anemia and carnitine supplementation in hemodialyzed patients. *Kidney Int. Suppl.* **1999**, *69*, S93–S106.
206. Thomas, S.; Fischer, F.P.; Mettang, T.; Pauli-Magnus, C.; Weber, J.; Kuhlmann, U. Effects of L-carnitine on leukocyte function and viability in hemodialysis patients: A double-blind randomized trial. *Am. J. Kidney Dis.* **1999**, *34*, 678–687.

207. Chazot, C.; Blanc, C.; Hurot, J.M.; Charra, B.; Jean, G.; Laurent, G. Nutritional effects of carnitine supplementation in hemodialysis patients. *Clin. Nephrol.* **2003**, *59*, 24–30.
208. Vaux, E.C.; Taylor, D.J.; Altmann, P.; Rajagopalan, B.; Graham, K.; Cooper, R.; Bonomo, Y.; Styles, P. Effects of carnitine supplementation on muscle metabolism by the use of magnetic resonance spectroscopy and near-infrared spectroscopy in end-stage renal disease. *Nephron Clin. Pract.* **2004**, *97*, c41–c48.
209. Steiber, A.L.; Davis, A.T.; Spry, L.; Strong, J.; Buss, M.L.; Ratkiewicz, M.M.; Weatherspoon, L.J. Carnitine treatment improved quality-of-life measure in a sample of Midwestern hemodialysis patients. *J. Parenter. Enteral. Nutr.* **2006**, *30*, 10–15.
210. Signorelli, S.S.; Fatuzzo, P.; Rapisarda, F.; Neri, S.; Ferrante, M.; Oliveri Conti, G.; Fallico, R.; di Pino, L.; Pennisi, G.; Celotta, G.; *et al.* A randomised, controlled clinical trial evaluating changes in therapeutic efficacy and oxidative parameters after treatment with propionyl L-carnitine in patients with peripheral arterial disease requiring haemodialysis. *Drugs Aging* **2006**, *23*, 263–270.
211. Rathod, R.; Baig, M.S.; Khandelwal, P.N.; Kulkarni, S.G.; Gade, P.R.; Siddiqui, S. Results of a single blind, randomized, placebo-controlled clinical trial to study the effect of intravenous L-carnitine supplementation on health-related quality of life in Indian patients on maintenance hemodialysis. *Indian J. Med. Sci.* **2006**, *60*, 143–153.
212. Duranay, M.; Akay, H.; Yilmaz, F.M.; Senes, M.; Tekeli, N.; Yücel, D. Effects of L-carnitine infusions on inflammatory and nutritional markers in haemodialysis patients. *Nephrol. Dial. Transplant.* **2006**, *21*, 3211–3214.
213. Verrina, E.; Caruso, U.; Calevo, M.G.; Emma, F.; Sorino, P.; de Palo, T.; Lavoratti, G.; Turrini Dertenois, L.; Cassanello, M.; Cerone, R.; *et al.* Effect of carnitine supplementation on lipid profile and anemia in children on chronic dialysis. *Pediatr. Nephrol.* **2007**, *22*, 727–733.
214. Fatouros, I.G.; Douroudos, I.; Panagoutsos, S.; Pasadakis, P.; Nikolaidis, M.G.; Chatzinikolaou, A.; Sovatzidis, A.; Michailidis.; Y.; Jamurtas, A.Z.; Mandalidis, D.; *et al.* Effects of L-carnitine on oxidative stress responses in patients with renal disease. *Med. Sci. Sports Exerc.* **2010**, *42*, 1809–1818.
215. Hakeshzadeh, F.; Tabibi, H.; Ahmadinejad, M.; Malakoutian, T.; Hedayati, M. Effects of L-carnitine supplement on plasma coagulation and anticoagulation factors in hemodialysis patients. *Renal Failure* **2010**, *32*, 1109–1114.
216. Tabibi, H.; Hakeshzadeh, F.; Hedayati, M.; Malakoutian, T. Effects of L-carnitine supplement on serum amyloid A and vascular inflammation markers in hemodialysis patients: A randomized controlled trial. *J. Ren. Nutr.* **2011**, *21*, 485–491.
217. Suchitra, M.M.; Ashalatha, V.L.; Sailaja, E.; Rao, A.M.; Reddy, V.S.; Bitla, A.R.; Sivakumar, V.; Rao, P.V. The effect of L-carnitine supplementation on lipid parameters, inflammatory and nutritional markers in maintenance hemodialysis patients. *Saudi J. Kidney Dis. Transpl.* **2011**, *22*, 1155–1159.
218. Naini, A.E.; Sadeghi, M.; Mortazavi, M.; Moghadasi, M.; Harandi, A.A. Oral carnitine supplementation for dyslipidemia in chronic hemodialysis patients. *Saudi J. Kidney Dis. Transpl.* **2012**; *23*, 484–488.

219. Sgambat, K.; Frank, L.; Ellini, A.; Sable, C.; Moudgil, A. Carnitine supplementation improves cardiac strain rate in children on chronic hemodialysis. *Pediatr. Nephrol.* **2012**, *27*, 1381–1387.

220. Mercadal, L.; Coudert, M.; Vassault, A.; Pieroni, L.; Debure, A.; Ouziala, M.; Depreneuf, H.; Fumeron, C.; Servais, A.; Bassilios, N.; *et al.* L-Carnitine treatment in incident hemodialysis patients: The multicenter, randomized, double-blinded, placebo-controlled CARNIDIAL trial. *Clin. J. Am. Soc. Nephrol.* **2012**, *7*, 1836–1842.

221. Mingrone, G.; Greco, A.V.; Capristo, E.; Benedetti, G.; Giancaterini, A.; de Gaetano, A.; Gasbarrini, G. L-Carnitine improves glucose disposal in type 2 diabetic patients. *J. Am. Coll. Nutr.* **1999**, *18*, 77–82.

222. Giancaterini, A.; de Gaetano, A.; Mingrone, G.; Gniuli, D.; Liverani, E.; Capristo, E.; Greco, A.V. Acetyl-L-carnitine infusion increases glucose disposal in type 2 diabetic patients. *Metabolism* **2000**, *49*, 704–708.

223. De Grandis, D.; Minardi, C. Acetyl-L-carnitine (levacecarnine) in the treatment of diabetic neuropathy. A long-term, randomised, double-blind, placebo-controlled study. *Drugs R&D* **2002**, *3*, 223–231.

224. Derosa, G.; Cicero, A.F.; Gaddi, A.; Mugellini, A.; Ciccarelli, L.; Fogari, R. The effect of L-carnitine on plasma lipoprotein(a) levels in hypercholesterolemic patients with type 2 diabetes mellitus. *Clin. Ther.* **2003**, *25*, 1429–1439.

225. Ragozzino, G.; Mattera, E.; Madrid, E.; Salomone, P.; Fasano, C.; Gioia, F.; Acerra, G.; del Guercio, R.; Federico, P. Effects of propionyl-carnitine in patients with type 2 diabetes and peripheral vascular disease: Results of a pilot trial. *Drugs R&D* **2004**, *5*, 185–190.

226. Sima, A.A.; Calvani, M.; Mehra, M.; Amato, A. Acetyl-L-Carnitine Study Group. Acetyl-L-carnitine improves pain, nerve regeneration, and vibratory perception in patients with chronic diabetic neuropathy: An analysis of two randomized placebo-controlled trials. *Diabetes Care* **2005**, *28*, 89–94.

227. Rahbar, A.R.; Shakerhosseini, R.; Saadat, N.; Taleban, F.; Pordal, A.; Gollestan, B. Effect of L-carnitine on plasma glycemic and lipidemic profile in patients with type II diabetes mellitus. *Eur. J. Clin. Nutr.* **2005**, *59*, 592–596.

228. Solfrizzi, V.; Capurso, C.; Colacicco, A.M.; D'Introno, A.; Fontana, C.; Capurso, S.A.; Torres, F.; Gadaleta, A.M.; Koverech, A.; Capurso, A.; *et al.* Efficacy and tolerability of combined treatment with L-carnitine and simvastatin in lowering lipoprotein(a) serum levels in patients with type 2 diabetes mellitus. *Atherosclerosis* **2006**, *188*, 455–461.

229. Signorelli, S.S; Neri, S.; di Pino, L.; Marchese, G.; Ferrante, M.; Oliveri Conti, G.; Fallico, R.; Celotta, G.; Pennisi, G.; Anzaldi, M. Effect of PLC on functional parameters and oxidative profile in type 2 diabetes-associated PAD. *Diabetes Res. Clin. Pract.* **2006**, *72*, 231–237.

230. Malaguarnera, M.; Vacante, M.; Avitabile, T.; Malaguarnera, M.; Cammalleri, L.; Motta, M. L-Carnitine supplementation reduces oxidized LDL cholesterol in patients with diabetes. *Am. J. Clin. Nutr.* **2009**, *89*, 71–76.

231. Ruggenenti, P.; Cattaneo, D.; Loriga, G.; Ledda, F.; Motterlini, N.; Gherardi, G.; Orisio, S.; Remuzzi, G. Ameliorating hypertension and insulin resistance in subjects at increased cardiovascular risk: Effects of acetyl-L-carnitine therapy. *Hypertension* **2009**, *54*, 567–574.

232. Molfino, A.; Cascino, A.; Conte, C.; Ramaccini, C.; Rossi Fanelli, F.; Laviano, A. Caloric restriction and L-carnitine administration improves insulin sensitivity in patients with impaired glucose metabolism. *J. Parenter. Enteral. Nutr.* **2010**, *34*, 295–299.
233. Uzun, N.; Sarikaya, S.; Uluduz, D.; Aydin, A. Peripheric and automatic neuropathy in children with type 1 diabetes mellitus: The effect of L-carnitine treatment on the peripheral and autonomic nervous system. *Electromyogr. Clin. Neurophysiol.* **2005**, *45*, 343–351.
234. Loffredo, L.; Pignatelli, P.; Cangemi, R.; Andreozzi, P.; Panico, M.A.; Meloni, V.; Violi, F. Imbalance between nitric oxide generation and oxidative stress in patients with peripheral arterial disease: Effect of an antioxidant treatment. *J. Vasc. Surg.* **2006**, *44*, 525–530.
235. De Marchi, S.; Zecchetto, S.; Rigoni, A.; Prior, M.; Fondrieschi, L.; Scuro, A.; Rulfo, F.; Arosio, E. Propionyl-L-carnitine improves endothelial function, microcirculation and pain management in critical limb ischemia. *Cardiovasc. Drugs Ther.* **2012**, *26*, 401–408.
236. Goldenberg, N.A.; Krantz, M.J.; Hiatt, W.R. L-Carnitine plus cilostazol *versus* cilostazol alone for the treatment of claudication in patients with peripheral artery disease: A multicenter, randomized, double-blind, placebo-controlled trial. *Vasc. Med.* **2012**, *17*, 145–154.
237. Xu, X.Q.; Jing, Z.C.; Jiang, X.; Zhao, Q.H.; He, J.; Dai, L.Z.; Wu, W.H.; Li, Y.; Yao, J. Clinical efficacy of intravenous L-carnitine in patients with right-sided heart failure induced by pulmonary arterial hypertension. *Zhonghua Xin Xue Guan Bing Za Zhi* **2010**, *38*, 152–155.
238. Serati, A.R.; Motamedi, M.R.; Emami, S.; Varedi, P.; Movahed, M.R. L-Carnitine treatment in patients with mild diastolic heart failure is associated with improvement in diastolic function and symptoms. *Cardiology* **2010**, *116*, 178–182.
239. Lango, R.; Smoleński, R.T.; Rogowski, J.; Siebert, J.; Wujtewicz, M.; Słomińska, E.M.; Lysiak-Szydłowska, W.; Yacoub, M.H. Propionyl-L-carnitine improves hemodynamics and metabolic markers of cardiac perfusion during coronary surgery in diabetic patients. *Cardiovasc. Drugs Ther.* **2005**, *19*, 267–275.
240. Xiang, D.; Sun, Z.; Xia, J.; Dong, N.; Du, X.; Chen, X. Effect of L-carnitine on cardiomyocyte apoptosis and cardiac function in patients undergoing heart valve replacement operation. *J. Huazhong Univ. Sci. Technol. Med. Sci.* **2005**, *25*, 501–504.
241. Łapiński, T.W.; Grzeszczuk, A. The impact of carnitine on serum ammonia concentration and lipid metabolism in patients with alcoholic liver cirrhosis. *Pol. Merkur. Lekarski* **2003**, *15*, 38–41.
242. Lim, C.Y.; Jun, D.W.; Jang, S.S.; Cho, W.K.; Chae, J.D.; Jun, J.H. Effects of carnitine on peripheral blood mitochondrial DNA copy number and liver function in non-alcoholic fatty liver disease. *Korean J. Gastroenterol.* **2010**, *55*, 384–389.
243. Malaguarnera, M.; Pistone, G.; Astuto, M.; dell'Arte, S.; Finocchiaro, G.; lo Giudice, E.; Pennisi, G. L-Carnitine in the treatment of mild or moderate hepatic encephalopathy. *Dig. Dis.* **2003**, *21*, 271–275.
244. Malaguarnera, M.; Vacante, M.; Giordano, M.; Motta, M.; Bertino, G.; Pennisi, M.; Neri, S.; Malaguarnera, M.; Li Volti, G.; Galvano, F. L-Carnitine supplementation improves hematological pattern in patients affected by HCV treated with Peg interferon-α 2b plus ribavirin. *World J. Gastroenterol.* **2011**, *17*, 4414–4420.

245. Neri, S.; Pistone, G.; Saraceno, B.; Pennisi, G.; Luca, S.; Malaguarnera, M. L-Carnitine decreases severity and type of fatigue induced by interferon-alpha in the treatment of patients with hepatitis C. *Neuropsychobiology* **2003**, *47*, 94–97.
246. Malaguarnera, M.; Vacante, M.; Giordano, M.; Pennisi, G.; Bella, R.; Rampello, L.; Malaguarnera, M.; Li Volti, G.; Galvano, F. Oral acetyl-L-carnitine therapy reduces fatigue in overt hepatic encephalopathy: A randomized, double-blind, placebo-controlled study. *Am. J. Clin. Nutr.* **2011**, *93*, 799–808.
247. Tomassini, V.; Pozzilli, C.; Onesti, E.; Pasqualetti, P.; Marinelli, F.; Pisani, A.; Fieschi, C. Comparison of the effects of acetyl L-carnitine and amantadine for the treatment of fatigue in multiple sclerosis: Results of a pilot, randomised, double-blind, crossover trial. *J. Neurol. Sci.* **2004**, *218*, 103–108.
248. Gavrilova, S.I.; Kalyn, IaB.; Kolykhalov, I.V.; Roshchina, I.F.; Selezneva, N.D. Acetyl-L-carnitine (carnicetine) in the treatment of early stages of Alzheimer's disease and vascular dementia. *Zh. Nevrol. Psikhiatr. Im. S. S. Korsakova* **2011**, *111*, 16–22.
249. Tarighat Esfanjani, A.; Mahdavi, R.; Ebrahimi Mameghani, M.; Talebi, M.; Nikniaz, Z.; Safaiyan, A. The effects of magnesium, L-carnitine, and concurrent magnesium-L-carnitine supplementation in migraine prophylaxis. *Biol. Trace Elem. Res.* **2012**, *150*, 42–48.
250. Pistone, G.; Marino, A.; Leotta, C.; dell'Arte, S.; Finocchiaro, G.; Malaguarnera, M. Levocarnitine administration in elderly subjects with rapid muscle fatigue: Effect on body composition, lipid profile and fatigue. *Drugs Aging* **2003**, *20*, 761–767.
251. Miyagawa, T.; Kawamura, H.; Obuchi, M.; Ikesaki, A.; Ozaki, A.; Tokunaga, K.; Inoue, Y.; Honda, M. Effects of oral L-carnitine administration in narcolepsy patients: A randomized, double-blind, cross-over and placebo-controlled trial. *PLoS One* **2013**, *8*, e53707.
252. Rossini, M.; di Munno, O.; Valentini, G.; Bianchi, G.; Biasi, G.; Cacace, E.; Malesci, D.; la Montagna, G.; Viapiana, O.; Adami, S. Double-blind, multicenter trial comparing acetyl L-carnitine with placebo in the treatment of fibromyalgia patients. *Clin. Exp. Rheumatol.* **2007**, *25*, 182–188.
253. Vermeulen, R.C.; Scholte, H.R. Exploratory open label, randomized study of acetyl- and propionyl-carnitine in chronic fatigue syndrome. *Psychosom. Med.* **2004**, *66*, 276–282.
254. Beghi, E.; Pupillo, E.; Bonito, V.; Buzzi, P.; Caponnetto, C.; Chiò, A.; Corbo, M.; Giannini, F.; Inghilleri, M.; Bella, V.L.; *et al.* Randomized double-blind placebo-controlled trial of acetyl-L-carnitine for ALS. *Amyotroph. Lateral Scler. Frontotemporal Degener.* **2013**, *14*, 397–405.
255. Cruciani, R.A.; Dvorkin, E.; Homel, P.; Malamud, S.; Culliney, B.; Lapin, J.; Portenoy, R.K.; Esteban-Cruciani, N. Safety, tolerability and symptom outcomes associated with L-carnitine supplementation in patients with cancer, fatigue, and carnitine deficiency: A phase I/II study. *J. Pain Symptom Manag.* **2006**, *32*, 551–559.
256. Cruciani, R.A.; Dvorkin, E.; Homel, P.; Culliney, B.; Malamud, S.; Lapin, J.; Portenoy, R.K.; Esteban-Cruciani, N. L-Carnitine supplementation in patients with advanced cancer and carnitine deficiency: A double-blind, placebo-controlled study. *J. Pain Symptom Manag.* **2009**, *37*, 622–631.

257. Cruciani, R.A.; Zhang, J.J.; Manola, J.; Cella, D.; Ansari, B.; Fisch, M.J. L-Carnitine supplementation for the management of fatigue in patients with cancer: An eastern cooperative oncology group phase III, randomized, double-blind, placebo-controlled trial. *J. Clin. Oncol.* **2012**, *30*, 3864–3869.
258. Mantovani, G. Randomised phase III clinical trial of 5 different arms of treatment on 332 patients with cancer cachexia. *Eur. Rev. Med. Pharmacol. Sci.* **2010**, *14*, 292–301.
259. Madeddu, C.; Dessì, M.; Panzone, F.; Serpe, R.; Antoni, G.; Cau, M.C.; Montaldo, L.; Mela, Q.; Mura, M.; Astara, G.; *et al.* Randomized phase III clinical trial of a combined treatment with carnitine + celecoxib ± megestrol acetate for patients with cancer-related anorexia/cachexia syndrome. *Clin. Nutr.* **2012**, *31*, 176–182.
260. Kraft, M.; Kraft, K.; Gärtner, S.; Mayerle, J.; Simon, P.; Weber, E.; Schütte, K.; Stieler, J.; Koula-Jenik, H.; Holzhauer, P.; *et al.* L-Carnitine-supplementation in advanced pancreatic cancer (CARPAN) a randomized multicentre trial. *Nutr. J.* **2012**, *11*, 52.
261. Hershman, D.L.; Unger, J.M.; Crew, K.D.; Minasian, L.M.; Awad, D.; Moinpour, C.M.; Hansen, L.; Lew, D.L.; Greenlee, H.; Fehrenbacher, L.; *et al.* Randomized double-blind placebo-controlled trial of acetyl-L-carnitine for the prevention of taxane-induced neuropathy in women undergoing adjuvant breast cancer therapy. *J. Clin. Oncol.* **2013**, *31*, 2627–2633.
262. Schöls, L.; Zange, J.; Abele, M.; Schillings, M.; Skipka, G.; Kuntz-Hehner, S.; van Beekvelt, M.C.; Colier, W.N.; Müller, K.; Klockgether, T.; *et al.* L-Carnitine and creatine in Friedreich's ataxia. A randomized, placebo-controlled crossover trial. *J. Neural Transm.* **2005**, *112*, 789–796.
263. El-Beshlawy, A.; Ragab, L.; Fattah, A.A.; Ibrahim, I.Y.; Hamdy, M.; Makhlouf, A.; Aoun, E.; Hoffbrand, V.; Taher, A. Improvement of cardiac function in thalassemia major treated with L-carnitine. *Acta Haematol.* **2004**, *111*, 143–148.
264. Karimi, M.; Mohammadi, F.; Behmanesh, F.; Samani, S.M.; Borzouee, M.; Amoozgar, H.; Haghpanah, S. Effect of combination therapy of hydroxyurea with L-carnitine and magnesium chloride on hematologic parameters and cardiac function of patients with beta-thalassemia intermedia. *Eur. J. Haematol.* **2010**, *84*, 52–58.
265. Ellaway, C.J.; Peat, J.; Williams, K.; Leonard, H.; Christodoulou, J. Medium-term open label trial of L-carnitine in Rett syndrome. *Brain Dev.* **2001**, *23*, S85–S89.
266. Torrioli, M.G.; Vernacotola, S.; Peruzzi, L.; Tabolacci, E.; Mila, M.; Militerni, R.; Musumeci, S.; Ramos, F.J.; Frontera, M.; Sorge, G.; *et al.* A double-blind, parallel, multicenter comparison of L-acetylcarnitine with placebo on the attention deficit hyperactivity disorder in fragile X syndrome boys. *Am. J. Med. Genet. A* **2008**, *146*, 803–812.
267. Geier, D.A.; Kern, J.K.; Davis, G.; King, P.G.; Adams, J.B.; Young, J.L.; Geier, M.R. A prospective double-blind, randomized clinical trial of levocarnitine to treat autism spectrum disorders. *Med. Sci. Monit.* **2011**, *17*, PI15–PI23.
268. Osio, M.; Muscia, F.; Zampini, L.; Nascimbene, C.; Mailland, E.; Cargnel, A.; Mariani, C. Acetyl-L-carnitine in the treatment of painful antiretroviral toxic neuropathy in human immunodeficiency virus patients: An open label study. *J. Peripher. Nerv. Syst.* **2006**, *11*, 72–76.
269. Youle, M.; Osio, M.; ALCAR Study Group. A double-blind, parallel-group, placebo-controlled, multicentre study of acetyl L-carnitine in the symptomatic treatment of antiretroviral toxic neuropathy in patients with HIV-1 infection. *HIV Med.* **2007**, *8*, 241–250.

270. Benedini, S.; Perseghin, G.; Terruzzi, I.; Scifo, P.; Invernizzi, P.L.; del Maschio, A.; Lazzarin, A.; Luzi, L. Effect of L-acetylcarnitine on body composition in HIV-related lipodystrophy. *Horm. Metab. Res.* **2009**, *41*, 840–845.
271. Pignatelli, P.; Tellan, G.; Marandola, M.; Carnevale, R.; Loffredo, L.; Schillizzi, M.; Proietti, M.; Violi, F.; Chirletti, P.; Delogu, G. Effect of L-carnitine on oxidative stress and platelet activation after major surgery. *Acta Anaesthesiol. Scand.* **2011**, *55*, 1022–1028.
272. Mikhailova, T.L.; Sishkova, E.; Poniewierka, E.; Zhidkov, K.P.; Bakulin, I.G.; Kupcinskas, L.; Lesniakowski, K.; Grinevich, V.B.; Malecka-Panas, E.; Ardizzone, S.; *et al.* Randomised clinical trial: The efficacy and safety of propionyl-L-carnitine therapy in patients with ulcerative colitis receiving stable oral treatment. *Aliment. Pharmacol. Ther.* **2011**, *34*, 1088–1097.
273. Mattiazzi, M.; Vijayvergiya, C.; Gajewski, C.D.; deVivo, D.C.; Lenaz, G.; Wiedmann, M.; Manfredi, G. The mtDNA *T8993G* (*NARP*) mutation results in an impairment of oxidative phosphorylation that can be improved by antioxidants. *Hum. Mol. Genet.* **2004**, *13*, 869–879.
274. Chinnery, P.F. Mitochondrial Disorders Overview. Synonyms: Mitochondrial Encephalomyopathies, Mitochondrial Myopathies, Oxidative Phosphorylation Disorders, Respiratory Chain Disorders. Available online: http://www.ncbi.nlm.nih.gov/books/NBK1224 (accessed on 3 November 2014).
275. Di Mauro, S.; Mancuso, M. Mitochondrial diseases: Therapeutic approaches. *Biosci. Rep.* **2007**, *27*, 125–137.
276. Smith, R.A.; Murphey, M.P. Mitochondria-targeted antioxidants as therapies. *Discov. Med.* **2011**, *11*, 106–114.
277. Lee, S.H.; Kim, M.J.; Kim, B.J.; Kim, S.R.; Chun, S.; Ryu, J.S.; Kim, G.S.; Lee, M.C.; Koh, J.M.; Chung, S.J. Homocysteine-lowering therapy or antioxidant therapy for bone loss in Parkinson's disease. *Mov. Disord.* **2010**, *25*, 332–340.
278. Pagano, G.; Aiello Talamanca, A.; Castello, G.; d'Ischia, M.; Pallardó, F.V.; Petrović, S.; Porto, B.; Tiano, L.; Zatterale, A. From clinical description to *in vitro* and animal studies, and backwards to patients: Oxidative stress and mitochondrial dysfunction in Fanconi anaemia. *Free Radic. Biol. Med.* **2013**, *58*, 118–125.
279. Seco-Cervera, M.; Spis, M.; García-Giménez, J.L. Oxidative stress and antioxidant response in fibroblasts from Werner and Atypical Werner Syndromes. *Aging* **2014**, *6*, 231–245.
280. Bhagavan, H.N.; Chopra, R.K. Plasma coenzyme Q10 response to oral ingestion of coenzyme Q10 formulations. *Mitochondrion* **2007**, *7*, S78–S88.
281. Roginsky, V.A.; Tashlitsky, V.N.; Skulachev, V.P. Chain-breaking antioxidant activity of reduced forms of mitochondria-targeted quinones, a novel type of geroprotectors. *Aging* **2009**, *1*, 481–489.
282. Teichert, J.; Hermann, R.; Ruus. P.; Preiss, R. Plasma kinetics, metabolism, and urinary excretion of alpha-lipoic acid following oral administration in healthy volunteers. *J. Clin. Pharmacol.* **2003**, *43*, 1257–1267.
283. Pettegrew, J.W.; Levine, J.; McClure, R.J. Acetyl-L-carnitine physical-chemical, metabolic and therapeutic properties: Relevance for its mode of action in Alzheimer'disease and geriatric depression. *Mol. Psychiatry* **2000**, *5*, 616–632.
284. Madiraju, P.; Pande, S.V.; Prentki, M.; Murthy Madiraju, S.R. Mitochondrial acetylcarnitine provides acetyl groups for nuclear histone acetylation. *Epigenetics* **2009**, *4*, 4–6

285. Rosca, M.G.; Lemieux, H.; Hoppel, C.L. Mitochondria in the ederly: Is acetylcarnitine a rejuvinator? *Adv. Drug Deliv. Rev.* **2009**, *61*, 1332–1342.
286. Breithaupt-Grögler, K.; Niebch, G.; Schneider, E.; Erb, K.; Hermann, R.; Blume, H.H.; Schug, B.S.; Belz, G.G. Dose-proportionality of oral thioctic acid—Coincidence of assessments via pooled plasma and individual data. *Eur. J. Pharm. Sci.* **1999**, *8*, 57–65.
287. Rebouche, C.J. Kinetics, pharmacokinetics, and regulation of L-carnitine and acetyl-L-carnitine metabolism. *Ann. N. Y. Acad. Sci.* **2004**, *1033*, 30–41.
288. Camp, K.M.; Lloyd-Puryear, M.A.; Yao, L.; Groft, S.C.; Parisi, M.A.; Mulberg, A.; Gopal-Srivastava, R.; Cederbaum, S.; Enns, G.M.; Ershow, A.G.; *et al.* Expanding research to provide an evidence base for nutritional interventions for the management of inborn errors of metabolism. *Mol. Genet. Metab.* **2013**, *109*, 319–328.

3

Skeletal Muscle Mitochondrial Energetic Efficiency and Aging

Raffaella Crescenzo, Francesca Bianco, Arianna Mazzoli, Antonia Giacco, Giovanna Liverini and Susanna Iossa *

Department of Biology, University of Naples "Federico II", Napoli I-80126, Italy;
E-Mails: rcrescen@unina.it (R.C.); francibianco@yahoo.it (F.B.); arimazzoli@hotmail.it (A.M.); antonia.giacco@hotmail.it (A.G.); liverini@unina.it (G.L.)

* Author to whom correspondence should be addressed; E-Mail: susiossa@unina.it

Academic Editor: Lars Olson

Abstract: Aging is associated with a progressive loss of maximal cell functionality, and mitochondria are considered a key factor in aging process, since they determine the ATP availability in the cells. Mitochondrial performance during aging in skeletal muscle is reported to be either decreased or unchanged. This heterogeneity of results could partly be due to the method used to assess mitochondrial performance. In addition, in skeletal muscle the mitochondrial population is heterogeneous, composed of subsarcolemmal and intermyofibrillar mitochondria. Therefore, the purpose of the present review is to summarize the results obtained on the functionality of the above mitochondrial populations during aging, taking into account that the mitochondrial performance depends on organelle number, organelle activity, and energetic efficiency of the mitochondrial machinery in synthesizing ATP from the oxidation of fuels.

Keywords: mitochondria; aging; skeletal muscle; proton leak

---

1. Introduction

Increasing age leads to a decline in cell functionality, generally termed as "aging" [1]. All tissue and organs of the body are involved in the phenomenon of aging, but the extent of the cellular impairment is greatly variable, since post-mitotic tissues are the most sensitive targets of the aging process [2]. Among the latter, skeletal muscle tissue is profoundly affected during aging, and its functional decline is characterized by a progressive atrophy, that becomes more severe with advancing age and that from a certain point onwards can lead to mobility impairment, increased risk of falls, and physical frailty [3,4]. The loss of function in skeletal muscle is also responsible for the development of age-associated insulin resistance and related metabolic disturbances [5–9]. For these reasons, understanding the mechanisms underlying aging in skeletal muscle is fundamental for promotion of health and mobility in the elderly. To this end, the most used animal model of human aging is represented by ad libitum fed caged rodents, that exhibit a sedentary lifestyle and unrestricted diet, and that develop an age-induced spontaneous obesity and insulin resistance already at 4 months of age [7].

Although the loss of skeletal muscle protein mass during aging could at least partially explain the decline in muscle performance, a role for mitochondria has been postulated in the aging process [10,11]. In fact, mitochondria have a major role in energetic homeostasis by determining ATP availability in the cells. A decrease in mitochondrial function could therefore cause an inability to meet cellular ATP demands, so that skeletal muscle cells lose their capacity to adapt to physiological stress imposed across the entire lifespan [12]. In addition, dysfunctional mitochondria could contribute to the development of age-induced insulin resistance [13], since mitochondrial oxidative capacity has been considered a good predictor of insulin sensitivity [14].

One important issue that should be taken into account when studying mitochondria in skeletal muscle is that the mitochondrial population is heterogeneous, composed of mitochondria located either beneath the sarcolemmal membrane (subsarcolemmal, SS) or between the myofibrils (intermyofibrillar, IMF) [15] (Figure 1). Since these two mitochondrial populations exhibit different energetic characteristics and can be differently affected by physiological stimuli [16], it is important that both are separately studied. Therefore, the purpose of the present review is to summarize the results obtained on the functionality of the above mitochondrial populations during aging, taking into account that the mitochondrial performance depends on organelle number, organelle activity, and energetic efficiency of the mitochondrial machinery in synthesizing ATP from the oxidation of fuels. A search in PubMed of relevant articles was conducted, by using query "skeletal muscle mitochondria and aging", "subsarcolemmal mitochondria and aging", and "intermyofibrillar mitochondria and aging", with the inclusion of related articles by the same groups.

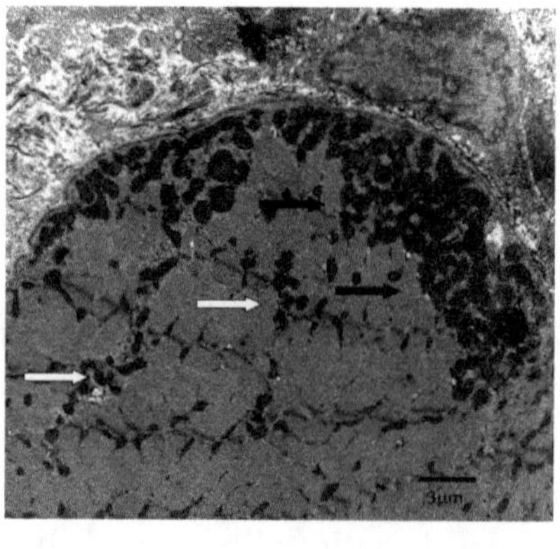

Figure 1. Heterogeneous mitochondrial populations in skeletal muscle cells. Scale bar = 3 μm. White arrows = Intermyofibrillar mitochondria; Black arrows = Subsarcolemmal mitochondria.

2. Age-Related Changes in Mitochondria

Although many studies have addressed the issue of mitochondrial performance attenuation during aging in skeletal muscle, the obtained results are contradictory, with some papers reporting impairment of mitochondria with increasing age [17–23] and others showing no age-induced change [24,25], underscoring the complexity of understanding in this area.

The discrepancy between the published data can partly be explained by differences in experimental approach. However, even when comparing the results obtained with similar methodological approach, divergent outcomes are evident. For example, among the studies that have measured the activity of enzymatic complexes, such as citrate synthase and electron transport chain complexes, to obtain indirect insights into the energy producing (respiratory) capacity of the mitochondria, some of them have reported an age-dependent decrease in aging muscle [26–28] while other studies found no variation [29] or reported a variable response in different muscles [30]. An alternative approach used to study mitochondrial alteration in aging muscle is assessing total mitochondrial content. Again, some studies reported that mitochondrial content is reduced in aging muscle [31,32] and others found no change [21,33–35]. There is however consensus on the finding that aging skeletal muscle has a blunted capacity for generation of new mitochondria in response to both endurance exercise training [36] and chronic electrical stimulation [37,38].

To our knowledge, studies carried out specifically on mitochondria located beneath the sarcolemmal membrane (SS) and between the myofibrils (IMF) from skeletal muscle during aging are scarce. In a pioneering research, Farrar *et al.* [39] found that ADP-stimulated respiration did not change with increasing age in IMF and SS mitochondria, while aging decreased the amount of IMF proteins. Chabi *et al.* [32] studied senescent (3 years) rats and have found that in SS and IMF mitochondria the capacity for ATP production was reduced, as a result of diminished mitochondrial content per gram of muscle. Drew *et al.* [40] found a decrease in ATP production between 12 and 26 months of age in SS mitochondria from gastrocnemius muscle in Fisher rats. Very few studies specifically addressed the differential regulation of SS and IMF mitochondria with aging in humans. These studies evaluated the SS and IMF mitochondrial content of skeletal muscle in young and old men and reported no age-dependent change in both mitochondrial populations [34,41]. From all the above results, it emerges a differential effect of aging on SS and IMF mitochondria, at least in animal models. More studies on humans are needed to validate the differential effect of aging on the two mitochondrial populations. In addition, studies on changes in mitochondrial function induced by aging or other physiological stimuli should be carried out on the two different mitochondrial populations existing in skeletal muscle.

3. Mitochondrial Energetic Efficiency during Aging

From the analysis of the published results on the differential effect of aging on IMF and SS mitochondria, no clear picture can be obtained, partly because of differences in the experimental approach used to evaluate mitochondrial function. The mitochondrial performance depends on organelle number, organelle activity, and energetic efficiency of the mitochondrial machinery in synthesizing ATP from the oxidation of fuels. In addition, it is well known that the amount of fuels oxidized by the cell is dictated mainly by ATP turnover rather than by mitochondrial oxidative capacity and therefore a decrease in mitochondrial capacity and/or number becomes more important when cells increase their metabolic activity, *i.e.*, during contraction, while an increased mitochondrial efficiency still alter the amount of oxidized fuels also at rest (Figure 2). The efficiency with which dietary calories are converted to ATP is determined by the degree of coupling of oxidative phosphorylation. If the respiratory chain is highly efficient at pumping protons out of the mitochondrial inner membrane, and the ATP synthesis is highly efficient at converting the proton flow through its proton channel into ATP (from ADP), then the mitochondria will generate maximum ATP and minimum heat per calorie. In contrast, if the efficiency of proton pumping is reduced and/or more protons are required to make each ATP molecule, then each calorie will yield less ATP.

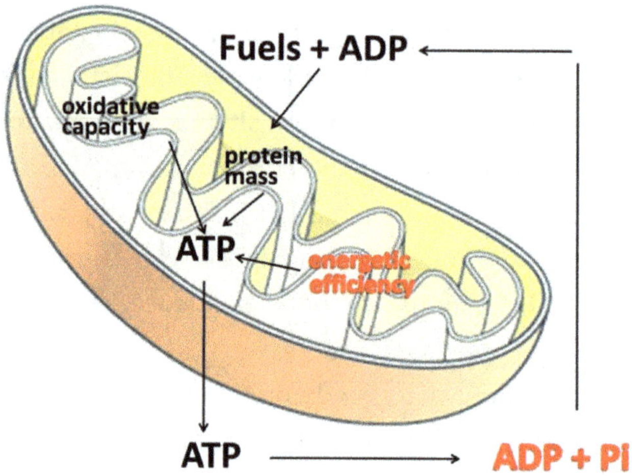

Figure 2. Factors affecting cellular fuel oxidation. The amount of burned fuels mainly depends on mitochondrial energetic efficiency and ATP turnover (in red). Mitochondrial mass and oxidative capacity are less important because mitochondria are thought to have a much greater capacity to generate ATP than what is usually required [42].

The main point of regulation of the oxidative phosphorylation efficiency [43] is represented by the degree of coupling between oxygen consumption and ATP synthesis, which is always lower than 1 and can vary according to the metabolic needs of the cell [44]. Among the factors, which affect mitochondrial degree of coupling, an important role is played by the permeability of the mitochondrial inner membrane to H^+ ions (leak). It is now well known that mitochondrial inner membrane exhibit a basal proton leak pathway, whose contribution to the basal metabolic rate in rats has been estimated to be about 20%–25% [45]. In addition, it is well known that fatty acids can act as natural uncouplers of oxidative phosphorylation, by generating a fatty acid-dependent proton leak pathway [46–48], which is a function of the amount of unbound fatty acids in the cell.

The issue of mitochondrial efficiency has been explored in humans *in vivo*, and it has been found that the effect of aging is fiber type-dependent. In fact, in mildly uncoupled fiber types (*i.e.*, tibialis anterior) no age effect is evident, while a substantial uncoupling takes place with aging in well coupled fiber types (*i.e.*, dorsal interosseous) [49]. Due to the *in vivo* conditions, these studies do not distinguish between SS and IMF mitochondria. Since aging has been show to selectively affect IMF but not SS mitochondria in heart [50], and taking into account the above considerations, a study was carried out on the putative changes in mitochondrial performance and efficiency during aging in SS and IMF mitochondria. In the transition from young adulthood (60 days) to adulthood (180 days), SS and IMF skeletal muscle mitochondria displayed an increase in the degree of coupling and efficiency, as well as a decreased fatty acid dependent proton leak [51]. The above modifications in mitochondrial performance occurred concomitantly with an increase in whole body lipids and plasma non-esterified fatty acids (NEFA) [51], suggesting a link between skeletal muscle increased mitochondrial efficiency and metabolic impairments.

These results were extended to the evaluation of how the progression of aging affects skeletal muscle mitochondrial function by measuring mitochondrial respiratory capacity and proton leak in SS and IMF mitochondria from adult (six months) and old (two years) rats [52]. A significant decrease in oxidative capacity was found in skeletal muscle homogenates, as well as in SS and IMF mitochondria from old rats. Oxidative capacity measured in the homogenate reflects the product of mitochondrial mass and activity, while oxidative capacity in isolated organelles depends only on mitochondrial activity, and therefore the similar age-induced decrease in oxidative capacity in homogenates and isolated mitochondria found in old rats could be indicative of a lack of changes in mitochondrial mass. A decreased mitochondrial mass has been found in senescent rats (36 months old) [32], thus suggesting that an impairment of mitochondrial biogenesis occurs in late aging and/or it takes place selectively in specific muscles, such as gastrocnemius, whose mitochondrial mass has been found decreased in old rats [20,53].

Both mitochondrial populations from old rats exhibited a significant decrease in proton leak [52], suggesting that with increasing age the efficiency of oxidative phosphorylation increases both in SS and IMF mitochondria. Similar results have been obtained *in vivo* in aged rat skeletal muscle, where a trend for a higher coupling efficiency was found [20]. Skeletal muscle cells in sedentary laboratory rats operate in conditions of low ADP availability, near to state 4 (when no ADP is available), with a high contribution of proton leak to total oxygen consumed by mitochondria [54], and therefore the decreased proton leak found in SS and IMF mitochondria from old rats is physiologically relevant. When mitochondria are more efficient, less substrates are oxidized to obtain ATP. Therefore, the increased mitochondrial coupling in skeletal muscle could contribute to the decreased energy expenditure that is evident even after the decrease in lean mass has been taken into account and that is at the basis of age-induced obesity, since skeletal muscle energy metabolism accounts for about 30% of whole body energy expenditure in resting conditions [45].

One interesting question is: What could be the impact of the increased mitochondrial coupling on ATP yield? Although it is very difficult to calculate the theoretical difference in ATP yield per calorie in highly efficient *vs.* inefficient proton pumping, a rough estimate could be obtained using a published estimate of the control value of proton leak on *P*/*O* ratio in skeletal muscle [55]. This value is reported to be −0.72 [55], so we can calculate that the 40% decrease in proton leak found in old rats [52] would result in a 29% increase in *P*/*O* ratio, and therefore in the amount of ATP obtained per unit of oxygen consumed. However, when mitochondria are more coupled, ATP is produced at a slower rate [44,56], that could be unable to meet cellular energy demands, especially during skeletal muscle contraction. In fact, in elderly subjects it has been found that a lower speed of ATP production is associated with higher fatigability [57].

Another deleterious consequence of reduced substrate burning in skeletal muscle could be intracellular triglyceride accumulation and lipotoxicity, since NEFA serum levels are significantly higher in older rats. In fact, in conditions of increased plasma NEFA, more fatty acids enter in the cells and if they are poorly oxidized, they accumulate intracellularly in myocytes mainly as long-chain fatty acyl-CoA, monoacylglcyerol, diacylglycerol, phosphatidic acid, triacylglycerol and ceramide. Among these fatty acid derivatives, high intramyocellular levels of diacylglycerol and ceramides are directly associated with insulin resistance [58]. The metabolic implications of increased mitochondrial energetic efficiency are summarized in Figure 3.

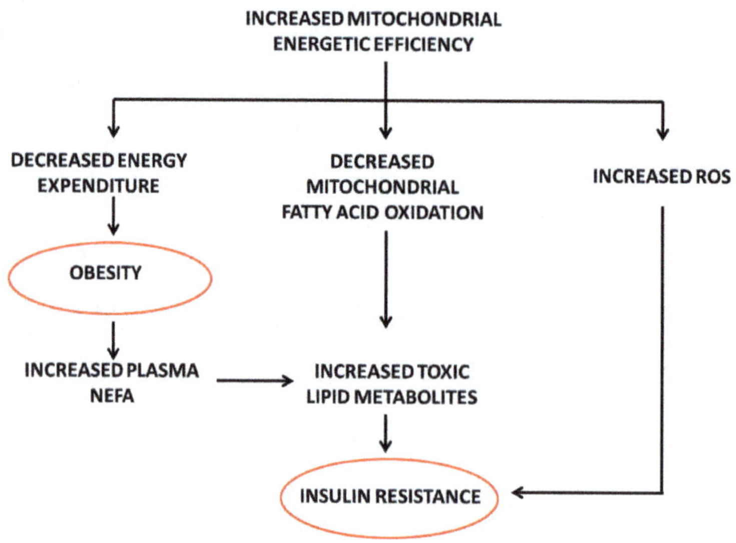

Figure 3. Summary of the metabolic implications of the increased mitochondrial energetic efficiency in skeletal muscle with aging. The red circles identify the pathological outcomes.

4. Oxidative Stress in Aging Mitochondria

Reactive oxygen species (ROS) production increases when mitochondrial potential is higher [59,60], and therefore increased mitochondrial energetic efficiency could induce a condition of increased oxidative stress. In fact, one of the postulated physiological roles for the uncoupling effect of fatty acids is to maintain mitochondrial membrane potential below the critical threshold for ROS production, especially *in situ*ations of low rates of ATP turnover, such as in resting skeletal muscle [61]. It has been proposed that ROS generation during the normal oxidative activity of mitochondria leads to damage of lipids, proteins and DNA, especially in postmitotic tissues, such as skeletal muscle, and that this oxidative damage is at the basis of the biological phenomenon of cellular aging [10,11]. Therefore, the increased coupling of mitochondrial oxidative phosphorylation found in 180 day-old rats [51] could led to the oxidative damage of skeletal muscle cell, detectable later in life. However, the increased uncoupling protein 3 (UCP3) protein content found in SS and IMF mitochondria from 180 days old rats could be involved in the protection from oxidative damage [62,63]. In fact, it has been proposed that UCP3 translocates fatty acid peroxides from the inner to the outer membrane leaflet, thus preserving macromolecules from being oxidized by very aggressive fatty acid peroxides [62,63], while its uncoupling effect is considered very low, due to its low content in skeletal muscle mitochondria [64]. Therefore, the up-regulation of UCP3 during aging could buffer oxidative damage, that otherwise could be even higher.

In 2 year-old rats, an increase in the degree of lipid peroxidation was found only in SS mitochondria, although the decrease in proton leak is the same in both mitochondrial populations [52], while

Chabi *et al.* [32] studied senescent (3 years) rats and found that ROS production was enhanced in SS and IMF mitochondria. The differential susceptibility to oxidative stress of SS and IMF mitochondria could depend on their different content of UCP3. In fact, a significant increase in UCP3 content was evident in SS and IMF mitochondria from old rats, but it was more marked in IMF mitochondria (about 10-fold) than in SS mitochondria (about 5-fold). Therefore, the lower increase in UCP3 content in SS mitochondria is probably the cause of the higher oxidative damage found in this mitochondrial population, while IMF mitochondria seem more protected by oxidative damage by the marked up-regulation of UCP3, so preserving the capacity to produce ATP for muscle contraction. One could speculate that, as aging proceeds, SS mitochondria undergo progressive oxidative damage with loss of functional activity. In fact, in senescent (30–36 month-old) animals, an increase in mitochondrial proton leak [65], a decrease in mitochondrial coupling [66] or a decrease in mitochondrial membrane potential in SS but not in IMF mitochondria has been found [32]. In conclusion, the increased coupling of SS mitochondria causes an increase of the oxidative damage in this mitochondrial population, that is located beneath the plasma membrane and provides ATP for membrane transports and signal transduction pathways [67]. On the other hand, IMF mitochondria, that provide ATP for muscle contraction, seems to be more protected from oxidative damage, and could thus increase their oxidative capacity and density in response to endurance training even in old age [68].

A decreased UCP3 content has been found in skeletal muscle mitochondria from old rats [31,69,70], but these results have been obtained using Fischer 344 rats, a rat strain that gain weight only moderately with age compared with other strains (*i.e.*, Sprague-Dawley, Wistar, Long Evans) [8]. Therefore, it can be hypothesized that the regulation of UCP3 with aging in skeletal muscle mitochondria is strain-dependent. Therefore, it is possible that age-induced oxidative damage in skeletal muscle and age-induced obesity are intimately linked. Further studies on the degree of obesity and oxidative damage induced by aging in different strains and species are needed to substantiate the hypothesis.

5. Conclusions

In the rat model of human obesity the progression of aging is accompanied by an increased efficiency of SS and IMF mitochondria but an increased oxidative damage occurs only in the SS population. Therefore, a differential susceptibility of SS and IMF mitochondria to aging-induced damage emerges, although more studies on humans are needed to validate the differential effect of aging on the two mitochondrial populations. These observations also indicate that studies on changes in mitochondrial function induced by aging or other physiological stimuli should be carried out on the two different mitochondrial populations existing in skeletal muscle.

Acknowledgments

The work was supported by University “Federico II” of Naples.

Author Contributions

All authors contributed to the writing of this review.

References

1. Figueiredo, P.A.; Mota, M.P.; Appell, H.J.; Duarte, J.A. The role of mitochondria in aging of skeletal muscle. *Biogerontology* **2008**, *9*, 67–84.
2. Kwong, L.K.; Sohal, R.S. Age-related changes in activities of mitochondrial electron transport complexes in various tissues of the mouse. *Arch. Biochem. Biophys.* **2000**, *373*, 16–22.
3. Cruz-Jentoft, J.; Landi, F.; Topinkova, E.; Michel, J.P. Understanding sarcopenia as a geriatric syndrome. *Curr. Opin. Clin. Nutr. Metab. Care* **2010**, *13*, 1–7.
4. Marzetti, E.; Calvani, R.; Cesari, M.; Buford, T.W.; Lorenzi, M.; Behnke, B.J.; Leeuwenburgh, C. Mitochondrial dysfunction and sarcopenia of aging: From signaling pathways to clinical trials. *Int. J. Biochem. Cell Biol.* **2013**, *45*, 2288–2301.
5. Shimokata, H.; Tobin, J.; Muller, D.C.; Elahi, D.; Coon, P.J.; Andres, R. Studies in the distribution of body fat: 1. Effects of age, sex and obesity. *J. Gerontol.* **1989**, *44*, M66–M73.
6. Reaven, G.M.; Reaven, E.P. Age, glucose intolerance, and non-insulin dependent diabetes mellitus. *J. Am. Geriatr. Soc.* **1985**, *33*, 286–290.
7. Barzilai, N.; Rossetti, L. Relationship between changes in body composition and insulin responsiveness in models of the aging rat. *Am. J. Physiol.* **1995**, *269*, E591–E597.
8. Larkin, L.M.; Reynolds, T.H.; Supiano, M.A.; Kahn, B.B.; Halter, J.B. Effect of aging and obesity on insulin responsiveness and glut-4 glucose transporter content in skeletal muscle of Fisher 344 × Brown Norway rats. *J. Gerontol.* **2001**, *56*, B486–B492.
9. Iossa, S.; Mollica, M.P.; Lionetti, L.; Crescenzo, R.; Botta, M.; Barletta, A.; Liverini, G. Acetyl-L-carnitine supplementation differently influences nutrient partitioning, serum leptin concentration and skeletal muscle mitochondrial respiration in young and old rats. *J. Nutr.* **2002**, *132*, 636–642.
10. Lenaz, G. Role of mitochondria in oxidative stress and ageing. *Biochim. Biophyis. Acta* **1998**, *1366*, 53–67.
11. Lenaz, G.; D'Aurelio, M.; Merlo Pich, M.; Genova, M.L.; Ventura, B.; Bovina, C.; Formiggini, G.; Parenti Castelli, G. Mitochondrial bioenergetics in aging. *Biochim. Biophys. Acta* **2000**, *1459*, 397–404.
12. Shigenaga, M.K.; Hagen, T.M.; Ames, B.N. Oxidative damage and mitochondrial decay in aging. *Proc. Natl. Acad. Sci. USA* **1994**, *91*, 10771–10778.
13. Petersen, K.F.; Befroy, D.; Dufour, S.; Dziura, J.; Arijan, C.; Rothman, D.L.; DiPietro, L.; Cline, G.W.; Shulman, G.I. Mitochondrial disfunction in the elderly: Possible role in insulin resistance. *Science* **2003**, *300*, 1140–1142.
14. Bruce, C.R.; Anderson, M.J.; Carey, A.L.; Newman, D.G.; Bonen, A.; Kriketos, A.D.; Cooney, G.J.; Hawley, J.A. Muscle oxidative capacity is a better predictor of insulin sensitivity than lipid status. *J. Clin. Endocrinol. Metab.* **2003**, *88*, 5444–5451.
15. Drew, B.; Leeuwenburgh, C. Ageing and subcellular distribution of mitochondria: Role of mitochondrial DNA deletions and energy production. *Acta Physiol. Scand.* **2004**, *182*, 333–341.

16. Mollica, M.P.; Lionetti, L.; Crescenzo, R.; D'Andrea, E.; Ferraro, M.; Liverini, G.; Iossa, S. Heterogeneous bioenergetic behaviour of subsarcolemmal and intermyofibrillar mitochondria in fed and fasted rats. *Cell. Mol. Life Sci.* **2006**, *63*, 358–366.
17. Figueiredo, P.A.; Powers, S.K.; Ferreira, R.M.; Appell, H.J.; Duarte, J.A. Aging impairs skeletal muscle mitochondrial bioenergetic function. *J. Gerontol. A Biol. Sci. Med. Sci.* **2009**, *64*, 21–33.
18. Trounce, I.; Byrne, E.; Marzuki, S. Decline in skeletal muscle mitochondrial respiratory chain function: Possible factor in ageing. *Lancet* **1989**, *1*, 637–639.
19. Conley, K.E.; Jubrias, S.A.; Esselman, P.C. Oxidative capacity and aging in human muscle. *J. Physiol.* **2000**, *526*, 203–210.
20. Gouspillou, G.; Bourdel-Marchasson, I.; Rouland, R.; Calmettes, G.; Biran, M.; Deschodt-Arsac, V.; Miraux, S.; Thiaudiere, E.; Pasdois, P.; Detaille, D.; *et al.* Mitochondrial energetics is impaired *in vivo* in aged skeletal muscle. *Aging Cell* **2014**, *13*, 39–48.
21. Gouspillou, G.; Sgarioto, N.; Kapchinsky, S.; Purves-Smith, F.; Norris, B.; Pion, C.H.; Barbat-Artigas, S.; Lemieux, F.; Taivassalo, T.; Morais. J.A.; *et al.* Increased sensitivity to mitochondrial permeability transition and myonuclear translocation of endonuclease G in atrophied muscle of physically active older humans. *FASEB J.* **2014**, *28*, 1621–1633.
22. Gouspillou, G.; Bourdel-Marchasson, I.; Rouland, R.; Calmettes, G.; Franconi, J.M.; Deschodt-Arsac, V.; Diolez, P. Alteration of mitochondrial oxidative phosphorylation in aged skeletal muscle involves modification of adenine nucleotide translocator. *Biochim. Biophys. Acta* **2010**, *1797*, 143–151.
23. Picard, M.; Ritchie, D.; Thomas, M.M.; Wright, K.J.; Hepple, R.T. Alterations in intrinsic mitochondrial function with aging are fiber type-specific and do not explain differential atrophy between muscles. *Aging Cell* **2011**, *10*, 1047–1055.
24. Rasmussen, U.F.; Krustrup, P.; Kjaer, M.; Rasmussen, H.N. Experimental evidence against the mitochondrial theory of aging. A study on isolated human skeletal muscle mitochondria. *Exp. Gerontol.* **2003**, *38*, 877–886.
25. Rasmussen, U.F.; Krustrup, P.; Kjaer, M.; Rasmussen, H.N. Human skeletal muscle mitochondrial metabolism in youth and senescence: No signs of functional changes of ATP formation and mitochondrial capacity. *Pflug. Arch.* **2003**, *446*, 270–278.
26. Desai, V.G.; Weindruch, R.; Hart, R.W.; Feuers, R.J. Influences of age and dietary restriction on gastrocnemius electron transport system activities in mice. *Arch. Biochem. Biophys.* **1996**, *333*, 145–151.
27. Hagen, J.L.; Krause, D.J.; Baker, D.J.; Fu, M.; Tarnopolsky, M.A.; Hepple, R.T. Skeletal muscle aging in F344BN F1-hybrid rats: I. Mitochondrial dysfunction contributes to the age-associated reduction in VO_{2max}. *J. Gerontol.* **2004**, *59*, 1099–1110.
28. Lombardi, A.; Silvestri, E.; Cioffi, F.; Senese, R.; Lanni, A.; Goglia, F.; de Lange, P.; Moreno, M. Defining the trascriptomic and proteomic profiles of rat ageing skeletal muscle by the use of a cDNA array, 2D- and Blue native-PAGE approach. *J. Proteomics* **2009**, *72*, 708–721.
29. Barrientos, A.; Casademont, J.; Rotig, A.; Miro, O.; Urbano-Marquez, A.; Rustin, P.; Cardellach, F. Absence of relationship between the level of electron transport chain activities and aging in human skeletal muscle. *Biochem. Biophys. Res. Commun.* **1996**, *229*, 536–539.

30. Lyons, C.N.; Mathieu-Costello, O.; Moyes, C.D. Regulation of skeletal muscle mitochondrial content during aging. *J. Gerontol.* **2006**, *61*, 3–13.
31. Kerner, J.; Turkaly, P.J.; Minkler, P.E.; Hoppel, C.L. Aging skeletal muscle mitochondria in the rat: Decreased uncoupling protein-3 content. *Am. J. Physiol.* **2001**, *281*, E1054–E1062.
32. Chabi, B.; Ljubicic, V.; Menzies, K.J.; Huang, J.H.; Saleem, A.; Hood, D.A. Mitochondrial function and apoptotic susceptibility in aging skeletal muscle. *Aging Cell* **2008**, *7*, 2–12.
33. Mathieu-Costello, O.; Ju, Y.; Trejo-Morales, M.; Cui, L. Greater capillary-fiber interface per fiber mitochondrial volume in skeletal muscles of old rats. *J. Appl. Physiol.* **2005**, *99*, 281–289.
34. Callahan, D.M.; Bedrin, N.G.; Subramanian, M.; Berking, J.; Ades, P.A.; Toth, M.J.; Miller, M.S. Age-related structural alterations in human skeletal muscle fibers and mitochondria are sex specific: Relationship to single-fiber function. *J. Appl. Physiol.* **2014**, *116*, 1582–1592.
35. Konopka, R.; Suer, M.K.; Wolff, C.A.; Harber, M.P. Markers of human skeletal muscle mitochondrial biogenesis and quality control: Effects of age and aerobic exercise training. *J. Gerontol.* **2014**, *69*, 371–378.
36. Betik, C.; Thomas, M.M.; Wright, K.J.; Riel, C.D.; Hepple, R.T. Exercise training from late middle age until senescence does not attenuate the declines in skeletal muscle aerobic function. *Am. J. Physiol.* **2009**, *297*, R744–R755.
37. Ljubicic, V.; Hood, D.A. Diminished contraction-induced intracellular signaling towards mitochondrial biogenesis in aged skeletal muscle. *Aging Cell* **2009**, *8*, 394–404.
38. Ljubicic, V.; Joseph, A.M.; Adhihetty, P.J.; Huang, J.H.; Saleem, A.; Uguccioni, G.; Hood, D.A. Molecular basis for an attenuated mitochondrial adaptive plasticity in aged skeletal muscle. *Aging* **2009**, *1*, 818–830.
39. Farrar, R.P.; Martin, T.P.; Ardies, C.M. The interaction of aging and endurance exercise upon the mitochondrial function of skeletal muscle. *J. Gerontol.* **1981**, *36*, 642–647.
40. Drew, B.; Phaneuf, S.; Dirks, A.; Selman, C.; Gredilla, R.; Lezza, A.; Barja, G.; Leeuwenburgh, C. Effects of aging and caloric restriction on mitochondrial energy production in gastrocnemius muscle and heart. *Am. J. Physiol.* **2003**, *284*, R474–R480.
41. Nielsen, J.; Suetaa, C.; Hvid, L.G.; Schroder, H.D.; Aagaard, P.; Orteblad, N. Subcellular localization-dependent decrements in skeletal muscle glycogen and mitochondria content following short-term disuse in young and old men. *Am. J. Physiol.* **2010**, *299*, E1053–E1060.
42. Holloszy, J.O. Skeletal muscle "mitochondrial deficiency" does not mediate insulin resistance. *Am. J. Clin. Nutr.* **2009**, *89*, 463S–466S.
43. Kadenbach, B. Intrinsic and extrinsic uncoupling of oxidative phosphorylation. *Biochim. Biophys. Acta* **2003**, *1604*, 77–94.
44. Stucki, J.W. The optimal efficiency and the economic degrees of coupling of oxidative phosphorylation. *Eur. J. Biochem.* **1980**, *109*, 269–283.
45. Rolfe, D.F.S.; Brown, G.C. Cellular energy utilisation and molecular origin of standard metabolic rate in mammals. *Physiol. Rev.* **1997**, *77*, 731–758.
46. Skulachev, V.P. Fatty acid circuit as a physiological mechanism of uncoupling of oxidative phosphorylation. *FEBS Lett.* **1991**, *294*, 158–162.

47. Jezek, P.; Engstova, H.; Zackova, M.; Vercesi, A.E.; Costa, A.D.T.; Arruda, P.; Garlid, K.D. Fatty acid cycling mechanism and mitochondrial uncoupling proteins. *Biochim. Biophys. Acta* **1998**, *1365*, 319–327.
48. Soboll, S.; Grundel, S.; Schwabe, U.; Scholtz, R. Influence of fatty acids on energy metabolism. 2. Kinetics of changes in metabolic rates and changes in subcellular adenine nucleotide contents and pH gradients following addition of octanoate and oleate in perfused rat liver. *Eur. J. Biochem.* **1984**, *141*, 231–236.
49. Amara, C.E.; Shankland, E.G.; Jubrias, S.A.; Marcinek, D.J.; Kushmerick, M.J.; Conley, K.E. Mild mitochondrial uncoupling impacts cellular aging in human muscles *in vivo*. *Proc. Natl. Acad. Sci. USA* **2007**, *104*, 1057–1062.
50. Fannin, S.W.; Lesnefsky, E.J.; Slabe, T.J.; Hassan, M.O.; Hoppel, C.L. Aging selectively decreases oxidative capacity in rat heart interfibrillar mitochondria. *Arch. Biochem. Biophys.* **1999**, *372*, 399–407.
51. Iossa, S.; Mollica, M.P.; Lionetti, L.; Crescenzo, R.; Tasso, R.; Liverini, G. A possible link between skeletal muscle mitochondrial efficiency and age-induced insulin resistance. *Diabetes* **2004**, *53*, 2861–2866.
52. Crescenzo, R.; Bianco, F.; Mazzoli, A.; Giacco, A.; Liverini, G.; Iossa, S. Alterations in proton leak, oxidative status and uncoupling protein 3 content in skeletal muscle subsarcolemmal and intermyofibrillar mitochondria in old rats. *BMC Geriatr.* **2014**, *14*, 79.
53. Koltai, E.; Hart, N.; Taylor, A.W.; Goto, S.; Ngo, J.K.; Davies, K.J.A.; Radak, Z. Age-associated declines in mitochondrial biogenesis and protein quality control factors are minimized by exercise training. *Am. J. Physiol.* **2012**, *303*, R127–R134.
54. Hafner, R.P.; Brown, G.C.; Brand, M.D. Analysis of the control of respiration rate, phosphorylation rate, proton leak rate and protonmotive force in isolated mitochondria using the "top-down" approach of metabolic control theory. *Eur. J. Biochem.* **1990**, *188*, 313–319.
55. Brand, M.D.; Chien, L.F.; Ainscow, E.K.; Rolfe, D.F.S.; Porter, R.K. The causes and functions of mitochondrial proton leak. *Biochim. Biophys. Acta* **1994**, *1187*, 132–139.
56. Cairns, C.B.; Walther, J.; Harken, A.H.; Banerjee, A. Mitochondrial oxidative phosphorylation thermodynamic efficiencies reflect physiological organ roles. *Am. J. Physiol.* **1998**, *274*, R1376–R1383.
57. Santanasto, A.J.; Glynn, N.W.; Jubrias, S.A.; Conley, K.E.; Boudreau, R.M.; Amati, F.; Mackey, D.C.; Simonsick, E.M.; Strotmeyer, E.S.; Coen, P.M.; *et al.* Skeletal muscle mitochondrial function and fatigability in older adults. *J. Gerontol. A Biol. Sci. Med. Sci.* **2015**, doi:10.1093/gerona/glu134.
58. Stannard, S.R.; Johnson, N.A. Insulin resistance and elevated triglyceride in muscle: More important for survival than thrifty genes? *J. Physiol.* **2003**, *554*, 595–607.
59. Korshunov, S.S.; Skulachev, V.P.; Starkov, A.A. High protonic potential actuates a mechanism of production of reactive oxygen species in mitochondria. *FEBS Lett.* **1997**, *416*, 15–18.
60. Mailloux, R.J.; Harper, M.E. Uncoupling proteins and the control of mitochondrial reactive oxygen species production. *Free Radic. Biol. Med.* **2011**, *51*, 1106–1115.
61. Brand, M.D. Uncoupling to survive? The role of mitochondrial inefficiency in ageing. *Exp. Gerontol.* **2000**, *35*, 811–820.

62. Lombardi, A.; Busiello, R.A.; Napolitano, L.; Cioffi, F.; Moreno, M.; de Lange, P.; Silvestri, E.; Lanni, A.; Goglia, F. UCP3 translocates lipid hydroperoxide and mediates lipid hydroperoxide-dependent mitochondrial uncoupling. *J. Biol. Chem.* **2010**, *285*, 16599–16605.
63. Goglia, F.; Skulachev, V.P. A function for novel uncoupling proteins: Antioxidant defence of mitochondrial matrix by translocating fatty acid peroxides from the inner to the outer membrane leaflet. *FASEB J.* **2003**, *17*, 1585–1591.
64. Divakaruni, A.S.; Brand, M.D. The regulation and physiology of mitochondrial proton leak. *Physiology* **2011**, *26*, 192–205.
65. Lal, S.B.; Ramsey, J.J.; Monemdjou, S.; Weindruch, R.; Harper, M.E. Effects of caloric restriction on skeletal muscle mitochondrial proton leak in aging rats. *J. Gerontol.* **2001**, *56*, B116–B122.
66. Marcinek, D.J.; Schenkman, K.A.; Ciesielski, W.A.; Lee, D.; Conley, K.E. Reduced mitochondrial coupling *in vivo* alters cellular energetic in aged mouse skeletal muscle. *J. Physiol.* **2005**, *569*, 467–473.
67. Hood, D. Plasticity in skeletal, cardiac, and smooth muscle: Contractile activity-induced mitochondrial biogenesis in skeletal muscle. *J. Appl. Physiol.* **2001**, *90*, 1137–1157.
68. Beyer, R.E.; Starnes, J.W.; Edington, D.W.; Lipton, R.J.; Compton, R.T.; Kwasman, M.A. Exercise-induced reversal of age-related declines of oxidative reactions, mitochondrial yield, and flavins in skeletal muscle of the rat. *Mech. Ageing Dev.* **1984**, *24*, 309–323.
69. Bevilacqua, L.; Ramsey, J.J.; Hagopian, K.; Weindruch, R.; Harper, M.E. Long-term caloric restriction increases UCP3 content but decreases proton leak and reactive oxygen species production in rat skeletal muscle mitochondria. *Am. J. Physiol.* **2005**, *289*, E429–E438.
70. Barazzoni, R.; Nair, K.S. Changes in uncoupling protein-2 and -3 expression in aging rat skeletal muscle, liver, and heart. *Am. J. Physiol.* **2001**, *280*, E413–E419.

4

Synthesis, Radiolabelling and *In Vitro* Characterization of the Gallium-68-, Yttrium-90- and Lutetium-177-Labelled PSMA Ligand, CHX-A''-DTPA-DUPA-Pep

Benjamin Baur [1,†,*], Christoph Solbach [1,†], Elena Andreolli [1,2], Gordon Winter [1], Hans-Jürgen Machulla [1] and Sven N. Reske [1]

1 Clinic for Nuclear Medicine, University of Ulm, Albert-Einstein-Allee, 89081 Ulm, Germany; E-Mails: christoph.solbach@uniklinik-ulm.de (C.S.); elena.andreolli@gmail.com (E.A.); gordon.winter@uniklinik-ulm.de (G.W.); hans.machulla@uniklinik-ulm.de (H.-J.M.); sven.reske@uniklinik-ulm.de (S.N.R.)

2 Department of Health Sciences, University of Milano-Bicocca, Milan 20900, Italy

† These authors contributed equally to this work.

* Author to whom correspondence should be addressed; E-Mail: benjamin.baur@uniklinik-ulm.de;

Abstract: Since prostate-specific membrane antigen (PSMA) has been identified as a diagnostic target for prostate cancer, many urea-based small PSMA-targeting molecules were developed. First, the clinical application of these Ga-68 labelled compounds in positron emission tomography (PET) showed their diagnostic potential. Besides, the therapy of prostate cancer is a demanding field, and the use of radiometals with PSMA bearing ligands is a valid approach. In this work, we describe the synthesis of a new PSMA ligand, CHX-A''-DTPA-DUPA-Pep, the subsequent labelling with Ga-68, Lu-177 and Y-90 and the first *in vitro* characterization. In cell investigations with PSMA-positive LNCaP C4-2 cells, K_D values of $\leq 14.67 \pm 1.95$ nM were determined, indicating high biological activities towards PSMA. Radiosyntheses with Ga-68, Lu-177 and Y-90 were developed under mild reaction conditions (room temperature, moderate pH of 5.5 and 7.4, respectively) and resulted in nearly quantitative radiochemical yields within 5 min.

Keywords: PSMA; prostate-specific membrane antigen; PET; positron emission tomography; prostate cancer; DUPA; Ga-68; Y-90; Lu-177; radionuclide therapy

---

1. Introduction

Prostate cancer (PCa) is still one of the leading causes of cancer deaths among men. Despite using novel therapeutic approaches, mortality from metastasizing prostate cancer is still high [1]. Therefore, the early diagnosis of prostate cancer to prevent tumor dissemination is highly desirable. Furthermore, effective treatment strategies for disseminated prostate cancer are urgently needed. Comprehensibly, targeting of PCa or its metastases is a demanding task in the field of molecular imaging with positron emission tomography (PET) and for targeted internal radiation therapy.

Prostate-specific membrane antigen (PSMA) is a peptidase that catalyzes the hydrolysis of *N*-acetyl-L-aspartyl-L-glutamate (NAAG) into the corresponding *N*-acetyl-L-aspartate (NAA) and L-glutamate [2]. The use of PSMA as a target for diagnostic and therapeutic agents is a highly valid approach. Compared to healthy human prostate tissue, in almost all PCa tumors, the expression of PSMA is 10–80 fold higher.

Besides, the PSMA levels are increased in the neovasculature of other solid tumors, as well [3–6]. Therefore, selective addressing of PSMA with small molecules labelled with a positron emitting radionuclide is a considerable approach for the diagnosis of prostate cancer with PET.

Based on the chemical structure of NAAG, several glutamate-urea-glutamate-based peptides bearing a 2-[3-(1,3-dicarboxypropyl)-ureido]pentanedioic acid (DUPA) moiety were developed in the last few years [7]. These molecules showed high affinity and specific binding to PSMA, as demonstrated in binding studies using PSMA expressing LNCaP cell lines. Beside the slightly modified chemical structure, these molecules differ mainly in the selection of the chelator for the complexation of the desired radionuclide [7–9].

Due to their proven excellent affinity to PSMA, it is desirable to develop radionuclide-based therapeutic strategies using adapted PSMA targeting. In radiometal therapy approaches, the application of Y-90 and Lu-177 is favored. The use of Y-90 ($E_{\beta max}$: 2.3 MeV, $t_{½}$: 64 h) is more appropriate in the treatment of larger tumor lesions, while Lu-177 ($E_{\beta max}$: 0.5 MeV, $t_{½}$: 6.7 d) is more suitable for the treatment of smaller lesions and metastases, accompanied by a minimization of kidney dose in comparison to the application of Y-90 labelled peptides [10]. Moreover, due to the contemporary beta- and gamma-emission, Lu-177 is a useful diagnostic tool for scintigraphy of tumoral uptake [11].

For both diagnostic and therapeutic application, 1,4,7,10-tetraazacyclododecane-*N,N',N'',N'''*-tetraacetic acid (DOTA) is the mostly used chelator for the complexation of radiometals, like Ga-68, Lu-177 and Y-90, to small molecules [12,13]. However, DOTA can show some undesirable characteristics for therapy. Possible immunogenicity in humans has been published, as well as unfavorable kinetics in the complexation of radiometals [14,15]. Furthermore, labelling reactions using DOTA as the chelating agent are usually carried out at high temperatures under acidic conditions and long reaction times [16]. The preferred labelling procedure for peptides should consist of a simple, fast and quantitative labelling step at room temperature and neutral pH to avoid decomposition. Therefore, the use of alternative chelators is a demanding approach. In addition, the chelator must provide sufficient stability *in vivo*.

The preceding developments based on the chelator, diethylenetriaminepentaacetic acid (DTPA), lead to the promising cyclohexyl substituted analogue, cyclohexyl-diethylene triamine pentaacetic acid (CHX-A''-DTPA) [17]. It showed high stability *in vivo*, and radiolabelling with Y-90 was achieved under mild conditions (pH = 6 at room temperature) [17]. Thus, CHX-A''-DTPA is a highly promising alternative to the mostly used DOTA. Previous work using the PSMA ligand (5S,8S,22S,26S)-1-amino-5,8-dibenzyl-4,7,10,19,24-pentaoxo-3,6,9,18,23,25-hexaazaoctacosane-22,26,28-tricarboxylic acid (DUPA-Pep), conjugated with DOTA, revealed a high PSMA-affinity with a K_D of 21.6 ± 0.4 nM measured in a PSMA-positive LNCaP C4-2 cell assay [18]. Furthermore, the first clinical application of [^{68}Ga]Ga-DOTA-DUPA-Pep showed the feasibility of *in vivo* targeting. [^{68}Ga]Ga-HBED-PSMA showed promising clinical results for prostate cancer patients so far [19,20].

Thus, in the present study, we describe the synthesis of CHX-A"-DTPA-DUPA-Pep (scheme 1), subsequent labelling with Ga-68, Y-90 and Lu-177, and the investigation of its biological activity as a new potential PSMA ligand for the diagnosis and therapy of prostate cancer.

Scheme 1. Synthesis of cyclohexyldiethylenetriamine pentaacetic acid (5*S*,8*S*,22*S*,26*S*)-1-amino-5,8-dibenzyl-4,7,10,19,24-pentaoxo-3,6,9,18,23,25-hexaazaoctacosane-22,26,28-tricarboxylic acid trifluoroacetate (CHX-A"-DTPA-DUPA-Pep).

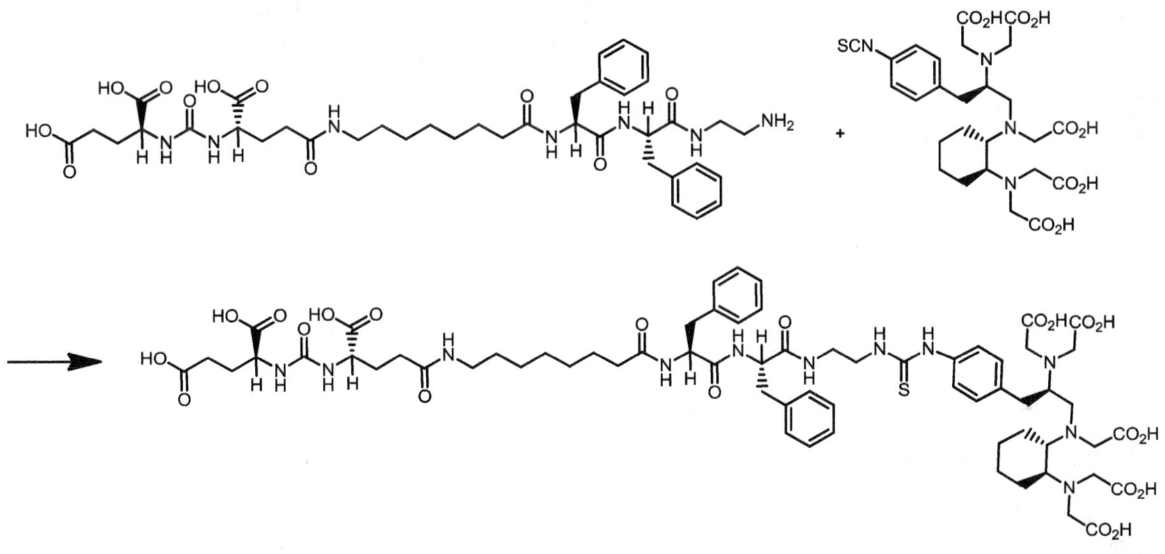

2. Experimental

[^{90}Y]YCl_3 and [^{177}Lu]$LuCl_4$, in 0.04 M HCl were purchased from Eckert and Ziegler and Isotope Technologies, respectively. The ^{68}Ge/^{68}Ga generator was obtained from IDB Holland BV. All chemicals and solvents were purchased from Aldrich, Fluka, Applichem, Waters and Merck and used without further purification. (5*S*,8*S*,22*S*,26*S*)-1-Amino-5,8-dibenzyl-4,7,10,19,24-pentaoxo-3,6,9,18,23,25-hexaazaoctacosane-22,26,28-tricarboxylic acid trifluoroacetate ("DUPA-Pep" trifluoroacetate) was obtained from ABX (Germany). *N*-[(*R*)-2-Amino-3-(*p*-isothiocyanato-phenyl) propyl]-trans-(*S,S*)-cyclohexane-1,2-diamine-*N,N,N'-,N'',N''*-pentaacetic acid (*p*-SCN-Bn-CHX-A"-DTPA) was achieved from Macrocyclics (Dallas, TX, USA). Diethylenetriaminepentaacetic acid (≥99%) was obtained from Fluka (Buchs, Switzerland). Solvents for analytics were of HPLC-grade purity. Chemicals for organic synthesis were of >98% purity. Chemicals for radiolabelling were of trace metal® grade. An analytical assay was carried out by thin-layer chromatography (TLC) with Silica gel 60 RP-18 F_{254}S (on alumina sheets of 5 × 7.5 cm, Merck) as the stationary phase, visualized by UV-detection for preparative synthesis and by phosphorimager analysis (FLA3000. Raytest, Straubenhardt, Germany) for radiosynthesis. In addition, a radio high-pressure liquid chromatography system (radio-HPLC, P680 Dionex, Idstein, Germany), equipped with NaI(TI)-scintillation detector (GABI, Raytest) and a UV-Vis detector (UVD 170U, Dionex), was used for the identification of the labelled peptide. For the control of radiochemical yields (RCYs), a γ-counter (Cobra^{TM} II, Packard Instrument, Canberra, Australia) was applied. ^{1}H- and ^{13}C- nuclear magnetic resonance spectra were recorded by a 400 MHz Bruker Avance spectrometer (Bruker, Karlsruhe, Germany) at ambient temperature. MS spectra were measured using a Triple Stage Quadrupole TSQ 70 Finnigan-MAT-mass spectrometer (Finnigan-MAT, Bremen, Germany).

2.1. Organic Synthesis of CHX-A"-DTPA-DUPA-Pep

DUPA-Pep trifluoroacetate (10 mg, 12.5 µmol) was dissolved in DMSO (1 mL), and DIPEA (15 µL, 90 µmol) and *p*-SCN-Bn-CHX-A"-DTPA·3HCl (Macrocyclics, Dallas, TX., USA) (7 mg, 11 nmol) were added. The reaction was carried out for 16 h. The product was purified by preparative HPLC consisting of a C18 column (Gemini 5µ C18, 250 × 21 mm, Phenomenex, Aschaffenburg, Germany) as the stationary phase and by use of gradient elution technique (gradient profile: 0–3 min 20% MeCN in H_2O (0.1% TFA), 3–15 min from 20% to 40% MeCN in H_2O (0.1% TFA), 15–30 min 40% MeCN in H_2O (0.1% TFA); a flow rate of 7 mL/min, UV detection at 220 and 254 nm) as the mobile phase gave CHX-A"-DTPA-DUPA-Pep (12.5 mg, 8.6 µmol, 72%) as a colorless solid after lyophilization with a purity of >98% (HPLC). ^{1}H NMR (400 MHz, DMSO-d_6) δ = 1.2 (m, 14H); 1.7 (m, 4H); 2,0 (m, 8 H); 2.2 (m, 2H); 2.7 (m, 1H); 2.8 (m, 1H); 3.0 (m, 6H); 3.2 (m, 4H); 3.4 (m, 4H); 3.6 (m, 7H); 4.1 (m, 5H); 4.5 (m, 3H); 6.4 (m, 2H); 7.3 (m, 14H); 7.8 (m, 2H); 8.0 (m, 3H); 9.6 (s, 1H); 12.6 (s, 5H). ^{13}C NMR (300 MHz, DMSO-d_6) δ = 25.1 (2CH_2); 26.3 (CH_2); 27.5 (2*CH_2); 28.3 (2*CH_2); 28.3 (2*CH_2); 28.4 (CH_2); 28.5 (CH_2); 29.1 (CH_2); 29.8 (CH_2); 31.6 (CH_2); 35.1 (CH_2); 37,0 (CH_2); 37.2 (2*CH_2); 37.7 (CH_2); 37.9 (2*CH_2); 38.5 (3*CH_2); 48.3 (2*CH_2); 51.6 (2*CH); 52.1 (CH); 53.6 (2*CH); 53.9 (2*CH); 126.1 (CH (phenylalanine)); 126.3 (CH (phenylalanine)); 127.9 (2*C2 phenyl(desferal)); 128.0 (2*C3 phenyl(desferal)); 129.1 (4*CH (phenylalanine)); 129.2 (4*CH (phenylalanine)); 137.5 (C1, C6 phenyl(desferal)); 137.9 (2*C (phenylalanine)); 155.3 (2*C = O); 157.2 (HN-(C = O)-NH); 158.2 (2*C = O); 171.0 (2*COOH); 171.1 (COOH); 172.1 (2*COOH); 173.7 (COOH); 174.1 (2*COOH); 174.2 (C = S). ESI-HRMS (*m*/*z*): MALDI calcd. for $C_{65}H_{89}N_{11}O_{21}S$, 1391,60; found, $[M + H]^+$1392.60177.

2.2. Radionuclide Production and Radiochemistry

2.2.1. Production of [^{68}Ga]Ga-Chloride

Ga-68 ($t_{½}$ = 67.7 min, $β^+$: 89%, $E_{β^+ max}$, 1.9 MeV; EC: 11%, $E_{γ\ max}$: 4.0 MeV) was prepared as described by Meyer *et al.* [21]. Briefly, Ga-68 was eluted from a TiO_2-based ^{68}Ge/^{68}Ga generator with *ca.* 2 mL of 1 M HCl and collected into a vial containing 1 mL of 9.5 M HCl. The resulting solution was loaded onto a strong anion exchange resin (100 mg Dowex 1 × 8), and the activity was retained as the $[GaCl_4]^-$ complex. After elution with 200 µL of water, the pH of the solution was neutralized by adding the appropriate volume of 2 M Na_2CO_3.

2.2.2. Radiosynthesis of [^{68}Ga]Ga-CHX-A"-DTPA-DUPA-Pep

The Ga-68 activity in neutral aqueous solution (13 µL; 9.3–9.6 MBq) was added to a freshly prepared solution of CHX-A"-DTPA-DUPA-Pep, dissolved in DMSO (2 mg/mL) and 0.25 M HEPES buffer (pH 7.4, 500 µL), resulting in a total volume of 514–563 µL and a final pH of 7.4. Different amounts of peptide (1, 10, 25, 50 and 100 µg; 0.7, 7.2, 18, 36 and 72 nM) were used to study the dependence of RCY on peptide concentration and reaction time. The reaction was carried out at room temperature for 60 min. At different reaction times (1, 5, 10, 30 and 60 min), 1-µL aliquots were withdrawn and analyzed by radio-TLC on RP-18 thin layer plates as the stationary phase and 0.1% TFA/MeOH 30/70 *v*/*v* as the mobile phase.

2.2.3. Radiosynthesis of [^{90}Y]Y-CHX-A"-DTPA-DUPA-Pep

Y-90 activity in 0.04 M HCl solution (20 µL; 3.7–3.8 MBq) was added to a freshly prepared solution of CHX-A"-DTPA-DUPA-Pep, dissolved in DMSO (2 mg/mL) and 0.5 M NH_4OAc buffer (pH 5.5, 500 µL), resulting in a total volume of 533–570 µL and a final pH of 5.5. Different amounts of peptide (25, 50 and 100 µg; 18, 36 and 72 nM) were used to study the dependence of RCY on peptide concentration and reaction time. The reaction was carried out at room temperature for 60 min. At different reaction times (5, 10, 30 and 60 min), 1-µL aliquots were withdrawn and analyzed by radio-TLC on RP-18 thin layer plates as the stationary phase and 0.1% TFA/MeOH 30/70 *v*/*v* as the mobile phase.

2.2.4. Radiosynthesis of [^{177}Lu]Lu-CHX-A"-DTPA-DUPA-Pep

Lu-177 activity in 0.04 M HCl solution (5 µL; 6.3–6.7 MBq) was added to a freshly prepared solution of CHX-A"-DTPA-DUPA-Pep, dissolved in DMSO (2 mg/mL) and 0.5 M NH_4OAc buffer (pH 5.5, 500 µL), resulting in to total volume of 510–555 µL and a final pH of 5.5. Different amounts of peptide (10, 25, 50 and 100 µg; 7.2, 18, 36 and 72 nM) were used to study the dependence of RCY on peptide concentration and reaction time. The reaction was carried out at room temperature for 60 min. At different reaction times (5, 10, 30 and 60 min), 1-µL aliquots were withdrawn and analyzed by radio-TLC on RP-18 thin layer plates as the stationary phase and 0.1% TFA/MeOH 30/70 *v*/*v* as the mobile phase.

2.2.5. Radio-TLC Analysis

RCYs of the labelling reactions described above were determined in dependency on reaction time and peptide concentration. One microliter was withdrawn from every reaction mixture at different definite time points for radio-TLC analysis. RP-18 Silica gel plates were used as the stationary phase and 0.1% TFA/MeOH 30/70 *v*/*v* as the mobile phase, in which the free radiometal remains at the baseline (R_f: 0.01). The R_f value of CHX-A"-DTPA-DUPA-Pep was found to be 0.73 (λ: 254 nm), while the R_f value of radio labelled compound was found to be *ca.* 0.78 (phosphorimager). Quantitative assays of radioactive spots were carried out by phosphorimager to determine the amount of radio labelled chelate and free radiometal.

2.2.6. Radio-HPLC Analysis

Radiochemical purity was determined by gradient radio-HPLC analysis. A C18 column (Phenomenex Gemini, 5 μm, C18, 250 × 4.6 mm) was used as stationary phase. Gradient elution technique was performed by the use of Solvent A (H_2O, 0.1% TFA) and Solvent B (MeCN, 0.1% TFA) as the mobile phase (gradient profile: 85% A at 5 min; 85%–30% A from 5 to 14 min, 30%–85% A from 14 to 16 min; flow rate: 2 mL/min). The retention time of [^{68}Ga]Ga-CHX-A''-DTPA-DUPA-Pep was 9.4 min (Figure 1).

Figure 1. Exemplary chromatograms of [^{68}Ga]Ga-CHX-A''-DTPA-DUPA-Pep in the radioactivity channel (upper signal; the first peak shows the total activity injected detector bypass for recovery rate calculation) and CHX-A''-DTPA-DUPA-Pep in the UV-channel (bottom signal; λ: 254 nm).

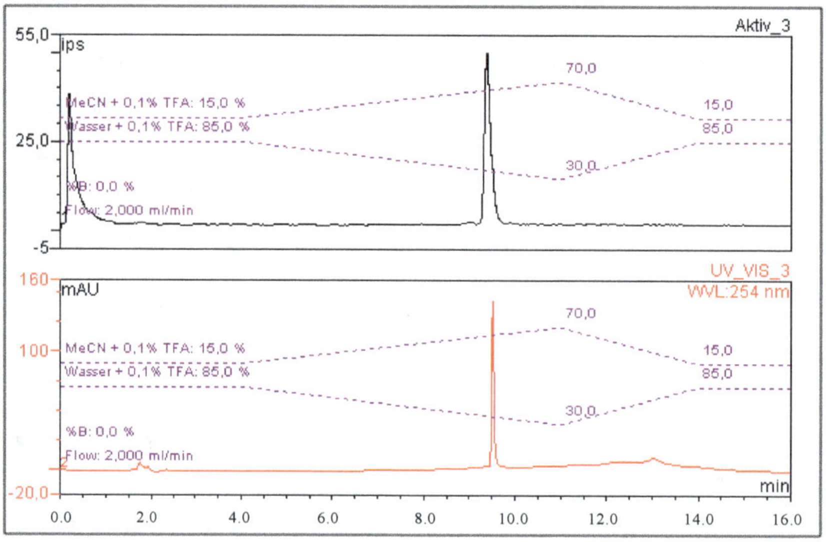

2.3. Stability of Labelled CHX-A''-DTPA-DUPA-Pep

The stability of the labelled peptide in human serum and phosphate buffered saline (PBS) was verified via radio-HPLC after 30 min, 2, 4 and 8 h for [^{68}Ga]Ga-CHX-A''-DTPA-DUPA-Pep and by radio-TLC on RP-18 thin layer plates as the stationary phase and 0.1% TFA/MeOH 30/70 *v*/*v* as the mobile phase after 1 h, 3 h, 6 h, 24 h, 48 h and 72 h for [^{177}Lu]Lu-CHX-A''-DTPA-DUPA-Pep and [^{90}Y]Y-CHX-A''-DTPA-DUPA-Pep. The radiochemical stability of the labelled compounds was determined in serum and PBS buffer at 37 °C. The total volume was 1.2 mL (980 μL of serum or buffer and 220 μL of product solution). In stability tests, 12 MBq of Ga-68 and 4.5 MBq of each radionuclide, Lu-177 and Y-90, were applied. Two hundred microliters of the mixture were withdrawn at each time point, and the samples were analyzed by radio-HPLC. Radio-TLC analyses were performed without further reprocessing.

The PBS samples were injected into a radio-HPLC system without further work-up. Serum samples of the Y-90 and Lu-177 labelled compound were tested without further work-up by radio-TLC. Serum samples for radio-HPLC-investigations underwent a serum protein precipitation work-up procedure before injection. Briefly: an equal volume of EtOH (200 μL) was added to the samples. Subsequently, the samples were centrifuged for 10 min at 10,000 rpm. Ultrafiltration of the supernatant was performed using an Amicon ultra centrifugal filter (3 KDa NMWL, Millipore, Darmstadt, Germany) for 10 min at 10,000 rpm. The filtrate was analyzed by radio-HPLC.

Exchange experiments with 500-fold excess of DTPA and additionally with hydrolysed *p*-SCN-Bn-CHX-A"-DTPA were performed for each of the three radiolabeled CHX-A"-DTPA-DUPA-Pep compounds. Briefly, after radiolabelling with Ga-68, Y-90 and Lu-177, a 500-fold excess of DTPA or hydrolysed *p*-SCN-Bn-CHX-A"-DTPA (in 500 µL of 0.5 M NH_4OAc buffer (pH 5.5)) was added to the sealed reaction mixture stored at RT. Aliquots were withdrawn at two time points (30 min and 240 min) and analyzed by radio-TLC procedure (RP-18 Silica gel plates as the stationary phase and 0.1% TFA/MeOH 30/70 *v*/*v* as the mobile phase).

2.4. Cell Culture and Analysis of Binding Specificity of CHX-A"-DTPA-DUPA-Pep

The PSMA-positive cell line, LNCaP C4-2, was obtained from ViroMed Laboratories (Minnetonka, MN, USA) and grown in T-media (Dulbecco's Modified Eagle's Medium (DMEM), high glucose, Gibco) and 20% Ham's F-12 (Biochrom, Berlin, Germany) supplemented with 5% fetal bovine serum, 5 µg/mL insulin (Sigma, St. Louis, MO., USA), 13.65 pg/mL triiodothyronine (Sigma), 5 µg/mL apo-transferrin (Sigma), 0.244 µg/mL D-biotin (Sigma) and 25 µg/mL adenine (Sigma). As a negative control, PC-3 cells (DSMZ, ACC 465) were maintained in DMEM and 10% Ham's F12 with 10% FBS, 1% penicillin/streptomycin and 2 mM glutamine. All cell lines were incubated at 37 °C under constant humidity and an atmosphere of 5% CO_2.

To determine the binding coefficient (K_D) of the radio-labelled peptide, 5×10^5 cells/well were grown in coated 12-well plates in 1 mL of medium for 48 h. The cells were washed twice with PBS, and 900 µL fresh media were added. Radiolabelled peptide was added, resulting in final concentrations of 480, 240, 120, 60, 30, 15 and 7.5 nM. In parallel, the PSMA-inhibitor (2-PMPA) was applied in a final concentration of 30 µM to determine unspecific binding. All samples were prepared in triplicate. Following 60 min of incubation at 37 °C, cells were washed twice to remove unbound activity and afterwards lysed in 1 mL of 0.5 M NaOH. Activity was measured in a gamma counter (COBRA™ II, Packard Instrument). Aliquots of the solution added to the cells were also measured for the calculation of the cellular uptake as %ID. Data were analyzed using GraphPad Prism 5.02 (one site, total and non-specific binding evaluation).

2.5. Statistical Aspects

All experiments were conducted with $n \geq 3$. Data are expressed as the mean ± SD.

3. Results and Discussion

3.1. Organic Synthesis of CHX-A"-DTPA-DUPA-Pep

CHX-A"-DTPA-DUPA-Pep synthesis was successfully achieved by coupling DUPA-Pep and p-SCN-Bn-CHX-A"-DTPA in the presence of DIPEA as the base. After preparative HPLC purification and lyophilization of the purified fraction, the peptide was obtained as a white powder in yields of 72% and a purity of ≥98% (HPLC).

3.2. Radiochemistry

Radiolabelling with Ga-68 was performed in HEPES buffer (pH = 7.4) at room temperature by adding the prepared solution of [^{68}Ga]GaCl_3 to the peptide reaction mixture. The radiochemical yields were evaluated in dependence on the reaction time and the amount of peptide via radio-TLC (Table 1, Figure 2). Starting from 18 nM (25 µg) of CHX-A"-DTPA-DUPA-Pep, the RCY lay over 95% after 30 min of reaction time. Increasing the amount of peptide, to 72 nM (100 µg), a RCY >95% was obtained after 1 min of reaction time. Due to nearly quantitative labelling, the product solution was applied without further purification.

Table 1. Dependence of the radiochemical yield (RCY) of [^{68}Ga]Ga-CHX-A"-DTPA-DUPA-Pep on the peptide amount (concentration) and the reaction time at room temperature.

Amount (µg)	RCY % (1 min)	RCY % (5 min)	RCY % (10 min)	RCY % (30 min)	RCY % (60 min)
1	0.2 ± 0 .1	0.2 ± 0.0	0.3 ± 0.1	0.5 ± 0.1	0.6 ± 0.1
10	3.7 ± 0.3	14.0 ± 0.3	21.9 ± 0.8	41.3 ± 4.0	56.5 ± 5.8
25	29.0 ± 1.5	70.0 ± 1.3	85.6 ± 0.8	96.6 ± 0.3	97.1 ± 0.6
50	69.4 ± 0.9	95.4 ± 0.4	96.6 ± 0.6	96.6 ± 0.2	97.2 ± 0.7
100	95.1 ± 0.3	95.8 ± 0.1	96.1 ± 0.1	97.3 ± 0.1	97.7 ± 0.6

Figure 2. Dependence of the RCY of [^{68}Ga]Ga-CHX-A"-DTPA-DUPA-Pep on the peptide amount (concentration) and reaction time at room temperature.

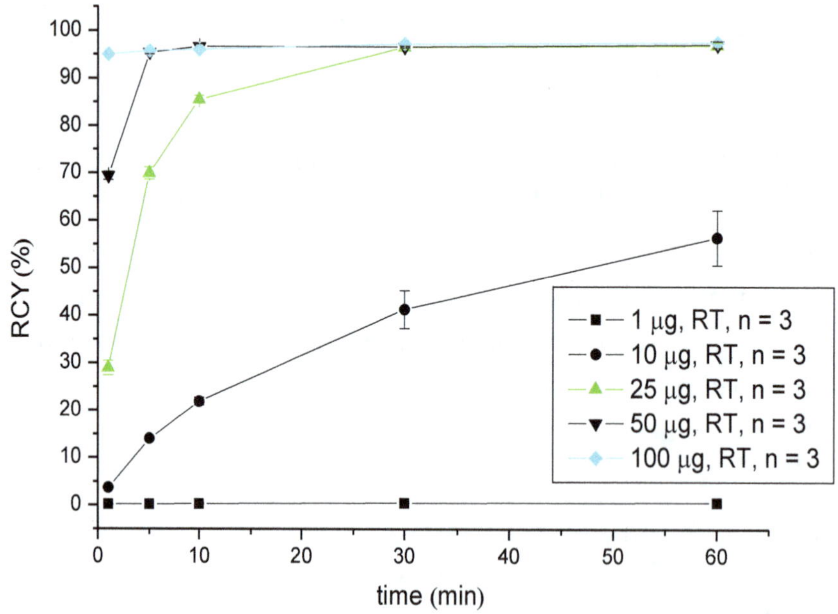

Regarding radiometals for radiotherapy (Lu-177 and Y-90), the radiolabelling was performed at pH 5.5 in 0.5 M NH_4OAc buffer and showed RCYs >95% for low amounts of peptide (25 µg (18 nM) in the case of ^{90}Y-labelling and 10 µg (7.2 nM) in the case of ^{177}Lu-labelling) after 5 min of reaction time. A practically quantitative RCY of >99% was obtained in both radiosyntheses after 30 min of reaction time when the highest amount of peptide (100 µg; 72 nM) was used (Tables 2 and 3, Figures 3 and 4).

Table 2. Dependence of the RCY of [^{90}Y]Y-CHX-A"-DTPA-DUPA-Pep on peptide amount (concentration) and reaction time at room temperature.

Amount (μg)	RCY % (5 min)	RCY % (10 min)	RCY % (30 min)	RCY % (60 min)
25	98.2 ± 0.2	97.3 ± 1.0	98.7 ± 0.2	97.9 ± 0.1
50	98.6 ± 0.1	98.2 ± 0.2	98.5 ± 0.3	99.0 ± 0.1
100	98.4 ± 0.4	98.2 ± 0.2	99.3 ± 0.2	99.0 ± 0.1

Table 3. Dependence of the RCY of [^{177}Lu]Lu-CHX-A"-DTPA-DUPA-Pep on peptide amount (concentration) and reaction time at room temperature.

Amount (μg)	RCY % (5 min)	RCY % (10 min)	RCY % (30 min)	RCY % (60 min)
10	96.2 ± 1.0	95.2 ± 0.8	98.6 ± 0.4	96.7 ± 0.3
25	95.7 ± 0.5	97.2 ± 0.7	93.4 ± 0.4	96.1 ± 1.1
50	95.4 ± 0.1	96.7 ± 0.2	98.5 ± 0.4	98.4 ± 0.7
100	98.4 ± 0.1	98.6 ± 0.3	99.3 ± 0.2	99.6 ± 0.2

Figure 3. Dependence of the RCY of [^{90}Y]Y-CHX-A"-DTPA-DUPA-Pep on peptide amount (concentration) and reaction time at room temperature.

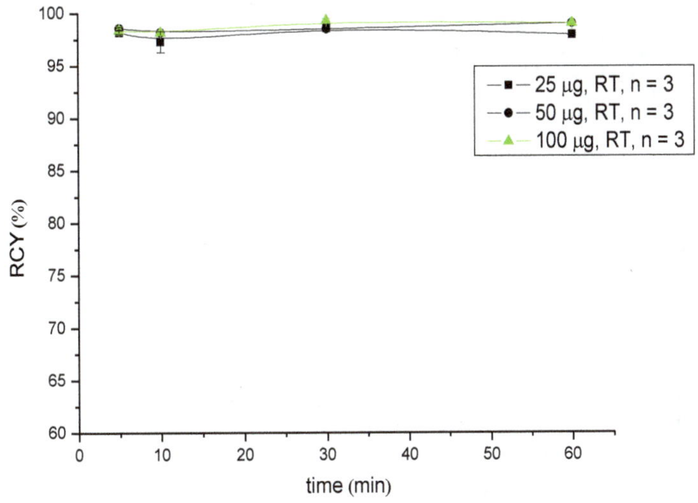

Figure 4. Dependence of the RCY of [^{177}Lu]Lu-CHX-A"-DTPA-DUPA-Pep on peptide amount (concentration) and reaction time at room temperature.

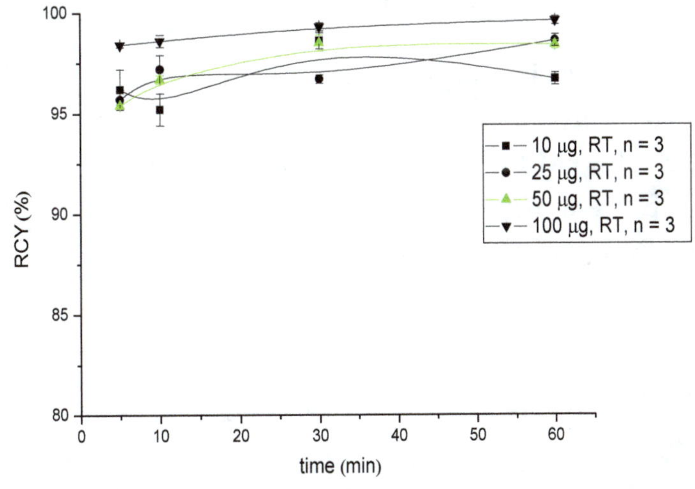

3.3. Stability Investigations

A high incorporation level and high complex stability is strongly desired in therapeutic applications, due to the severe myelotoxicity associated with Y-90 and the long half-life of Lu-177 (6.74 d). Thus, in order to verify the stability of the Y-90 and Lu-177 complexes, stability studies were performed in human serum and PBS at 37 °C. Additionally, the stability of the Ga-68 labelled compound was investigated. Radio-TLC analysis of [^{177}Lu]Lu- and [^{90}Y]Y-CHX-A"-DTPA-DUPA-Pep after 1 h, 3 h, 6 h, 24 h, 48 h and 72 h confirmed the stability of the radiolabelled product. Moreover, the incubation of the [^{68}Ga]Ga-CHX-A"-DTPA-DUPA-Pep for 30 min, 1 h, 2 h, 4 h and 8 h at 37 °C resulted in no detectable changes, as shown by radio-HPLC (Figure 5). Exchange experiments with a 500-fold excess of DTPA and with a 500-fold excess of hydrolysed *p*-SCN-Bn-CHX-A"-DTPA confirmed the stability of the investigated radiolabelled complexes. No radiometal labelled chelator (DTPA or hydrolysed *p*-SCN-Bn-CHX-A"-DTPA) was observed.

Figure 5. Stability study of [^{68}Ga]Ga-CHX-A"-DTPA-DUPA-Pep in human serum: overlaid radio-HPLC chromatograms after 8 h (1), 4 h (2), 2 h (3) and 30 min (4). The first peak shows the total activity injected (the detector bypass for the recovery rate calculation).

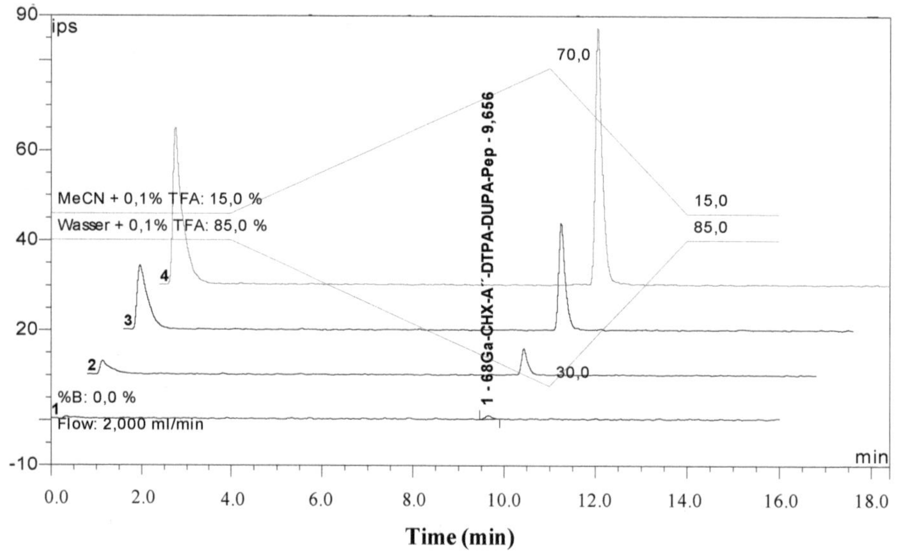

3.4. Determination of In Vitro Binding Specificity

The binding coefficient was investigated for all radiolabelled CHX-A"-DTPA-DUPA-Pep derivatives. In the PSMA-positive cell line, LNCaP C4-2, K_D values of 14.67 ± 1.95 nM for ^{177}Lu-labelled peptide, 8.0 ± 1.1 nM for Y-90 labelled peptide and 12.09 ± 0.66 nM for ^{68}Ga-labelled CHX-A"-DTPA-DUPA-Pep were determined. No specific binding could be observed for the PSMA-negative cell line, PC-3.

4. Conclusions

The synthesis of the DTPA-derivative bearing the PSMA targeting ligand, CHX-A"-DTPA-DUPA-Pep, was straightforward and efficient.

The radiolabelling of the new PSMA ligand with diverse radiometals, for both diagnosis and therapy, was investigated, and optimized conditions were developed.

Labelling with Ga-68 was performed at room temperature under neutral conditions. Significant differences in RCY were observed. Radio labelling with Ga-68 succeeds in a short reaction time with high radiochemical yields; ≥95% when 50 µg (36 nM) of the peptide was used. Additionally, the labelling of CHX-A"-DTPA-DUPA-Pep with Y-90 and Lu-177 was successfully developed. In both cases, the RCY was >95% after 5 min at room temperature using 25 µg (18 nM) peptide Y-90 and 10 µg (7.2 nM) Lu-177, respectively. In conclusion, RCYs over 95% up to quantitative yields were obtained for each radionuclide at room temperature at a moderate pH value.

In cell uptake experiments with the PSMA-positive cell line, LNCaP C4-2, the biological activity of CHX-A"-DTPA-DUPA-Pep was tested exemplarily for the Ga-68- and Lu-177- radiolabelled peptide. The obtained K_D values of 14.67 ± 1.95 nM for the Ga-68 labelled peptide, 8.0 ± 1.1 nM for the Y-90 labelled peptide and 12.09 ± 0.66 nM for the Lu-177 labelled peptide demonstrate high biological activity towards PSMA. The stability of the labelled peptide was confirmed in both human serum and PBS buffer. Additionally performed exchange experiments by use of DTPA and hydrolysed *p*-SCN-Bn-CHX-A"-DTPA confirmed the complex stability. The application of CHX-A"-DTPA as a chelator allows the choice of the suited radionuclide with respect to the selected (clinical) application. In the next step, it has to be elucidated if Ga-68, Y-90 and Lu-177 labelled CHX-A"-DTPA-DUPA-Pep may be useful ligands for the diagnosis and therapy of prostate cancer.

Acknowledgments

We gratefully acknowledge the funding of the University of Milano-Bicocca (Milan, Italy) through a PhD fellowship in Biomedical Technology.

Author Contributions

B.B. and C.S. coordinated the study, supervised the experiments and wrote the manuscript. B.B. and E.A. performed the chemical, the radiochemical syntheses and the stability experiments. G.W. performed the *in vitro* measurements. H.-J.M. and S.R. gave advice in the interpretation of the data and reviewed the manuscript. All authors approved the final manuscript.

References

1. Siegel, R.; Ward, E.; Brawley, O.; Jemal, A. Cancer statistics, 2011: The impact of eliminating socioeconomic and racial disparities on premature cancer deaths. *CA Cancer J. Clin.* **2011**, *61*, 212–236.
2. Robinson, M.B.; Blakely, R.D.; Couto, R.; Coyle, J.T. Hydrolysis of the brain dipeptide n-acetyl-l-aspartyl-l-glutamate. Identification and characterization of a novel n-acetylated alpha-linked acidic dipeptidase activity from rat brain. *J. Biol. Chem.* **1987**, *262*, 14498–14506.

3. Olson, W.C.; Heston, W.D.; Rajasekaran, A.K. Clinical trials of cancer therapies targeting prostate-specific membrane antigen. *Rev. Recent Clin. Trials* **2007**, *2*, 182–190.
4. Hao, G.; Zhou, J.; Guo, Y.; Long, M.A.; Anthony, T.; Stanfield, J.; Hsieh, J.T.; Sun, X. A cell permeable peptide analog as a potential-specific pet imaging probe for prostate cancer detection. *Amino Acids* **2011**, *41*, 1093–1101.
5. Su, S.L.; Huang, I.P.; Fair, W.R.; Powell, C.T.; Heston, W.D. Alternatively spliced variants of prostate-specific membrane antigen rna: Ratio of expression as a potential measurement of progression. *Cancer Res.* **1995**, *55*, 1441–1443.
6. Wernicke, A.G.; Varma, S.; Greenwood, E.A.; Christos, P.J.; Chao, K.S.C.; Liu, H.; Bander, N.H.; Shin, S.J. Prostate-specific membrane antigen expression in tumor-associated vasculature of breast cancers. *APMIS* **2013**, doi:10.1111/apm.12195.
7. Kularatne, S.A.; Venkatesh, C.; Santhapuram, H.K.; Wang, K.; Vaitilingam, B.; Henne, W.A.; Low, P.S. Synthesis and biological analysis of prostate-specific membrane antigen-targeted anticancer prodrugs. *J. Med. Chem.* **2010**, *53*, 7767–7777.
8. Malik, N.; Machulla, H.J.; Solbach, C.; Winter, G.; Reske, S.N.; Zlatopolskiy, B. Radiosynthesis of a new psma targeting ligand ([18f]fpy-dupa-pep). *Appl. Radiat. Isot.* **2011**, *69*, 1014–1018.
9. Eder, M.; Schafer, M.; Bauder-Wust, U.; Hull, W.E.; Wangler, C.; Mier, W.; Haberkorn, U.; Eisenhut, M. 68ga-complex lipophilicity and the targeting property of a urea-based psma inhibitor for pet imaging. *Bioconjug. Chem.* **2012**, *23*, 688–697.
10. Stabin, M.G.; Siegel, J.A. Physical models and dose factors for use in internal dose assessment. *Health Phys.* **2003**, *85*, 294–310.
11. Van Essen, M.; Krenning, E.P.; de Jong, M.; Valkema, R.; Kwekkeboom, D.J. Peptide receptor radionuclide therapy with radiolabelled somatostatin analogues in patients with somatostatin receptor positive tumours. *Acta Oncol.* **2007**, *46*, 723–734.
12. Kwekkeboom, D.J.; Teunissen, J.J.; Bakker, W.H.; Kooij, P.P.; de Herder, W.W.; Feelders, R.A.; van Eijck, C.H.; Esser, J.P.; Kam, B.L.; Krenning, E.P. Radiolabeled somatostatin analog [177lu-dota0,tyr3]octreotate in patients with endocrine gastroenteropancreatic tumors. *J. Clin. Oncol.* **2005**, *23*, 2754–2762.
13. Otte, A.; Herrmann, R.; Heppeler, A.; Behe, M.; Jermann, E.; Powell, P.; Maecke, H.R.; Muller, J. Yttrium-90 dotatoc: First clinical results. *Eur. J. Nucl. Med.* **1999**, *26*, 1439–1447.
14. Kosmas, C.; Snook, D.; Gooden, C.S.; Courtenayluck, N.S.; Mccall, M.J.; Meares, C.F.; Epenetos, A.A. Development of humoral immune-responses against a macrocyclic chelating agent (dota) in cancer-patients receiving radioimmunoconjugates for imaging and therapy. *Cancer Res.* **1992**, *52*, 904–911.
15. Kodama, M.; Koike, T.; Mahatma, A.B.; Kimura, E. Thermodynamic and kinetic studies of lanthanide complexes of 1,4,7,10,13-pentaazacyclopentadecane-n,n',n'',n''',n''''-pentaacetic acid and 1,4,7,10,13,16-hexaazacyclooctadecane-n,n',n'',n''',n'''',n'''''-hexaacetic acid. *Inorg. Chem.* **1991**, *30*, 1270–1273.
16. Wadas, T.J.; Wong, E.H.; Weisman, G.R.; Anderson, C.J. Coordinating radiometals of copper, gallium, indium, yttrium, and zirconium for pet and spect imaging of disease. *Chem. Rev.* **2010**, *110*, 2858–2902.

17. Kobayashi, H.; Wu, C.; Yoo, T.M.; Sun, B.F.; Drumm, D.; Pastan, I.; Paik, C.H.; Gansow, O.A.; Carrasquillo, J.A.; Brechbiel, M.W. Evaluation of the *in vivo* biodistribution of yttrium-labeled isomers of chx-dtpa-conjugated monoclonal antibodies. *J. Nucl. Med.* **1998**, *39*, 829–836.
18. Winter, G.; Zlatopolskiy, B.; Kull, T.; Bertram, J.; Genze, F.; Cudek, G.; Machulla, H.-J.; Reske, S. 68ga-dota-dupa-pep as a new peptide conjugate for molecular imaging of prostate carcinoma. *J. Nucl. Med. Meet. Abstr.* **2011**, *52*, 1597.
19. Reske, S.N.; Winter, G.; Baur, B.; Machulla, H.J.; Kull, T. Comment on afshar-oromieh *et al.*: Pet imaging with a [68ga]gallium-labelled psma ligand for the diagnosis of prostate cancer: Biodistribution in humans and first evaluation of tumour lesions. *Eur. J. Nucl. Med. Mol. Imaging.* **2013**, *40*, 969–970.
20. Afshar-Oromieh, A.; Malcher, A.; Eder, M.; Eisenhut, M.; Linhart, H.G.; Hadaschik, B.A.; Holland-Letz, T.; Giesel, F.L.; Kratochwil, C.; Haufe, S.; *et al.* Pet imaging with a [68ga] gallium-labelled psma ligand for the diagnosis of prostate cancer: Biodistribution in humans and first evaluation of tumour lesions. *Eur. J. Nucl. Med .Mol. Imaging* **2013**, *40*, 486–495.
21. Meyer, G.J.; Macke, H.; Schuhmacher, J.; Knapp, W.H.; Hofmann, M. 68a-labelled dota-derivatised peptide ligands. *Eur. J. Nucl. Med. Mol. Imaging* **2004**, *31*, 1097–1104.

5

Genes and Pathways Involved in Adult Onset Disorders Featuring Muscle Mitochondrial DNA Instability

Naghia Ahmed, Dario Ronchi * and Giacomo Pietro Comi *

Neurology Unit, IRCCS Foundation Ca' Granda Ospedale Maggiore Policlinico, Dino Ferrari Centre, Department of Pathophysiology and Transplantation, Università degli Studi di Milano, via Francesco Sforza 35, Milan 20122, Italy; E-Mail: naghia.ahmed89@gmail.com

* Authors to whom correspondence should be addressed;
E-Mails: dario.ronchi@gmail.com (D.R.); giacomo.comi@unimi.it (G.P.C.)

Academic Editors: Jaime M. Ross and Giuseppe Coppotelli

Abstract: Replication and maintenance of mtDNA entirely relies on a set of proteins encoded by the nuclear genome, which include members of the core replicative machinery, proteins involved in the homeostasis of mitochondrial dNTPs pools or deputed to the control of mitochondrial dynamics and morphology. Mutations in their coding genes have been observed in familial and sporadic forms of pediatric and adult-onset clinical phenotypes featuring mtDNA instability. The list of defects involved in these disorders has recently expanded, including mutations in the exo-/endo-nuclease flap-processing proteins MGME1 and DNA2, supporting the notion that an enzymatic DNA repair system actively takes place in mitochondria. The results obtained in the last few years acknowledge the contribution of next-generation sequencing methods in the identification of new disease loci in small groups of patients and even single probands. Although heterogeneous, these genes can be conveniently classified according to the pathway to which they belong. The definition of the molecular and biochemical features of these pathways might be helpful for fundamental knowledge of these disorders, to accelerate genetic diagnosis of patients and the development of rational therapies. In this review, we discuss the molecular findings disclosed in adult patients with muscle pathology hallmarked by mtDNA instability.

Keywords: mtDNA maintenance; multiple mtDNA deletions; adult mitochondrial disorders; progressive external ophthalmoplegia; mitochondrial myopathy

---

1. Introduction

Mitochondrial disorders display heterogeneous clinical presentations in terms of age at onset, progression and symptoms. This variability reflects their complex pathogenesis, which might affect structural proteins and enzymes involved in oxidative metabolism, finally leading to the mitochondrial dysfunction observed in multiple cell types and tissues [1].

Serum abnormalities (*i.e.*, increased lactate levels) or evidence of respiratory chain impairment in affected tissues (*i.e.*, isolated or combined complex deficiency) are often indicative of inadequate mitochondrial respiration, although they are not observed in many primary mitochondrial disorders [2]. Moreover, they lack specificity or should be tested in inaccessible tissues, jeopardizing their efficacy in a clinical setting. Therefore, clinical variability and the unavailability of reliable biomarkers delay the diagnosis in early- and late-onset mitochondrial disorders.

In adult non-syndromic clinical presentations, the involvement of skeletal muscle is mainly exhibited by external ophthalmoplegia, isolated or accompanied by limb weakness [3]. In these patients, histological studies on muscle biopsy usually reveal the presence of fibers not reacting to cytochrome c oxidase staining (COX-negative fibers) or showing signs of compensatory mitochondrial proliferation (ragged-red fibers) [4]. Southern blot analysis of muscle-extracted mitochondrial DNA (mtDNA) can detect reduced mtDNA content and multiple deletions, reflecting mtDNA instability [5]. Although subtle mtDNA deletions can accumulate in the muscle of patients harboring mtDNA point mutations [6], they are largely considered hallmarks of adult disorders due to defects in nuclear genes. Therefore, the disclosure of these alterations is helpful to identify a specific subgroup of mitochondrial disorders (multiple mtDNA deletions syndromes), but the large number of imputable genes hampers the chance of a prompt molecular diagnosis, especially in sporadic cases.

In 2000, missense mutations in *SLC25A4* (encoding for the mitochondrial translocator *ANT1*) have been identified as the first molecular defects resulting in impaired mtDNA maintenance in human [7]. Since then, the hunt for molecular defects underlying multiple mtDNA deletion syndromes has demonstrated to be a very active field of research contributing to improving the diagnostic yield of these conditions. At the same time, these studies expanded our knowledge of the proteins involved in the replication and repair of the mitochondrial genome. Some of these proteins were found to have multiple, sometimes redundant activities, while others seem to have unique functions [8].

In this review, we overview the genes hosting mutations linked with adult-onset phenotypes featuring the accumulation of mtDNA deletions in skeletal muscle. The limited number of patients harboring mutations in the most recently-described genes prevents the definition of clear genotype-phenotype correlations, which are still elusive, even for the most frequent genetic defects. However, the systematic collection of clinical and molecular data of diagnosed patients might help to disclose features shared by subjects harboring mutations in the same gene or pathway.

2. Molecular Features in Adult Mitochondrial Disorders Featuring Muscle mtDNA Instability

Disturbances of mtDNA homeostasis result in clinical presentations affecting every stage of life. Infantile and pediatric forms are mostly associated with a strong reduction of mtDNA content in affected tissues; therefore, they are collectively termed mtDNA depletion syndromes [9]. The low residual mtDNA levels cannot sustain a proper respiratory chain assembly: the resulting dysfunction mainly impacts complexes containing mtDNA-encoded subunits and spares complex II, which is entirely encoded by nuclear DNA.

In adults, large-scale deletions of mtDNA spontaneously accumulate in post-mitotic tissues with ageing [6]. The position and extension of such deletions within the mitochondrial genome may be heterogeneous, and in most cases, their accumulation does not result in clinical presentations, although a biochemical defect could be observed by histochemical or enzymatic studies. These slight alterations might also be involved in muscle weakness naturally occurring with ageing. On the opposite side, in primary mitochondrial disorders due to impaired mtDNA maintenance, deletions tend to accumulate in post-mitotic tissues (*i.e.*, muscle) and result in clinical phenotypes, even in relatively young subjects [10]. These alterations are easily detectable on muscle-extracted DNA specimens using Southern blot analysis and standard or quantitative PCR protocols. The co-existence of a quantitative (depletion) and qualitative (deletions) alteration of mtDNA in the same patient is not excluded.

Both pediatric and adult-onset presentations due to mtDNA instability originate from molecular defects in nuclear genes. This is relevant for human health, since such disorders are transmitted according to Mendelian inheritance. Infantile- or juvenile-onset mtDNA depletion syndromes are associated with recessive mutations, resulting in the loss of function of the encoded enzymes. Adult presentations are genetically more heterogeneous, presenting both recessive and dominant types of transmission, these two forms co-existing even for the same gene [11] (Figure 1).

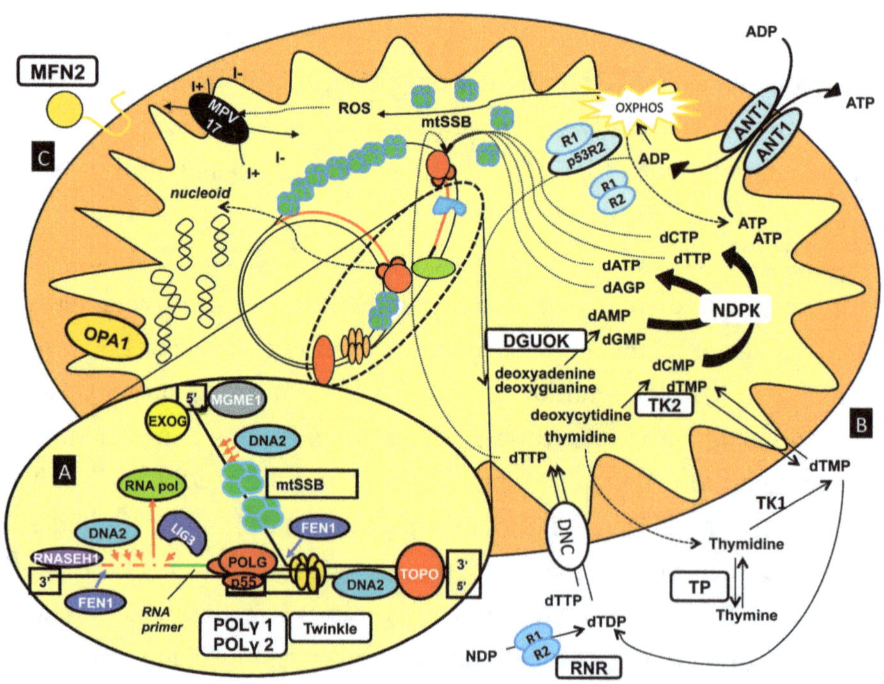

Figure 1. Schematic overview of the proteins and pathways involved in mtDNA maintenance. Zooming in on the mtDNA allows the identification of factors involved in mtDNA replication and repair (**A**) including: POLG, Twinkle, DNA2, MGME1; on the right (**B**) are the proteins assumed to affect the import and metabolism of the mitochondrial dNTP precursors; on the left (**C**) are the supposed localization of key factors ruling mitochondrial dynamics (MFN2, OPA1).

The number of genes involved in this class of disorders has rapidly increased in the last few years, mostly due to the results achieved using next-generation sequencing protocols in familial and sporadic cases. Although the encoded proteins might display multiple catalytic activities and the functional characterization is incomplete for most of them, a (provisional) classification of the genes involved in human mtDNA instability disorders can be attempted (Table 1):

(1) Genes encoding for key proteins of the core mtDNA replication machinery: *POLG* [12], *POLG2* [13], *PEO1* [14];
(2) Genes encoding for proteins involved in mtDNA repair and maintenance: *DNA2* [15], *MGME1* [16];
(3) Genes encoding for proteins preserving the mitochondrial nucleotide pool: *TP* [17], *TK2* [18], *DGUOK* [19], *SLC25A4* [7], *RRM2B* [20], *SUCLA2* [21], *SUCLG1* [22], *ABAT* [23];
(4) Genes encoding for proteins involved in mitochondrial dynamics and remodelling of mitochondrial membranes: *OPA1* [24], *MFN2* [25], *FBXL4* [26,27].

Although mutations of some genes have been so far restricted to infantile (*i.e.*, *FBXL4*) or adult (*i.e.*, *MGME1*) presentations, novel evidence supports the concept of a genetic overlap between these forms, with mutations in the same gene resulting in striking differences in age of onset, tissues affected, progression and outcome. The reason for this heterogeneity is currently unknown. The hypothesis that pediatric presentations could be associated with deleterious mutations, strongly impacting on enzyme activity of the encoded protein, while variants disclosed in adult patients could preserve higher levels of residual activity, is only partially supported by experimental data [28]. The low levels of expression and basal activity of TK2 observed in skeletal muscle might explain the tissue selectivity of TK2 dysfunction, but they are not useful to understand why age of onset and progression are so different between infantile and adult forms [29]. Moreover, the number of pediatric and adult patients sharing the same molecular defects is still low, making genotype-phenotype correlations hazardous or not obvious. It is likely that on-going projects of massive parallel sequencing in novel probands featuring mtDNA instability will further expand this genetic overlap.

The proposed classification does not include all of the genetic defects featuring muscle mtDNA instability. For example, although mutations in *SPG7*, encoding paraplegin, classically result in neurodegenerative diseases, such as hereditary spastic paraplegia [30] and optic neuropathies [31], they have been also described in patients with complex phenotypes showing prominent spastic ataxia with mitochondrial muscle pathology hallmarked by mtDNA deletions [32]. Similarly, mutations in *CHCHD10* have been originally described in familial amyotrophic lateral sclerosis with fronto-temporal dementia [33]. Muscle biopsies of affected subjects revealed respiratory chain dysfunction, COX-negative fibers and a large amount of mtDNA deletions. Following this observation, muscle mitochondrial defects, without signs of mtDNA instability, have been observed in other patients presenting pure motor neuron phenotypes [34,35] or isolated mitochondrial myopathy [36]. The role of CHCHD10 is not completely understood: some evidence indicates that it might take part in mitochondrial dynamics and cristae remodelling [37].

In the following paragraphs, we will focus on genes associated with adult presentations featuring primary accumulation of mtDNA deletions in muscle.

Table 1. Genes involved in disorders featuring mitochondrial DNA instability.

Pathway	Gene	Locus	Encoded Protein	Transmission	Onset	mtDNA Defects	Tissues Mainly Affected	Clinical Phenotypes
mtDNA replication	*POLG*	15q25	DNA polymerase gamma, catalytic subunit	AD, AR	Adult	dels	muscle	PEO, MM
				AR	Infantile	depl	liver	MDS, ME, AS
					Adult	depl	cerebellum	MIRAS
	POLG2	17q	DNA polymerase gamma, accessory subunit	AD	Adult	dels	muscle	PEO
	PEO1	10q24	Twinkle	AD	Adult	dels	muscle	PEO
				AR	Infantile	depl	liver	MDS
					Infantile	depl	brain	IOSCA, ME
mtDNA repair	*DNA2*	10q21.3–q22.1	DNA replication helicase/nuclease 2	AD	Adult	dels	muscle	PEO, MM
	MGME1	20p11.23	Mitochondrial genome maintenance exonuclease 1	AR	Adult	dels/depl	muscle	PEO, MM
dNTPs pools maintenance	*SLC25A4*	4q35	Adenine nucleotide translocator	AD	Adult	dels	muscle	PEO
	TYMP	22q13	Thymidine phosphorylase	AR	Late childhood Adolescence	dels/depl	muscle	MNGIE
	TK2	16q22–q23.1	Thymidine kinase 2	AR	Early childhood	depl	muscle	MDS
				AR	Adult	dels	muscle	PEO, MM
	DGUOK	2p13	Deoxyguanosine kinase	AR	Neonatal Infantile	depl	liver/muscle	MDS
				AR	Adult	dels	muscle	PEO
	RRM2B	8q23.1	Ribonucleotide reductase M2 B	AR	Infantile	depl	muscle	MDS
				AR	Adult	depl	muscle	MNGIE
				AD	Adult	dels	muscle	PEO
	SUCLA2	13q12.2–q13.1	Succinyl-CoA ligase, beta subunit	AR	Early childhood	depl	muscle	MDS
	SUCLG1	2p11.2	Succinyl-CoA ligase, alpha subunit	AR	Neonatal Infantile	depl	muscle/liver	MDS
	ABAT	16p13.2	4-aminobutyrate aminotransferase	AR	Infantile	depl	brain/muscle	MDS
Mitochondrial dynamics	*OPA1*	3q28–q29	Mitochondrial dynamin-like GTPase	AD	Adult	dels	muscle	OA plus
	MFN2	1p36.22	Mitofusin 2	AR	Adult	dels	muscle	OA plus
	MPV17	2p23.2	Mpv17 mitochondrial inner membrane protein	AR	Neonatal Infantile	depl	liver	MDS
				AR	Adult	dels	brain	ME
	FBXL4	6q16.1	F-box and leucine-rich repeat (LRR) protein	AR	Neonatal Infantile	depl	brain/muscle	ME/

Abbreviations: mitochondrial DNA (mtDNA), autosomal dominant (AD), autosomal recessive (AR), multiple mtDNA deletions (dels), mtDNA depletion (depl), progressive external ophthalmoplegia (PEO), mitochondrial myopathy (MM), mtDNA depletion syndrome (MDS), Alpers' syndrome (AS), mitochondrial recessive ataxia syndrome (MIRAS), mitochondrial encephalopathy (ME), optic atrophy (OA), mitochondrial neurogastrointestinal encephalopathy (MNGIE), infantile-onset spinocerebellar ataxia (IOSCA).

2.1. Genes Encoding for Members of the mtDNA Replication Machinery

The minimum replicative apparatus (replisome) of mtDNA includes the polymerase γ (POLG), the only DNA polymerase active within mitochondria of animal cells, the hexameric helicase Twinkle and the single-strand binding proteins (mtSSBPs). Other enzymatic activities, likely essential for mtDNA replication, include POLRMT, which supplies primers to start replication at the origin of the heavy strand, RNASEH1, which removes primers used during the synthesis of lagging strand, and TOP1MT, a 72-kDa topoisomerase, required to relax negative supercoils [38].

2.1.1. *POLG*

The mammalian mitochondrial polymerase γ is a 250-kDa heterotrimer composed of a 140-kDa catalytic α subunit (encoded by human *POLG*) and two 55-kDa accessory β subunits (encoded by human *POLG2*) [39]. It is synthesized as a precursor containing an amino-terminal leader sequence, which targets the protein towards mitochondria, where the precursor is cleaved. The catalytic subunit contains a carboxy terminal polymerase domain (*pol*) and an amino terminal 3′–5′ exonuclease domain (*exo*) with proofreading activity, separated by a linker region [40]. The intrinsic exonuclease 3′–5′ function greatly improves the replication fidelity, and it is necessary to suppress mtDNA deletions between direct repetitions [41].

Mutations in α subunits result in pediatric- and juvenile-onset presentations such as Alpers–Huttenlocher syndrome (severe infantile-onset encephalopathy with epilepsy associated with hepatic failure), SANDO (sensory ataxic neuropathy, dysarthria and ophthalmoplegia) and MIRAS (mitochondrial recessive ataxia syndrome) [42]. The analysis of muscle mtDNA reveals reduced mtDNA content and multiple deletions. *POLG* mutations also constitute the most frequent cause of familiar (dominant or recessive) and sporadic chronic external ophthalmoplegia [12]. In recessive forms, peripheral neuropathy is frequent and might occur decades before ptosis; the onset is precocious compared to dominant forms [43]. Beside muscle, *POLG* mutations often strike the adult central nervous system with presentations including cognitive impairment, dementia and obsessive disorders [44]. *POLG* defects have been documented in different clinical presentations sharing ataxia [45]. Parkinsonism accompanying progressive external ophthalmoplegia (PEO), ptosis and neuropathy was also found to segregate with *POLG* mutations in a few families [44,46,47]. Therefore, *POLG* defects should be regarded as a secondary genetic cause of Parkinson's disease, with affected patients presenting earlier onset and variable response to levodopa. Few reports also described premature ovarian failure with Parkinsonism and PEO in women harboring *POLG* variants [48,49].

The A467T mutation is the most common pathogenetic substitution in *POLG*, and it is estimated to occur in 36% of the mutated alleles. Its frequency varies between 0.2% and 1% in the general European asymptomatic population [50]. Almost all of the dominant *POLG* mutations associated with PEO are mapped on the polymerase domain of the enzyme. Mutated enzymes compete against wild-type proteins for binding to the replicative fork [51]. Experiments in *S. cerevisiae* displayed a similar behavior between human *POLG* mutations and the corresponding substitutions of the orthologue *mip1*: the severity of yeast phenotypes correlated with the age of onset of human presentations [52].

2.1.2. *POLG2*

The p55 β (accessory) subunits of polymerase γ are encoded by *POLG2*. They constitute a homodimer that binds asymmetrically to the catalytic portion of the holoenzyme (encoded by *POLG*). The proximal subunit strengthens DNA binding, while the distal subunit facilitates the nucleotide incorporation. The disruption of the interaction between the p55 accessory subunit and the p140 catalytic part might promote the stalling of the replication fork and produce mtDNA deletions. Furthermore, POLG2 has a role in nucleoids' structure [38,53].

Few pathogenetic mutations have been so far identified in *POLG2* (G451E, G416A, c.1207_1208ins24, R369G), the patients presenting autosomal dominant ptosis and PEO with onset in the third to fourth decade, mild proximal muscle weakness with exercise intolerance and other neurologic or systemic symptoms, similarly to *POLG*-mutated PEO patients [13,54,55].

2.1.3. *PEO1*

Also known as *C10orf2*, *PEO1* encodes for a hexameric helicase of the RecA-type superfamily, named Twinkle. It has structural similarities with the gp4 protein of T7 phage primase/helicase. The linker region, important for subunit interactions and the establishment of the functional hexamer, is known to be a mutational hotspot for dominant PEO [14]. Apart from 5′–3′ helicase activity, Twinkle is essential for the maintenance and the regulation of mtDNA copy number [56].

Mutations that suppress Twinkle helicase activity also compromise mtDNA replication and transcription, since the progression of the replicative fork is hampered, causing the accumulation of replication intermediates [57]. The presence of a cluster of mutations in small highly-conserved regions supports the proposed negative-dominant behavior of Twinkle mutations on mtDNA maintenance and transcription [43]. As a consequence, *PEO1* mutations are inherited according to an autosomal-dominant fashion. Heterozygous mutations might also arise *de novo* in sporadic patients [58]. Fratter and colleagues studied a cohort of 33 mutated patients (26 probands) presenting either missense mutations or in-frame duplications [59].

PEO1-mutated patients show variable onset ranging from the second to the eighth decade. Clinical features might include ptosis, ophthalmoparesis, proximal hyposthenia, ataxia, peripheral neuropathy, bulbar signs, cardiomyopathy, endocrine disorders, cataract and depression or avoiding personality tracts. A syndromic presentation, including sensory neuropathy with ataxia, dysarthria and ophthalmoparesis (SANDO), can also occur [60]. As for *POLG* mutations, *PEO1* defects have been found to segregate with Parkinsonism and additional syndromic features in dominant PEO families [61,62]. Recessive mutations in *PEO1* are less frequent and result in severe pediatric presentations, including IOSCA (infantile-onset spinocerebellar ataxia), mostly occurring in Finnish patients [63], and a hepatocerebral form of mtDNA depletion syndrome [64].

2.2. Genes Encoding for Factors Involved in mtDNA Repair

The accumulation of DNA damage is thought to play a critical role in the aging process, as well as mitochondrial disorders [65]. It was thought for many years that mitochondria lacked an enzymatic DNA repair system comparable to that in the nuclear compartment. However, it is now well established that DNA repair actively takes place in mitochondria, preserving genomic integrity through oxidative DNA damage processing, base excision repair (BER) pathways and further, still uncharacterized, mechanisms [66]. It is likely that the enzymatic activities required for repairing specific mtDNA damage might be also helpful in the processing of replication intermediates, narrowing the functional border between these two groups of proteins involved in mtDNA maintenance. The BER pathway engages the recognition and removal of deaminated, oxidized and alkylated DNA bases. Following the recognition of the damaged base, a group of endo/exonucleases catalyzes the consecutive steps of this orchestrated process. The nuclear BER has been largely explored, as well as the consequences of its dysfunction for human health [67]. Mammalian mitochondria also possess BER activities, promoted by a set of enzymes presenting double localization (nuclear and mitochondrial) or exclusively targeted to the mitochondrial compartment [68]. Interestingly, clinical presentations have been recently associated with altered mtDNA maintenance due to defective mitochondrial endo-/exo-nuclease activities.

2.2.1. *DNA2*

This gene has been recently considered as a candidate for the molecular screening of familiar or sporadic cases with PEO and/or myopathy accompanied by muscle accumulation of mtDNA deletions [15]. Human *DNA2* encodes for a member of the nuclease/helicase family. It was found in mammalian mitochondria, where it participates in the removal of RNA primers during mtDNA replication [69], but exclusive mitochondrial localization is debated [70]. The encoded protein interacts with polymerase γ and stimulates its catalytic activity [69]. As a member of the nuclease/helicase family, human DNA2 contains conserved nuclease, ATPase and helicase domains.

Experimental studies have demonstrated a role for DNA2 in the mitochondrial long patch base excision repair pathway primed by mtDNA defects due to oxidation, alkylation and hydrolysis. In particular, the nuclease activity is involved in the processing of flap intermediates occurring during the removal of damaged bases in BER. Inside mitochondria, this task is accomplished with the partnership of the flap structure-specific endonuclease 1 (FEN1). DNA2 seems also involved in the processing of Okazaki fragments, as demonstrated in yeast [71]. The function of DNA2 as a DNA helicase is less demonstrated. In yeast, helicase activity facilitates the formation of flap intermediates *in vitro* [72], but it was found to be dispensable *in vivo* [73]. Similarly, the mutations identified in patients and localized within nuclease and ATPase domains also impair helicase activity, as shown by *in vitro* studies [15]. Therefore, the helicase activity of DNA2 is also likely involved in mtDNA maintenance. At present, Twinkle remains the only established replicative helicase inside mitochondria. However, DNA2 might partially support Twinkle, since it co-localizes with Twinkle in nucleoids and its mitochondrial recruitment is induced by *PEO1* mutations [74].

A *DNA2* homozygous out of frame truncating mutation has been identified as a genetic cause of Seckel syndrome [75], a disorder characterized by *in utero* and postnatal growth retardation, intellectual disability, microcephaly, facial dysmorphisms and, rarely, cardiac malformations. Enhanced senescence with a marked increase of damaged nuclear DNA has been observed in patients' fibroblasts, supporting the hypothesis that DNA2 repair activity is affected in this disorder [75]. While these findings acknowledge a nuclear role for human DNA2, the striking differences between pediatric and adult presentations are less clear. The heterozygous mutations disclosed in adult patients impair

DNA2 activity less severely with respect to the homozygous truncating mutation linked with Seckel syndrome. Alternative nuclear and mitochondrial factors, replacing defective DNA2 and their differential expression among tissues, might also modulate the phenotype.

2.2.2. *MGME1*

This gene, previously known as *C20orf72*, encodes a RecB-type exonuclease belonging to the PD-(D/E)XK nuclease superfamily [76]. Cell fractioning experiments showed that the encoded polypeptide is targeted to mitochondria, where it is involved in mtDNA maintenance, promoting the turnover of the replication intermediates 7S. MGME1 cuts DNA, but not RNA or DNA-RNA hybrids; it seems to require free 5′-ends to exert its function (it does not work on circular DNA), and it has a greater affinity for single DNA than double-strand DNA *in vitro*. As in the case of DNA2, MGME1 also interacts with pol γ, therefore contributing a further 5′–3′ exonuclease activity to the mitochondrial machinery assembled at the replication fork [77].

Recessive mutations within the *MGME1* coding sequence have been observed in multiple patients from two families and a sporadic patient. Clinical presentation includes external ophthalmoplegia, muscle weakness and a progressive respiratory impairment, requiring assisted ventilation [16].

2.3. Genes Encoding for Proteins Maintaining the Mitochondrial dNTP Pool

As for the nuclear genome, mtDNA also require a balanced pool of dNTPs for an effective replication (and repair). In replicative tissues, cytosolic dNTPs, used in nuclear DNA synthesis, can be also imported into mitochondria through dedicated transporters, sustaining mtDNA maintenance. In non-replicative cells, the cytosolic *de novo* synthesis of dNTPs is downregulated, and the mtDNA replication, which persists in post-mitotic cells, relies on a set of intra-mitochondrial reactions to preserve the proportions of deoxyribonucleotides (dNTPs salvage pathway) [78]. Recessive mutations in key enzymes of this pathway might impair the supply of substrates for the mitochondrial DNA polymerase, resulting in defective mtDNA synthesis or increased replicative errors and genomic instability. In quiescent cells, *de novo* dNTP synthesis still partially supports the replication of mtDNA and the repair of the nuclear genome. Therefore, other proteins involved in nucleoside transport and metabolism might be important for mtDNA maintenance, as observed for the cytosolic ribonucleotide reductase (see below) [79]. Besides the obvious role in the supply of the "building blocks" for DNA synthesis, the deregulation of nucleotide levels might also have more complex consequences, overall conditioning the mitochondrial mass in the cell [80].

2.3.1. *TYMP*

Thymidine phosphorylase (TP, encoded by *TYMP*, also known as *ECGF1*) acts as a homodimer catalyzing the phosphorylation of thymidine phosphate to thymidine and 2-deoxy-D-ribose 2-phosphate. The direct reaction is important for nucleosides catabolism, while the reverse reaction participates in the pyrimidine salvage pathway.

Recessive mutations of this gene have been identified as the cause of mitochondrial neurogastrointestinal encephalomyopathy (MNGIE) [17]. These loss-of-function defects result in the reduction (or substantial absence) of thymidine phosphorylase activity and the toxic accumulation of nucleotides in plasma [17]. The imbalance of the dNTP pool increases the rate of mtDNA point mutations, and both mtDNA depletion and deletions are observed at Southern blot analysis of muscle mtDNA. Clinical manifestations depend on residual TP activity. Since *TP* is a homodimer enzyme, the presence of a mutant allele induces 25% wild-type enzyme synthesis: asymptomatic heterozygotes have 25%–35% of enzymatic residual activity, and this threshold represents a target for therapeutic strategies, like liver transplantation, that results in being be six-fold more efficient as a TP source than bone marrow [81]. Neuroradiological studies in *TYMP*-mutated patients often show leukoencephalopathy [82]. Brain MRI has been suggested as a tool for differential diagnosis of *POLG* involvement in recessive PEO presentations [43].

2.3.2. *TK2*

This gene encodes for the thymidine phosphorylase type 2, the enzyme that catalyze the rate-limiting step of deoxypirimidine phosphorylation within the mitochondrial dNTPs salvage pathway. The gene is upregulated in non-replicating cells, where the encoded product safeguards pool availability.

TK2 deficiency has been classically associated with the myopathic form of mtDNA depletion syndrome, featuring an infantile-onset severe phenotype with motor regression and early death due to diaphragmatic paralysis [18]. Recessive mutations have been also reported in presentations resembling spinal muscular atrophy [83]. Mutations in the same gene have been more recently observed in adult phenotypes with ptosis, external ophthalmoplegia, slowly-progressive proximal muscle weakness, muscular atrophy and dysarthria [84,85]. In adult patients with PEO due to *TK2* mutations, multiple deletions, but not depletion, have been found in muscle. The low basal TK2 activity in muscles could explain why most of the clinical features related to TK2 dysfunction affect this tissue [29]. The reason for the different presentations in pediatric and adult subjects is more elusive: indeed, biochemical studies performed in fibroblasts from adult patients with mild myopathy revealed very low TK2 activity levels, not dissimilar to those observed in pediatric cases.

2.3.3. *DGUOK*

DGUOK encodes for the mitochondrial deoxyguanosine kinase that catalyzes the first reaction in the purine salvage pathway inside mitochondrial matrix. Recessive mutations have been described in pediatric cases showing severe encephalohepatopathy with severe reduction of liver mtDNA content and premature death, unless the patients undergo a liver transplant [19]. Recently, the massive parallel sequencing of genes encoding for established or predicted mitochondrial proteins (MitoExome) has revealed compound heterozygous mutations even in adult sporadic and familial forms of mitochondrial myopathy [28]. Muscle analysis in mutated patients disclosed mtDNA deletions with variable mtDNA content and significant deficiency of DGUOK protein levels and activity. The main clinical features of adult forms include: muscle weakness, PEO, hearing loss and bulbar signs. One of the probands reported had previously undergone liver transplant. Notably, two siblings harboring *DGUOK* mutations displayed an atypical phenotype characterized by slowly progressive, predominantly distal, upper and lower limb muscle weakness, mild dysphonia and dysphagia, similarly to SMA-like presentations due to TK2 deficiency. These findings clearly demonstrate that the effects of *DGUOK* and *TK2* mutations are not limited to those tissues that are primarily affected in terms of severity and age of involvement.

2.3.4. *SLC25A4*

This gene encodes for the ANT1 member of the ADP/ATP translocator family, localized in the inner mitochondrial membrane. ANT1 predominates in post-mitotic tissues, including skeletal muscle, heart and brain. It acts as a homodimeric gate-channel operating the ADP/ATP exchange between mitochondria and cytoplasm. Since ANT1 senses the adenosine concentration in the two compartments, it is a part of the signaling pathway coupling cellular energy consumption and mitochondrial respiratory chain activity [86]. Moreover, ANT1 is a structural element of the mitochondrial permeability transition pores (MPTP) and has a role in the intrinsic apoptotic pathway [87].

Heterozygous *SLC25A4* mutations have been reported in patients with PEO and ptosis with or without generalized muscle weakness [7]. Cardiac involvement, including cardiomyopathy, disarrangement of myofibers, inflammation linked to heart disease and ischemic attacks [88], has been repeatedly described. Neurosensory hypoacusia, thyroid gout and dementia without affective aspects have also been reported [43]. Mutations in the same gene also result in Sengers syndrome, a multisystemic disorder featuring congenital cataract, hypertrophic cardiomyopathy, mitochondrial myopathy and lactic acidosis. Two forms are known for Sengers syndrome: a neonatal form with poor prognosis and a late-onset presentation with lifespan observed until the third decade [89].

2.3.5. *RRM2B*

This gene maps on chromosome 8q22.1–q23.3 and encodes the minor subunit (p53R2) of ribonucleotide reductase (RNR), the tetrameric enzyme catalyzing *de novo* dNTPs synthesis from the reduction of the corresponding ribonucleosides diphosphate. *RRM2B* expression is tightly regulated by the oncosuppressor p53. In post-mitotic cells, ribonucleotide reductase contributes to mitochondrial dNTP supply [90], in parallel with the salvage pathway.

Recessive mutations within the *RRM2B* coding sequence have been associated with the infantile encephalomyopathic form of mtDNA depletion syndrome [20]. In 2009, linkage analysis was used to drive the discovery of a heterozygous non-sense *RRM2B* mutation in affected members of a big North American family of European origin presenting adult-onset dominant PEO with muscle accumulation of mtDNA deletions. The same mutation in a second family was associated with additional clinical features, including: ataxic gait, hypoacusia, reduction of tendon reflexes and psychiatric disorders [91]. Kearns–Sayre syndrome (KSS) and MNGIE are rarer manifestations. *RRM2B* mutations were observed as the third most common molecular cause of PEO in a large cohort of British adult patients, following *POLG* and *PEO1* [92]. The mutated or truncated protein p53R2 might compete against wild-type protein for the binding of heterodimeric RNR, exerting a dominant negative effect [91], a pathogenic mechanism compatible with defects disclosed in both pediatric and adult subjects.

2.4. Genes Encoding for Protein Involved in Mitochondrial Dynamics and Remodeling

It is now well established that intracellular mitochondria are organized in dynamic networks and physiologically undergo remodelling cycling between fission and fusion events [93]. This behavior is crucial for the turnover of aged mitochondria and their intracellular transport, but also compensates the defects accumulating in the organelle as bypass products of oxidative metabolism [94]. Together with mitochondrial protein quality control system, mitochondrial dynamics have emerged as an important field of investigation to address the pathogenesis of human neurodegeneration. Indeed, increased mitochondrial fragmentation due to the reduced ratio between fusion (*i.e.*, MFN2 and OPA1) and fission (*i.e.*, DLP1/DRP1 and FIS1) proteins has been observed in tissues and cell cultures obtained from patients with major neurodegenerative disorders, including Alzheimer's and Huntington's disease. Furthermore, Parkinson's disease has been associated with mutations in PARK2 and PINK1 proteins, which orchestrate mitochondrial protein quality control and the turnover of aged mitochondria by ubiquitination and degradation of MFN2 [95].

2.4.1. *OPA1*

The product encoded by this gene is a regulatory mitochondrial fusion protein localized in the inner membrane, showing GTPase activity [96]. OPA1 is a main regulator of cristae morphology, but also plays a key role in response to cellular cytotoxic insult generated by hyperactivation of *N*-methyl-D-aspartate (NMDA) receptors for an altered homeostasis of mitochondrial calcium. Indeed, the increased agonism of NMDA receptors induces mitochondrial fragmentation and ultrastructural defects of the inner mitochondrial membrane [97]. Conversely, *OPA1* overexpression is protective against cytotoxic glutamate response, with an increased survival of neurons and preserved mitochondrial morphology [98].

Mutations in *OPA1* are the most common cause of isolated optic atrophy with dominant inheritance (ADOA) [99,100], as well as "plus" phenotypes where optic atrophy is accompanied by syndromic features, such as neurosensory deafness and ophthalmoplegia. Muscle from *OPA1* plus patients showed signs of mitochondrial sufferance, including COX-negative fibers and accumulation of mtDNA deletions [24,101]. OPA1 defects not only impact retinal gangliar cells leading to the progressive degeneration of optic nerve, but also nervous and muscle tissues. Some studies have documented a compensatory increase of mtDNA copies in affected muscle fibers, independently of the severity of the disease [102]. The percentage of COX-negative fibers seems to be four-fold superior in the DOA plus form than in pure presentations.

Recently, *OPA1* sequencing in probands from two Italian families affected by dominant chronic PEO complicated by Parkinsonism and dementia revealed two heterozygous missense mutations affecting highly conserved amino acid positions in the GTPase domain [103]. Multiple mtDNA deletions were detected in available muscle biopsies.

2.4.2. *MFN2*

This gene encodes an outer membrane protein with GTPase activity involved in mitochondrial fusion. Molecular defects affecting *MFN2* are a major cause of the Charcot–Marie–Tooth axonal neuropathy type 2A, an autosomal dominant disease characterized by a pronounced impairment of motor and sensory neurons [104].

Mutations in *MFN2* have been recently disclosed in familial multisystemic disorder with optic atrophy beginning in early childhood, associated with axonal neuropathy and mitochondrial myopathy in adult life, a presentation resembling *OPA1* plus phenotype. Mitochondrial DNA deletions were found in muscle while patients' fibroblasts disclosed defective repair of stress-induced mtDNA damage, leading to respiratory chain impairment [25]. Notably, a detrimental effect on mtDNA replication has been also recently documented in fibroblasts from CMT2A patients, which displayed mtDNA depletion and a minimal amount of deleted molecules [105], suggesting a common pathogenetic mechanism resulting from *MFN2* dysfunction, irrespective of the clinical phenotype.

2.4.3. *MPV17*

This gene encodes for an inner mitochondrial membrane protein whose function has remained obscure for several years. The encoded product is a member of the family of integral membrane proteins comprising PXMP2, MPV17, MP-L and FKSG24 (MPV17L2) in mammals. Studies on vertebrates suggested a role in nucleotide trafficking between mitochondria and cytoplasm [106]. Recently, MPV17 was identified as a non-selective ion channel, responsible for modulation of mitochondrial membrane potential, which is influenced by several conditions in the organelle, such as redox state and protein phosphorylation [107].

Mutations in *MPV17* have been firstly described in a severe juvenile-onset hepatoencephalopathy with major reduction of liver mtDNA content [108]. Coming to adult presentations, recessive *MPV17* mutations have been described in three patients presenting adult-onset multi-systemic disorders with neuropathy and leukoencephalopathy and featuring multiple mtDNA deletions in muscle [109–111].

3. Genotype-Phenotype Correlations

Mitochondrial disorders display a spectacular clinical heterogeneity. Presentations featuring muscle mtDNA instability, which constitute a tiny subgroup of the spectrum of mitochondrial disorders, do not elude this general consideration. Indeed, the overlapping of clinical features is frequent with heterogeneous genetic defects resulting in undistinguishable phenotypes (Figure 2).

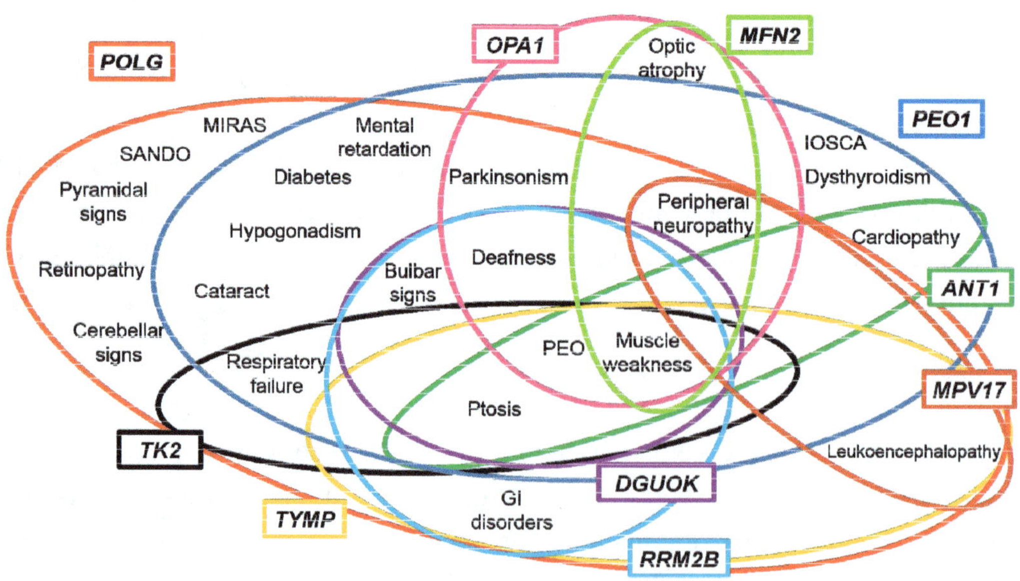

Figure 2. Representation of clinical phenotypes related to nuclear genes involved in mtDNA maintenance. Abbreviations: progressive external ophthalmoplegia (PEO), mitochondrial recessive ataxia syndrome (MIRAS), infantile-onset spinocerebellar ataxia (IOSCA), sensory ataxic neuropathy, dysarthria, and ophthalmoparesis (SANDO).

Previous studies reporting a cohort of patients accumulating multiple mtDNA deletions in muscle have attempted genotype-phenotype correlations, with modest results, even considering only forms due to mutations in the same gene [43]. The elucidation of such correlations might be useful to speed molecular diagnosis. The number of candidate genes for molecular tests overcomes the resources of standard laboratories, even if the application of next-generation sequencing in a clinical setting is expected to improve the diagnostic yield, reducing the cost of genetic testing [112]. Despite these issues, some remarks emerge from previous studies. Patients harboring *POLG* mutations display severer clinical features and age-related penetrance, while *ANT1* mutations are associated with milder phenotypes. *PEO1* patients show clinical presentations with intermediate severity of symptoms [43]. Bulbar weakness, deafness and gastrointestinal symptoms are observed in *TYMP* [113] and *RRM2B* patients [92], favoring the analysis of these genes before more frequent defects, such as *POLG* and *PEO1*. Cardiac involvement, reflected by conduction defects and left hypertrophy, is more representative of ANT1 involvement rather than *RRM2B* [92].

Peripheral neuropathy, even if not exclusive, is often associated with *POLG* mutations; however, it also occurs in *PEO1* cases [60]. Recently, a diagnostic flow-chart has been suggested for *POLG*-related diseases in pauci-symptomatic patients affected by peripheral neuropathy and suspected mitochondrial impairment [114]. Peripheral neuropathy has been also frequently observed in ADOA plus patients harboring *MFN2* and *OPA1* mutations [115].

4. Towards a Candidate-Pathway Approach

Despite the relevant achievements gained approaching mitochondrial patients with panel or exome sequencing technologies [116,117], the proportion of subjects without a molecular diagnosis is still high. The number of phenotypes reflecting muscle mtDNA instability has increased, while the borders between clinical presentations are getting narrow. As for other Mendelian disorders, the application of next-generation sequencing techniques in a research setting led to the discovery of novel genetic defects [112]. Overall, in only a few years, next-generation sequencing protocols disclosed more genetic defects than any other strategy (linkage studies, gene-candidate approach) in the last two decades. Nevertheless, the number of established diagnosis only slightly increased with a global improvement in the diagnostic yield of around 10%.

In the institutions where panel sequencing is routinely used for diagnostic purposes, the potential reduction of the time-to-diagnosis is challenged by the increased number of variants detected and the consequent efforts required to obtain a reliable genetic and biochemical validation. Taking into account these considerations, a detailed clinical assessment and a more accurate (or extended) genotyping performance might be important, but not decisive advancements. Conversely, the identification of clinical and molecular features shared by patients presenting defects in the same functional pathway, accompanied by quantitative measurements of deleted mtDNA molecules in muscle samples, could be used to drive the selection of genes or pathways to be addressed in molecular testing, improving the diagnosis of single patients in a short timeframe.

Acknowledgments

The financial support of Fondazione Cariplo to Dario Ronchi (Project number: 2014-1010) is gratefully acknowledged.

References

1. Schapira, A.H. Mitochondrial diseases. *Lancet* **2012**, *379*, 1825–1834.
2. Delonlay, P.; Rötig, A.; Sarnat, H.B. Respiratory chain deficiencies. *Handb. Clin. Neurol.* **2013**, *113*, 1651–1666.
3. Yu-Wai-Man, C.; Smith, F.E.; Firbank, M.J.; Guthrie, G.; Guthrie, S.; Gorman, G.S.; Taylor, R.W.; Turnbull, D.M.; Griffiths, P.G.; Blamire, A.M.; *et al.* Extraocular muscle atrophy and central nervous system involvement in chronic progressive external ophthalmoplegia. *PLoS ONE* **2013**, *8*, e75048, doi:10.1371/journal.pone.0075048.
4. Sundaram, C.; Meena, A.K.; Uppin, M.S.; Govindaraj, P.; Vanniarajan, A.; Thangaraj, K.; Kaul, S.; Kekunnaya, R.; Murthy, J.M. Contribution of muscle biopsy and genetics to the diagnosis of chronic progressive external opthalmoplegia of mitochondrial origin. *J. Clin. Neurosci.* **2011**, *18*, 535–538.
5. Wong, L.J.; Boles, R.G. Mitochondrial DNA analysis in clinical laboratory diagnostics. *Clin. Chim. Acta* **2005**, *354*, 1–20.
6. Zhang, C.; Baumer, A.; Maxwell, R.J.; Linnane, A.W.; Nagley, P. Multiple mitochondrial DNA deletions in an elderly human individual. *FEBS* **1992**, *297*, 34–38.
7. Kaukonen, J.; Juselius, J.K.; Tiranti, V.; Kyttälä, A.; Zeviani, M.; Comi, G.P.; Keränen, S.; Peltonen, L.; Suomalainen, A. Role of adenine nucleotide translocator 1 in mtDNA maintenaince. *Science* **2000**, *289*, 782–785.
8. Holt, I.J.; Reyes, A. Human mitochondrial DNA replication. *Cold Spring Harb. Perspect. Biol.* **2012**, *4*, doi:10.1101/cshperspect.a012971.
9. Finsterer, J.; Ahting, U. Mitochondrial depletion syndromes in children and adults. *Can. J. Neurol. Sci.* **2013**, *40*, 635–644.
10. Mitochondrial DNA Deletion Syndromes. Available online: http://www.ncbi.nlm.nih.gov/books/NBK1203/PubMed PMID: 20301382 (accessed on 3 May 2011).
11. Spinazzola, A.; Zeviani, M. Mitochondrial diseases: A cross-talk between mitochondrial and nuclear genomes. *Adv. Exp. Med. Biol.* **2009**, *652*, 69–84.
12. Van Goethem, G.; Dermaut, B.; Löfgren, A.; Martin, J.J.; van Broeckhoven, C. Mutation of *POLG* is associated with progressive external ophthalmoplegia characterized by mtDNA deletions. *Nat. Genet.* **2001**, *28*, 211–212.
13. Longley, M.J.; Clark, S.; Yu, W.; Man, C.; Hudson, G.; Durham, S.E.; Taylor, R.W.; Nightingale, S.; Turnbull, D.M.; Copeland, W.C.; *et al.* Mutant *POLG*2 disrupts DNA polymerasegamma subunits and causes progressive external ophthalmoplegia. *Am. J. Hum. Genet.* **2006**, *78*, 1026–1034.

14. Spelbrink, J.N.; Li, F.Y.; Tiranti, V.; Nikali, K.; Yuan, Q.P.; Tariq, M.; Wanrooij, S.; Garrido, N.; Comi, G.; Morandi, L.; *et al.* Human mitochondrial DNA deletions associated with mutations in the gene encoding Twinkle, a phage T7 gene 4-like protein localized in mitochondria. *Nat. Genet.* **2001**, *28*, 223–231.

15. Ronchi, D.; di Fonzo, A.; Lin, W.; Bordoni, A.; Liu, C.; Fassone, E.; Pagliarani, S.; Rizzuti, M.; Zheng, L.; Filosto, M.; *et al.* Mutations in *DNA2* link progressive myopathy to mitochondrial DNA instability. *Am. J. Hum. Genet.* **2013**, *92*, 293–300.

16. Kornblum, C.; Nicholls, T.J.; Haack, T.B.; Schöler, S.; Peeva, V.; Danhauser, K.; Hallmann, K.; Zsurka, G.; Rorbach, J.; Iuso, A.; *et al.* Loss-of-function mutations in *MGME1* impair mtDNA replication and cause multisystemic mitochondrial disease. *Nat. Genet.* **2013**, *45*, 214–249.

17. Nishino, I.; Spinazzola, A.; Papadimitriou, A.; Hammans, S.; Steiner, I.; Hahn, C.D.; Connolly, A.M.; Verloes, A.; Guimarães, J.; Maillard, I.; *et al.* Mitochondrial neurogastrointestinal encephalomyopathy: An autosomal recessive disorder due to thymidine phosphorylase mutations. *Ann. Neurol.* **2000**, *47*, 792–800.

18. Saada, A.; Shaag, A.; Mandel, H.; Nevo, Y.; Eriksson, S.; Elpeleg, O. Mutant mitochondrial thymidine kinase in mitochondrial DNA depletion myopathy. *Nat. Genet.* **2001**, *29*, 342–344.

19. Mandel, H.; Szargel, R.; Labay, V.; Elpeleg, O.; Saada, A.; Shalata, A.; Anbinder, Y.; Berkowitz, D.; Hartman, C.; Barak, M.; *et al.* The deoxyguanosine kinase gene is mutated in individuals with depleted hepatocerebral mitochondrial DNA. *Nat. Genet.* **2001**, *29*, 337–341, Erratum in: *Nat. Genet.* 29 December 2001.

20. Bourdon, A.; Minai, L.; Serre, V.; Jais, J.P.; Sarzi, E.; Aubert, S.; Chrétien, D.; de Lonlay, P.; Paquis-Flucklinger, V.; Arakawa, H.; *et al.* Mutation of *RRM2B*, encoding p53-controlled ribonucleotide reductase (p53R2), causes severe mitochondrial DNA depletion. *Nat. Genet.* **2007**, *39*, 776–780.

21. Elpeleg, O.; Miller, C.; Hershkovitz, E.; Bitner-Glindzicz, M.; Bondi-Rubinstein, G.; Rahman, S.; Pagnamenta, A.; Eshhar, S.; Saada, A. Deficiency of the ADP-forming succinyl-CoA synthase activity is associated with encephalomyopathy and mitochondrial DNA depletion. *Am. J. Hum. Genet.* **2005**, *76*, 1081–1086.

22. Ostergaard, E.; Christensen, E.; Kristensen, E.; Mogensen, B.; Duno, M.; Shoubridge E.A.; Wibrand, F. Deficiency of the alpha subunit of succinate-coenzyme A ligase causes fatal infantile lactic acidosis with mitochondrial DNA depletion. *Am. J. Hum. Genet.* **2007**, *81*, 383–387.

23. Besse, A.; Wu, P.; Bruni, F.; Donti, T.; Graham, B.H.; Craigen, W.J.; McFarland, R.; Moretti, P.; Lalani, S.; Scott K.L.; *et al.* The GABA transaminase, *ABAT*, is essential for mitochondrial nucleoside metabolism. *Cell Metab.* **2015**, *21*, 417–427.

24. Amati-Bonneau, P.; Valentino, M.L.; Reynier, P.; Gallardo, M.E.; Bornstein, B.; Boissière, A.; Campos, Y.; Rivera, H.; de la Aleja, J.G.; Carroccia, R.; *et al.* *OPA1* mutations induce mitochondrial DNA instability and optic atrophy "plus" phenotypes. *Brain* **2008**, *131*, 338–351.

25. Rouzier, C.; Bannwarth, S.; Chaussenot, A.; Chevrollier, A.; Verschueren, A.; Bonello-Palot, N.; Fragaki, K.; Cano, A.; Pouget, J.; Pellissier, J.F.; *et al.* The *MFN2* gene is responsible for mitochondrial DNA instability and optic atrophy "plus" phenotype. *Brain* **2012**, *135*, 23–34.

26. Bonnen, P.E.; Yarham, J.W.; Besse, A.; Wu, P.; Faqeih, E.A.; Al-Asmari, A.M.; Saleh, M.A.; Eyaid, W.; Hadeel, A.; He, L.; *et al.* Mutations in *FBXL4* cause mitochondrial encephalopathy and a disorder of mitochondrial DNA maintenance. *Am. J. Hum. Genet.* **2003**, *93*, 471–481, Erratum in: *Am. J. Hum. Genet.* **2013**, *93*, 773.
27. Gai, X.; Ghezzi, D.; Johnson, M.A.; Biagosch, C.A.; Shamseldin, H.E.; Haack, T.B.; Reyes, A.; Tsukikawa, M.; Sheldon, C.A.; Srinivasan, S.; *et al.* Mutations in *FBXL4*, encoding a mitochondrial protein, cause early-onset mitochondrial encephalomyopathy. *Am. J. Hum. Genet.* **2013**, *93*, 482–495.
28. Ronchi, D.; Garone, C.; Bordoni, A.; Gutierrez Rios, P.; Calvo, S.E.; Ripolone, M.; Ranieri, M.; Rizzuti, M.; Villa, L.; Magri, F.; *et al.* Next-generation sequencing reveals *DGUOK* mutations in adult patients with mitochondrial DNA multiple deletions. *Brain* **2012**, *135*, 3404–3415.
29. Saada, A.; Shaag, A.; Elpeleg, O. mtDNA depletion myopathy: Elucidation of the tissue specificity in the mitochondrial thymidine kinase (*TK2*) deficiency. *Mol. Genet. Metab.* **2003**, *79*, 1–5.
30. Casari, G.; de Fusco, M.; Ciarmatori, S.; Zeviani, M.; Mora, M.; Fernandez, P.; de Michele, G.; Filla, A.; Cocozza, S.; Marconi, R.; *et al.* Spastic paraplegia and OXPHOS impairment caused by mutations in paraplegin, a nuclear-encoded mitochondrial metalloprotease. *Cell* **1998**, *93*, 973–983.
31. Klebe, S.; Depienne, C.; Gerber, S.; Challe, G.; Anheim, M.; Charles, P.; Fedirko, E.; Lejeune, E.; Cottineau, J.; Brusco, A.; *et al.* Spastic paraplegia gene 7 in patients with spasticity and/or optic neuropathy. *Brain* **2012**, *135*, 2980–2993.
32. Pfeffer, G.; Gorman, G.S.; Griffin, H.; Kurzawa-Akanbi, M.; Blakely, E.L.; Wilson, I.; Sitarz, K.; Moore, D.; Murphy, J.L.; Alston, C.L.; *et al.* Mutations in the SPG7 gene cause chronic progressive external ophthalmoplegia through disordered mitochondrial DNA maintenance. *Brain* **2014**, *137*, 1323–1336.
33. Bannwarth, S.; Ait-El-Mkadem, S.; Chaussenot, A.; Genin, E.C.; Lacas-Gervais, S.; Fragaki, K.; Berg-Alonso, L.; Yusuke Kageyama.; Serre, V.; Moore, D.G.; *et al.* A mitochondrial origin for frontotemporal dementia and amyotrophic lateral sclerosis through *CHCHD10* involvement. *Brain* **2014**, *137*, 2329–2345.
34. Ronchi, D.; Riboldi, G.; del Bo, R.; Ticozzi, N.; Scarlato, M.; Galimberti, D.; Corti, S.; Silani, V.; Bresolin, N.; Comi, G.P. *CHCHD10* mutations in Italian patients with sporadic amyotrophic lateral sclerosis. *Brain* **2015**, doi:10.1093/brain/awu384.
35. Penttilä, S.; Jokela, M.; Bouquin, H.; Saukkonen, A.M.; Toivanen, J.; Udd, B. Late onset spinal motor neuronopathy is caused by mutation in *CHCHD10*. *Ann. Neurol.* **2015**, *77*, 163–172.
36. Ajroud-Driss, S.; Fecto, F.; Ajroud, K.; Lalani, I.; Calvo, S.E.; Mootha, V.K.; Deng, H.X.; Siddique, N.; Tahmoush, A.J.; Heiman-Patterson, T.D.; *et al.* Mutation in the novel nuclear-encoded mitochondrial protein *CHCHD10* in a family with autosomal dominant mitochondrial myopathy. *Neurogenetics* **2015**, *16*, 1–9.
37. Cogliati, S.; Frezza, C.; Soriano, M.E.; Varanita, T.; Quintana-Cabrera, R.; Corrado, M.; Cipolat, S.; Costa, V.; Casarin, A.; Gomes, L.C.; *et al.* Mitochondrial cristae shape determines respiratory chain supercomplexes assembly and respiratory efficiency. *Cell* **2013**, *155*, 160–171.
38. Wanrooij, S.; Falkenberg, M. The human mitochondrial replication fork in health and disease. *Biochim. Biophys. Acta* **2010**, *1797*, 1378–1388.

39. Spelbrink, J.N.; Toivonen, J.M.; Hakkaart, G.A.; Kurkela, J.M.; Cooper, H.M.; Lehtinen, S.K.; Lecrenier, N.; Back, J.W.; Speijer, D.; Foury, F.; *et al. In vivo* functional analysis of the human mitochondrial DNA polymerase *POLG* expressed in cultured human cells. *J. Biol. Chem.* **2000**, *275*, 24818–24828.
40. Graziewicz, M.A.; Longley, M.J.; Bienstock, R.J.; Zeviani, M.; Copeland, W.C. Structure-function defects of human mitochondrial DNA polymerase in autosomal dominant progressive external ophthalmoplegia. *Nat. Struct. Mol. Biol.* **2004**, *11*, 770–776.
41. Del Bo, R.; Bordoni, A.; Sciacco, M.; di Fonzo, A.; Galbiati, S.; Crimi, M.; Bresolin, N.; Comi, G.P. Remarkable infidelity of polymerase gammaA associated with mutations in *POLG*1 exonuclease domain. *Neurology* **2003**, *61*, 903–908.
42. Isohanni, P.; Hakonen, A.H.; Euro, L.; Paetau, I.; Linnankivi, T.; Liukkonen, E.; Wallden, T.; Luostarinen, L.; Valanne, L.; Paetau, A.; *et al. POLG*1 manifestations in childhood. *Neurology* **2011**, *76*, 811–815.
43. Van Goethem, G.; Martin, J.J.; van Broeckhoven, C. Progressive External Ophtalmoplegia characterized by multiple deletions of mitochondrial DNA. *Neuromol. Med.* **2003**, *3*, 129–146.
44. Delgado-Alvarado, M.; de la Riva, P.; Jiménez-Urbieta, H.; Gago, B.; Gabilondo, A.; Bornstein, B.; Rodríguez-Oroz, M.C. Parkinsonism, cognitive deficit and behavioural disturbance caused by a novel mutation in the polymerase gamma gene. *J. Neurol. Sci.* **2015**, *350*, 93–97.
45. Synofzik, M.; Srulijes, K.; Godau, J.; Berg, D.; Schöls, L. Characterizing *POLG* ataxia: Clinics, electrophysiology and imaging. *Cerebellum* **2012**, *11*, 1002–1011.
46. Mancuso, M.; Filosto, M.; Oh, S.J.; di Mauro, S. A novel polymerase gamma mutation in a family with ophthalmoplegia, neuropathy, and Parkinsonism. *Arch. Neurol.* **2004**, *61*, 1777–1779.
47. Davidzon, G.; Greene, P.; Mancuso, M.; Klos, K.J.; Ahlskog, J.E.; Hirano, M.; di Mauro, S. Early-onset familial parkinsonism due to *POLG* mutations. *Ann. Neurol.* **2006**, *59*, 859–862.
48. Luoma, P.; Melberg, A.; Rinne, J.O.; Kaukonen, J.A.; Nupponen, N.N.; Chalmers, R.M.; Oldfors, A.; Rautakorpi, I.; Peltonen, L.; Majamaa, K.; *et al.* Parkinsonism, premature menopause, and mitochondrial DNA polymerase gamma mutations: Clinical and molecular genetic study. *Lancet* **2004**, *364*, 875–882.
49. Pagnamenta, A.T.; Taanman, J.W.; Wilson, C.J.; Anderson, N.E.; Marotta, R.; Duncan, A.J.; Bitner-Glindzicz, M.; Taylor, R.W.; Laskowski, A.; Thorburn, D.R.; *et al.* Dominant inheritance of premature ovarian failure associated with mutant mitochondrial DNA polymerase gamma. *Hum. Reprod.* **2006**, *21*, 2467–2473.
50. Neeve, V.C.M.; Samuels, D.C.; Bindoff, L.A.; van den Bosch, B.; van Goethem, G.; Smeet, H.; Lombès, A.; Jardel, C.; Hirano, M.; DiMauro, S.; *et al.* What is influencing the phenotype of the common homozygous polymerase γ mutation p.Ala467Thr? *Brain* **2012**, *135*, 3614–3626.
51. Stuart, G.R.; Santos, J.H.; Strand, M.K.; van Houten, B.; Copeland, W.C. Mitochondrial and nuclear DNA defects in Saccharomyces cerevisiae with mutations in DNA polymerase gamma associated with progressive external ophthalmoplegia. *Hum. Mol. Genet.* **2006**, *15*, 363–374.
52. Stumpf, J.D.; Bailey, C.M.; Spell, D.; Stillwagon, M.; Anderson, K.S.; Copeland, W.C. *mip1* containing mutations associated with mitochondrial disease causes mutagenesis and depletion of mtDNA in Saccharomyces cerevisiae. *Hum. Mol. Genet.* **2010**, *19*, 2123–2133.

53. Di Re, M.; Sembongi, H.; He, J.; Reyes, A.; Yasukawa, T.; Martinsson, P.; Bailey, L.J.; Goffart, S.; Boyd-Kirkup, J.D.; Wong, T.S.; *et al.* The accessory subunit of mitochondrial DNA polymerase gamma determines the DNA content of mitochondrial nucleoids in human cultured cells. *Nucleic Acids Res.* **2009**, *37*, 5701–5713.
54. Walter, M.C.; Czermin, B.; Muller-Ziermann, S.; Bulst, S.; Stewart, J.D.; Hudson, G.; Schneiderat, P.; Abicht, A.; Holinski-Feder, E.; Lochmüller, H.; *et al.* Late-onset ptosis and myopathy in a patient with a heterozygous insertion in *POLG2*. *J. Neurol.* **2010**, *257*, 1517–1523.
55. Young, M.J.; Longley, M.J.; Li, F.Y.; Kasiviswanathan, R.; Wong, L.J.; Copeland, W.C. Biochemical analysis of human *POLG2* variants associated with mitochondrial disease. *Hum. Mol. Genet.* **2011**, *20*, 3052–3066.
56. Tyynismaa, H.; Sembongi, H.; Bokori-Brown, M.; Granycome, C.; Ashley, N.; Poulton, J.; Jalanko, A.; Spelbrink, J.N.; Holt, I.J.; Suomalainen, A. Twinkle helicase is essential for mtDNA mainteinance and regulates mtDNA copy number. *Hum. Mol. Genet.* **2004**, *13*, 3219–3227.
57. Milenkovic, D.; Matic, S.; Kühl, I.; Ruzzenente, B.; Freyer, C.; Jemt, E.; Park, C.B.; Falkenberg, M.; Larsson, N.G. TWINKLE is an essential mitochondrial helicase required for synthesis of nascent D-loop strands and complete mtDNA replication. *Hum. Mol. Genet.* **2013**, *22*, 1983–1993.
58. Virgilio, R.; Ronchi, D.; Hadjigeorgiou, G.M.; Bordoni, A.; Saladino, F.; Moggio, M.; Adobbati, L.; Kafetsouli, D.; Tsironi, E.; Previtali, S.; *et al.* Novel Twinkle (*PEO1*) gene mutations in Mendelian progressive external ophthalmoplegia. *J. Neurol.* **2008**, *255*, 1384–1391.
59. Fratter, C.; Gorman, G.S.; Stewart, J.D.; Buddles, M.; Smith, C.; Evans, J.; Seller, A.; Poulton, J.; Roberts, M.; Hanna, M.G.; *et al.* The clinical, histochemical, and molecular spectrum of *PEO1* (Twinkle)-linked adPEO. *Neurology* **2010**, *74*, 1619–1626.
60. Hanisch, F.; Kornhuber, M.; Alston, C.L.; Taylor, R.W.; Deschauer, M.; Zierz, S. SANDO syndrome in a cohort of 107 patients with CPEO and mitochondrial DNA deletions. *J. Neurol. Neurosurg. Psychiatry* **2014**, *86*, 630–634.
61. Baloh, R.H.; Salavaggione, E.; Milbrandt, J.; Pestronk, A. Familial parkinsonism and ophthalmoplegia from a mutation in the mitochondrial DNA helicase twinkle. *Arch. Neurol.* **2007**, *64*, 998–1000.
62. Kiferle, L.; Orsucci, D.; Mancuso, M.; Lo Gerfo, A.; Petrozzi, L.; Siciliano, G.; Ceravolo, R.; Bonuccelli, U. Twinkle mutation in an Italian family with external progressive ophthalmoplegia and parkinsonism: A case report and an update on the state of art. *Neurosci. Lett.* **2013**, *27*, 1–4.
63. Nikali, K.; Suomalainen, A.; Saharinen, J.; Kuokkanen, M.; Spelbrink, J.N.; Lönnqvist, T.; Peltonen, L. Infantile onset spinocerebellar ataxia is caused by recessive mutations in mitochondrial proteins Twinkle and Twinky. *Hum. Mol. Genet.* **2005**, *14*, 2981–2990.
64. Sarzi, E.; Goffart, S.; Serre, V.; Chrétien, D.; Slama, A.; Munnich, A.; Spelbrink, J.N.; Rötig, A. Twinkle helicase (*PEO1*) gene mutation causes mitochondrial DNA depletion. *Ann. Neurol.* **2007**, *62*, 579–587.
65. Kazachkova, N.; Ramos, A.; Santos, C.; Lima, M. Mitochondrial DNA damage patterns and aging: Revising the evidences for humans and mice. *Aging Dis.* **2013**, *4*, 337–350.
66. Alexeyev, M.; Shokolenko, I.; Wilson, G.; LeDoux, S. The maintenance of mitochondrial DNA integrity—Critical analysis and update. *Cold Spring Harb. Perspect. Biol.* **2013**, *5*, doi:10.1101/cshperspect.a012641.

67. Akbari, M.; Morevati, M.; Croteau, D.; Bohr, V.A. The role of DNA base excision repair in brain homeostasis and diseases. *DNA Repair* **2015**, doi:10.1016/j.dnarep.2015.04.029.

68. Prakash, A.; Doublié, S. Base Excision Repair in mitochondria. *J. Cell. Biochem.* **2015**, *116*, 1490–1499.

69. Zheng, L.; Zhou, M.; Guo, Z.; Lu, H.; Qian, L.; Dai, H.; Qiu, J.; Yakubovskaya, E.; Bogenhagen, D.F.; Demple, B.; *et al.* Human *DNA2* is a mitochondrial nuclease/helicase for efficient processing of DNA replication and repair intermediates. *Mol. Cell* **2008**, *32*, 325–336.

70. Duxin, J.P.; Dao, B.; Martinsson, P.; Rajala, N.; Guittat, L.; Campbell, J.L.; Spelbrink, J.N.; Stewart, S.A. Human *DNA2* is a nuclear and mitochondrial DNA maintenance protein. *Mol. Cell Biol.* **2009**, *29*, 4274–4282.

71. Gloor, J.W.; Balakrishnan, L.; Campbell, J.L.; Bambara, R.A. Biochemical analyses indicate that binding and cleavage specificities define the ordered processing of human Okazaki fragments by *DNA2* and FEN1. *Nucleic Acids Res.* **2012**, *40*, 6774–6786.

72. Shim, E.Y.; Chung, W.H.; Nicolette, M.L.; Zhang, Y.; Davis, M.; Zhu, Z.; Paull, T.T.; Ira, G.; Lee, S.E. Saccharomyces cerevisiae Mre11/Rad50/Xrs2 and Ku proteins regulate association of Exo1 and *DNA2* with DNA breaks. *EMBO J.* **2010**, *29*, 3370–3380.

73. Budd, M.E.; Campbell, J.L. *DNA2* is involved in CA strand resection and nascent lagging strand completion at native yeast telomeres. *J. Biol. Chem.* **2013**, *288*, 29414–29429.

74. Ding, L.; Liu, Y. Borrowing nuclear DNA helicases to protect mitochondrial DNA. *Int. J. Mol. Sci.* **2015**, *16*, 10870–10887.

75. Shaheen, R.; Faqeih, E.; Ansari, S. Genomic analysis of primordial dwarfism reveals novel disease genes. *Genome Res.* **2014**, *24*, 291–299.

76. Steczkiewicz, K.; Muszewska, A.; Knizewski, L.; Rychlewski, L.; Ginalski, K. Sequence, structure and functional diversity of PD-(D/E)XK phosphodiesterase superfamily. *Nucleic Acids Res.* **2012**, *40*, 7016–7045.

77. Nicholls, T.J.; Zsurka, G.; Reeva, V.; Schoeler, S.; Szczesny, R.J.; Cysewski, D.; Reyes, A.; Kornblum, C.; Sciacco, M.; Moggio, M.; *et al.* Linear mtDNA fragments and unusual mtDNA rearrangements associate with pathological deficiency of *MGME1* exonuclease. *Hum. Mol. Genet.* **2014**, *23*, 6147–6162.

78. Rampazzo, C.; Ferraro, P.; Pontarin, G.; Fabris, S.; Reichard, P.; Bianchi, V. Mitochondrial deoxyribonucleotides, pool sizes, synthesis, and regulation. *J. Biol. Chem.* **2004**, *279*, 17019–17026.

79. Pontarin, G.; Fijolek, A.; Pizzo, P.; Ferraro, P.; Rampazzo, C.; Pozzan, T.; Thelander, L.; Reichard, P.A.; Bianchi, V. Ribonucleotide reduction is a cytosolic process in mammalian cells independently of DNA damage. *Proc. Natl. Acad. Sci. USA* **2008**, *105*, 17801–17806.

80. Jenuth, J.P.; Peterson, A.C.; Shoubridge, E.A. Tissue-specific selection for different mtDNA genotypes in heteroplasmic mice. *Nat. Genet.* **1997**, *16*, 93–95.

81. Boschetti, E.; D'Alessandro, R.; Bianco, F.; Carelli, V.; Cenacchi, G.; Pinna, A.D.; del Gaudio, M.; Rinaldi, R.; Stanghellini, V.; Pironi, L.; *et al.* Liver as a source of thimidine phosphorylase replacement in mitochondrial neurogastrointestinal encephalomyopathy. *PLoS ONE* **2014**, *9*, e96692, doi:10.1371/journal.pone.0096692.

82. Scarpelli, M.; Ricciardi, G.K.; Beltramello, A.; Zocca, I.; Calabria, F.; Russignan, A.; Zappini, F.; Cotelli, M.S.; Padovani, A.; Tomelleri, G.; *et al.* The role of brain MRI in mitochondrial neurogastrointestinal encephalomyopathy. *Neuroradiol. J.* **2013**, *26*, 520–530.

83. Mancuso, M.; Salviati, L.; Sacconi, S.; Otaegui, D.; Camaño, P.; Marina, A.; Bacman, S.; Moraes, C.T.; Carlo, J.R.; Garcia, M.; *et al.* Mitochondrial DNA depletion: Mutations in thymidine kinase gene with myopathy and SMA. *Neurology* **2002**, *59*, 1197–1202.

84. Tyynismaa, H.; Sun, R.; Ahola-Erkkilä, S.; Almusa, H.; Pöyhönen, R.; Korpela, M.; Honkaniemi, J.; Isohanni, P.; Paetau, A.; Wang, L.; *et al.* Thymidine kinase 2 mutations in autosomal recessive progressive external ophtalmoplegia with multiple mitochondrial DNA deletions. *Hum. Mol. Genet.* **2012**, *21*, 66–75.

85. Alston, C.L.; Schaefer, A.M.; Raman, P.; Solaroli, N.; Krishnan, K.J.; Blakely, E.L.; He, L.; Craig, K.; Roberts, M.; Vyas, A.; *et al.* Late-onset respiratory failure due to *TK2* mutations causing multiple mtDNA deletions. *Neurology* **2013**, *81*, 2051–2053.

86. Liu, Y.; Chen, X.J. Adenine nucleotide translocase, mitochondrial stress, and degenerative cell death. *Oxid. Med. Cell. Longev.* **2013**, *2013*, 146860, doi:10.1155/2013/146860.

87. Baines, C.P.; Song, C.X.; Zheng, Y.T.; Wang, G.W.; Zhang, J.; Wang, O.L.; Guo, Y.; Bolli, R.; Cardwell, E.M.; Ping, P. Protein kinase Cepsilon interacts with and inhibits thepermeability transition pore in cardiac mitochondria. *Circ. Res.* **2003**, *92*, 873–880.

88. Dörner, A.; Schultheiss, H.P. Adenine nucleotide translocase in the focus of cardiovascular diseases. *Trends Cardiovasc. Med.* **2007**, *17*, 284–290.

89. Jordens, E.Z.; Palmieri, L.; Huizing, M.; van den Heuvel, L.P.; Sengers, R.C.; Dörner, A.; Ruitenbeek, W.; Trijbels, F.J.; Valsson, J.; Sigfusson, G.; *et al.* Adenine nucleotide translocator 1 deficiency associated with Sengers syndrome. *Ann. Neurol.* **2002**, *52*, 95–99.

90. Tanaka, H.; Arakawa, H.; Yamaguchi, T.; Shiraishi, K.; Fukuda, S.; Matsui, K.; Takei, Y.; Nakamura, Y. A ribonucleotide reductase gene involved in a p53-dependent cell-cycle checkpoint for DNA damage. *Nature* **2000**, *404*, 42–49.

91. Tyynismaa, H.; Ylikallio, E.; Patel, M.; Molnar, M.J.; Haller, R.G.; Suomalainen, A. A heterozygous truncating mutation in *RRM2B* causes autosomal-dominant progressive external ophtalmoplegia with multiple mtDNA deletions. *Am. J. Hum. Genet.* **2009**, *85*, 290–295.

92. Pitceathly, R.D.S.; Smith, C.; Fratter, C.; Alston, C.L.; He, L.; Craig, K.; Blakely, E.L.; Evans, J.C.; Taylor, J.; Shabbir, Z.; *et al.* Adults with *RRM2B*-related mitochondrial disease have distinct clinical and molecular characteristics. *Brain* **2012**, *135*, 3392–3403.

93. Ranieri, M.; Brajkovic, S.; Riboldi, G.; Ronchi, D.; Rizzo, F.; Bresolin, N.; Corti, S.; Comi, G.P. Mitochondrial fusion proteins and human diseases. *Neurol. Res. Int.* **2013**, doi:10.1155/2013/293893.

94. Mishra, P.; Chan, D.C. Mitochondrial dynamics and inheritance during cell division, development and disease. *Nat. Rev. Mol. Cell Biol.* **2014**, *15*, 634–646.

95. Itoh, K.; Nakamura, K.; Iijima, M.; Sesaki, H. Mitochondrial dynamics in neurodegeneration. *Trends Cell Biol.* **2013**, *23*, 64–71.

96. Belenguer, P.; Pellegrini, L. The dynamin GTPase *OPA1*, more than mitochondria? *Biochim. Biophys. Acta* **2013**, *1833*, 176–183.

97. Jahani-Asl, A.; Pilon-Larose, K.; Xu, W.; MacLaurin, J.G.; Park, D.S.; McBride, H.M.; Slack, R.S. The mitochondrial inner membrane GTPase, optic atrophy 1 (*OPA1*), restores mitochondrial morphology and promotes neuronal survival following excitotoxicity. *J. Biol. Chem.* **2011**, *286*, 4772–4782.

98. Nguyen, D.; Alavi, M.V.; Kim, K.Y.; Kang, T.; Scott, R.T.; Noh, Y.H.; Lindsey, J.D.; Wissinger, B.; Ellisman, M.H.; Weinreb, R.N.; *et al.* A new vicious cycle involvingglutamate excitotoxicity, oxidative stress and mitochondrial dynamics. *Cell Death Dis.* **2011**, *2*, doi:10.1038/cddis.2011.117.

99. Alexander, C.; Votruba, M.; Pesch, U.E.; Thiselton, D.L.; Mayer, S.; Moore, A.; Rodriguez, M.; Kellner, U.; Leo-Kottler, B.; Auburger, G.; *et al. OPA1*, encoding a dynamin-related GTPase, is mutated in autosomal dominant optic atrophy linked to chromosome 3q28. *Nat. Genet.* **2000**, *26*, 211–215.

100. Delettre, C.; Lenaers, G.; Griffoin, J.M.; Gigarel, N.; Lorenzo, C.; Belenguer, P.; Pelloquin, L.; Grosgeorge, J.; Turc-Carel, C.; Perret, E.; *et al.* Nuclear gene *OPA1*, encoding a mitochondrial dynamin-related protein, is mutated in dominant optic atrophy. *Nat. Genet.* **2000**, *26*, 207–210.

101. Hudson, G.; Amati-Bonneau, P.; Blakely, E.L.; Stewart, J.D.; He, L.; Schaefer, A.M.; Griffiths, P.G.; Ahlqvist, K.; Suomalainen, A.; Reynier, P.; *et al.* Mutation of *OPA1* causes dominant optic atrophy with external ophthalmoplegia, ataxia, deafness and multiple mitochondrial DNA deletions: A novel disorder of mtDNA maintenance. *Brain* **2008**, *131*, 329–337.

102. Zanna, C.; Ghelli, A.; Porcelli, A.M.; Karbowski, M.; Youle, R.J.; Schimpf, S.; Wissinger, B.; Pinti, M.; Cossarizza, A.; Vidoni, S.; *et al. OPA1* mutations associated with dominant optic atrophy impair oxidative phosphorylation and mitochondrial fusion. *Brain* **2008**, *131*, 352–367.

103. Carelli, V.; Musumeci, O.; Caporali, L.; Zanna, C.; la Morgia, C.; del Dotto, V.; Porcelli, A.M.; Rugolo, M.; Valentino, M.L.; Iommarini, L.; *et al.* Syndromic parkinsonism and dementia associated with *OPA1* missense mutations. *Ann. Neurol.* **2015**, doi:10.1002/ana.24410.

104. Evgrafov, O.V.; Mersiyanova, I.; Irobi.; J.; van den Bosch, L.; Dierick, I.; Leung, C.L.; Schagina, O.; Verpoorten, N.; van Impe, K.; Fedotov, V.; *et al.* Mutant small heat-shock protein 27 causes axonal Charcot-Marie-Tooth disease and distal hereditary motor neuropathy. *Nat. Genet.* **2004**, *36*, 602–606.

105. Vielhaber, S.; Debska-Vielhaber, G.; Peeva, V.; Schoeler, S.; Kudin, A.P.; Minin, I.; Schreiber, S.; Dengler, R.; Kollewe, K.; Zuschratter, W.; *et al.* Mitofusin 2 mutations affect mitochondrial function by mitochondrial DNA depletion. *Acta Neuropathol.* **2013**, *125*, 245–256.

106. Löllgen, S.; Weiher, H. The role of Mpv17 protein mutations of which cause mitochondrial DNA depletion syndromes (MDDS): Lessons from homologs in different species. *Biol. Chem.* **2015**, *396*, 13–25.

107. Antonenkov, V.D.; Isomursu, A.; Mennerich, D.; Vapola, M.H.; Weiher, H.; Kietzmann, T.; Hiltunen, J.K. The human mtDNA depletion syndrome gene MPV17 encodes a non-selective channel that modulates membrane potential. *J. Biol. Chem.* **2015**, *290*, 13840–13861.

108. Spinazzola, A.; Viscomi, C.; Fernandez-Vizarra, E.; Carrara, F.; D'Adamo, P.; Calvo, S.; Marsano, R.M.; Donnini, C.; Weiher, H.; Strisciuglio, P.; *et al.* MPV17 encodes an inner mitochondrial membrane protein and is mutated in infantile hepatic mitochondrial DNA depletion. *Nat. Genet.* **2006**, *38*, 570–576.

109. Garone, C.; Rubio, J.C.; Calvo, S.E.; Naini, A.; Tanji, K.; di Mauro, S.; Mootha, V.K.; Hirano, M. MPV17 mutations causing adult-onset multisystemic disorder with multiple mitochondrial DNA deletions. *Arch. Neurol.* **2012**, *69*, 1649–1651.
110. Blakely, E.L.; Butterworth, A.; Hadden, R.D.M.; Bodi, I.; Landping, H.; McFarland, R.; Taylor, R.W. MPV17 mutation causes neuropathy and leukoencephalopathy with multiple mtDNA deletions in muscle. *Neuromuscol. Disord.* **2012**, *22*, 587–591.
111. Mendelsohn, B.A.; Mehta, N.; Hameed, B.; Pekmezci, M.; Packman, S.; Ralph, J. Adul Onset Fatal Neurohepatopathy in a Woman Caused by MPV17 Mutation. *JIMD Rep.* **2014**, *13*, 37–41.
112. Wortmann, S.B.; Koolen, D.A.; Smeitink, J.A.; van den Heuvel, L.; Rodenburg, R.J. Whole exome sequencing of suspected mitochondrial patients in clinical practice. *J. Inherit. Metab. Dis.* **2015**, *38*, 437–443.
113. Garone, C.; Tadesse, S.; Hirano, M. Clinical and genetic spectrum of mitochondrial neurogastrointestinal encephalomyopathy. *Brain* **2011**, *134*, 3326–3332.
114. Tchikviladzé, M.; Gilleron, M.; Maisonobe, T.; Galanaud, D.; Laforêt, P.; Durr, A.; Eymard, B.; Mochel, F.; Ogier, H.; Béhin, A.; *et al.* A diagnostic flow chart for *POLG*-related diseases based on signs sensitivity and specificity. *J. Neurol. Neurosurg. Psychiatry* **2015**, *86*, 646–654.
115. Pareyson, D.; Saveri, P.; Sagnelli, A.; Piscosquito, G. Mitochondrial dynamics and inherited peripheral nerve diseases. *Neurosci. Lett.* **2015**, *596*, 66–77.
116. Lieber, D.S.; Calvo, S.E.; Shanahan, K.; Slate, N.G.; Liu, S.; Hershman, S.G.; Gold, N.B.; Chapman, B.A.; Thorburn, D.R.; Berry, G.T.; *et al.* Targeted exome sequencing of suspected mitochondrial disorders. *Neurology* **2013**, *80*, 1762–1770.
117. Calvo, S.E.; Compton, A.G.; Hershman, S.G.; Lim, S.C.; Lieber, D.S.; Tucker, E.J.; Laskowski, A.; Garone, C.; Liu, S.; Jaffe, D.B.; *et al.* Molecular diagnosis of infantile mitochondrial disease with targeted next-generation sequencing. *Sci. Transl. Med.* **2012**, *4*, doi:10.1126/scitranslmed.3003310.

6

Abnormal Mitochondrial Function and Impaired Granulosa Cell Differentiation in Androgen Receptor Knockout Mice

Ruey-Sheng Wang [1], Heng-Yu Chang [2], Shu-Huei Kao [3], Cheng-Heng Kao [4], Yi-Chen Wu [1], Shuyuan Yeh [5], Chii-Reuy Tzeng [1,*] and Chawnshang Chang [5,*]

[1] Graduate Institute of Clinical Medicine & Department of Obstetrics and Gynecology, College of Medicine, Taipei Medical University, Taipei 110, Taiwan; E-Mails: d102094001@tmu.edu.tw (R.-S.W.); vvichern@yahoo.com.tw (Y.-C.W.)

[2] Department of Biochemistry and Molecular Cell Biology, College of Medicine, Taipei Medical University, Taipei 110, Taiwan; E-Mail: hychang@tmu.edu.tw

[3] School of Medical Laboratory Science and Biotechnology, College of Medical Science and Technology, Taipei Medical University, Taipei 110, Taiwan; E-Mail: kaosh@tmu.edu.tw

[4] Center of General Education, Chang Gung University, Taoyuan 333, Taiwan; E-Mail: kao@mail.cgu.edu.tw

[5] George H. Whipple Lab for Cancer Research, Departments of Pathology, Urology and Radiation Oncology, University of Rochester Medical Center, Rochester, NY 14642, USA; E-Mail: shuyuan_yeh@urmc.rochester.edu

* Authors to whom correspondence should be addressed; E-Mails: chang@urmc.rochester.edu (C.C.); tzengcr@tmu.edu.tw (C.-R.T.)

Academic Editors: Jaime M. Ross and Giuseppe Coppotelli

Abstract: In the ovary, the paracrine interactions between the oocyte and surrounded granulosa cells are critical for optimal oocyte quality and embryonic development. Mice lacking the androgen receptor ($AR^{-/-}$) were noted to have reduced fertility with abnormal ovarian function that might involve the promotion of preantral follicle growth and prevention of follicular atresia. However, the detailed mechanism of how AR in granulosa cells exerts its effects on oocyte quality is poorly understood. Comparing *in vitro* maturation rate of oocytes, we found oocytes collected from $AR^{-/-}$ mice have a significantly poor maturating rate with 60% reached metaphase II and 30% remained in germinal vesicle breakdown stage, whereas 95% of wild-type *AR* ($AR^{+/+}$) oocytes had reached metaphase II. Interestingly, we found these $AR^{-/-}$ female mice also had an increased frequency of morphological alterations in the mitochondria of granulosa cells with reduced ATP generation (0.18 ± 0.02 *vs.* 0.29 ± 0.02 μM/mg protein; $p < 0.05$) and aberrant mitochondrial biogenesis. Mechanism dissection found loss of AR led to a significant decrease in the expression of peroxisome proliferator-activated receptor γ (PPARγ) co-activator 1-β (PGC1-β) and its sequential downstream genes, nuclear respiratory factor 1 (NRF1) and mitochondrial transcription factor A (TFAM), in controlling mitochondrial biogenesis. These results indicate that AR may contribute to maintain oocyte quality and fertility via controlling the signals of PGC1-β-mediated mitochondrial biogenesis in granulosa cells.

Keywords: androgen receptor; granulosa cell; PGC1-β; mitochondria

---

1. Introduction

Androgens classically mediate their genomic effects via the androgen receptor (AR), a protein encoded by an X chromosome gene, which exerts its biological function through activation of target gene expression via a sequence of processes [1,2]. The earlier studies indicated that androgens act directly on the development of ovarian follicles via AR, apart from serving as a substrate of aromatase (Cyp19a1) for estrogen synthesis. Androgens have been implicated to have a role in promoting follicular development [3,4], by up-regulating follicle-stimulating hormone (FSH) receptor (FSHR) expression and augmenting FSH-stimulated follicular differentiation [5–8]. In addition, AR is expressed in all cell types of the ovarian follicle, including granulosa cells, theca cells and the oocytes [9–11].

By using Cre/LoxP system, we and others generated the global $AR^{-/-}$ [12–14] and granulosa cell-specific $AR^{-/-}$ female mice [15,16]. Those data showed that female mice lacking AR have a reduced fertility, fewer oocytes were recovered after superovulation with gonadotropins, and fewer corpora lutea were observed. However, ablation of AR in oocytes had no effect on female fertility [15]. All together, these animal examples suggested that androgen, functions through the AR in granulosa cells, plays a primary and essential role for normal follicle development and optimal fertility. Earlier studies have shown that AR antagonists slow down mouse follicle growth [3,17] and prevent primary to secondary follicle transition in bovine [18]. Moreover, granulosa cell-specific $AR^{-/-}$ female mice [15,16] have been found to contain more preantral follicles, more atretic follicles, with fewer antral follicles and corpora lutea in their ovaries. These animal examples suggest that androgen/AR signaling in granulosa cells regulates normal follicular growth, mainly by controlling preantral follicle growth and development to antral follicles, meanwhile preventing follicular atresia by inhibition of granulosa cell apoptosis. However, the definitive mechanisms of the subfertility and the potential pathways by which granulosa cell AR exerts its effects on follicle development and follicle atresia are still ill-defined.

Earlier studies suggested that mitochondria dysfunction in granulosa cells leads to poor oocyte quality and contribute to female subfertility [19–21]. In the current study, the prophase I arrested oocyte was collected from the ovary and the *in vitro* maturation rate of oocytes was evaluated as the preliminary indication for oocyte quality. In addition, oocyte maturation often required the energy sources from the surrounding nursing cells, such as granulose cells, as we reported here. Thus, the objective of this study was to determine whether lack of AR in granulosa cells has impacts on the oocyte quality by changing mitochondrial status and differentiation status in granulosa cells. We found alterations in mitochondrial morphology, biogenesis and metabolism in the granulosa cells of $AR^{-/-}$ mice, suggesting that ablation of AR leads to mitochondrial dysfunction. In addition, the results in this study demonstrated that pregnant mare's serum gonadotropin (PMSG)-induced granulosa cell differentiation was impaired in $AR^{-/-}$ ovaries as indicated by decreasing the expression of genes involved in the granulosa cell differentiation. These consequences may be interrelated to significantly reduce *in vitro* maturation rate of $AR^{-/-}$ oocytes.

2. Results

2.1. Confirmation of Knockout Androgen Receptor (AR) in the Ovaries in the $AR^{-/-}$ Mice

Three primers ("select", "2–3" and "2–9", for the relative position of each primer in the *AR* gene see Figure S1) were synthesized to amplify mouse genomic DNA to distinguish the floxed *AR*, $AR^{-/-}$ and $AR^{+/+}$ mice. We were able to identify $AR^{-/-}$ mice by using "select" and "2–9" primers to PCR amplify the 238-bp DNA (Figure 1, upper panel). In contrast, $AR^{+/+}$ mice can produce 580-bp DNA fragments by using "select" and "2–9" primers (Figure 1, upper panel). In this study, we were using mice with a ubiquitous deletion of the *AR*, so we further confirmed that *AR* was knocked out in ovaries by Western blot (Figure 1, lower panel).

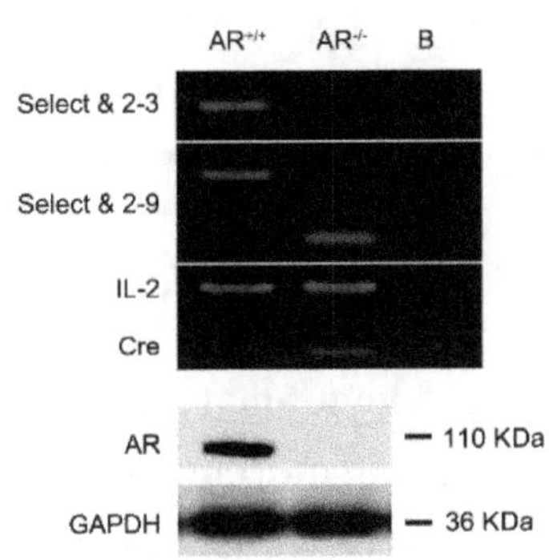

Figure 1. Genotyping of androgen receptor knockout ($AR^{-/-}$) female mice. We used two pairs of primers (i) "select" and "2–3"; and (ii) "select" and "2–9" to identify *AR* wild type ($AR^{+/+}$) and *AR* knockout ($AR^{-/-}$) female mice in our study. Genotyping was performed by using PCR on the genomic DNA isolated from the tails of 3-week-old mice. "Select" is a forward primer which is located in the intron 1 of *AR* gene with sequence 5'-GTTGAT ACCTTAACCTCTGC-3'. "2–3" is a reverse primer which is located at the 3' end of the exon 2 with the sequence 5'-CTTCAGCGGCTCTTTTGAAG-3'. "2–9" is a reverse primer which is located in intron 2 with the sequence 5'-CTTACATGTACTGTGAGAGG-3'. Using the "select" and "2–3" primers, we amplified a product with ~460 bp for $AR^{+/+}$ allele, and with no product for $AR^{-/-}$ allele. Using "select" and "2–9", we amplified a DNA fragment with 580-bp, which represents $AR^{+/+}$ allele and ~270 bp, which represents $AR^{-/-}$ allele (**upper panel**); The expression of Cre and internal control interleukin 2 (IL-2) were confirmed by PCR (bottom of upper panel). The lane marked B indicates blank control. Western blot shown that $AR^{-/-}$ ovaries do not express AR protein (**lower panel**).

2.2. Reduced Oocyte in Vitro Maturation Rate in $AR^{-/-}$ Mice

The *in vitro* oocyte maturation rate was examined to determine the potential AR roles in the oocyte maturation. The oocytes were collected from 4.5 weeks old $AR^{-/-}$ and $AR^{+/+}$ female mice that were previously treated with PMSG for 48 h. The result revealed that 95% of $AR^{+/+}$ oocytes had reached to metaphase II, whereas a significantly lower maturation rate (60%) was observed in the $AR^{-/-}$ oocytes (Figure 2, right panel). In the meantime, about 30% of $AR^{-/-}$ oocytes were arrested at prophase I and remained at germinal vesicle breakdown (GVBD) stage (Figure 2, left panel).

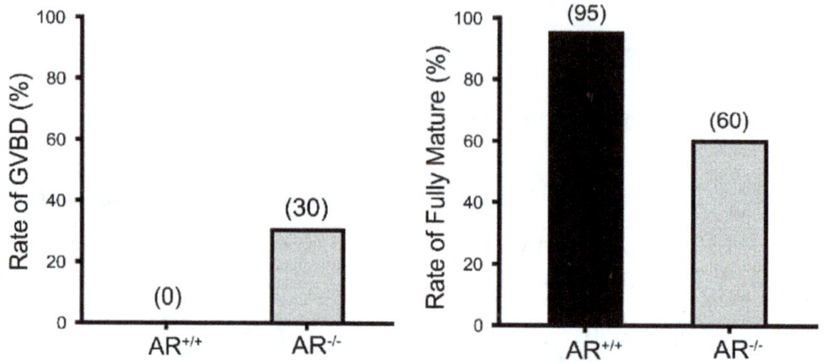

Figure 2. Reduced *in vitro* maturation rate of $AR^{-/-}$ oocytes. Histogram shows the *in vitro* maturation rate of $AR^{+/+}$ and $AR^{-/-}$ oocytes. After 18 h of *in vitro* culture, fewer $AR^{-/-}$ oocytes could reach to metaphase II (60%) compared with $AR^{+/+}$ oocytes (95%) (**right panel**); In the meantime, about 30% of $AR^{-/-}$ oocytes were remained in germinal vesicle breakdown (GVBD) stage (**left panel**), $n = 100$ follicles per genotype.

2.3. Change of Ovarian Follicle Morphology in $AR^{-/-}$ Mice

In order to elucidate whether poorer oocyte maturation rate in $AR^{-/-}$ mice was linked with morphological change of ovarian follicles, the morphology of ovarian follicles were checked with HE staining. Figure 3A shows follicle morphology after superovulation treatment. The largest follicles had reached to Pedersen class 6 (incipient antral) or Pedersen class 7 (early antral) stages in $AR^{+/+}$ ovaries, whereas in $AR^{-/-}$ ovaries, the majority of follicles did not show antrum formation in this time period. In agreement with our previous data [12], $AR^{-/-}$ ovaries contained considerably few antral follicles as compared to $AR^{+/+}$ littermates (Figure 3B).

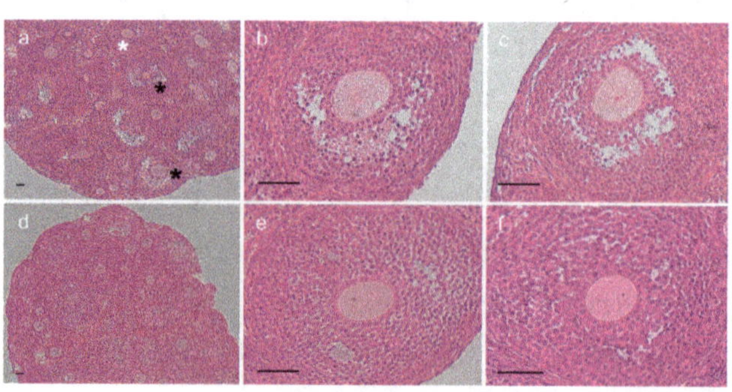

(A)

Figure 3. *Cont.*

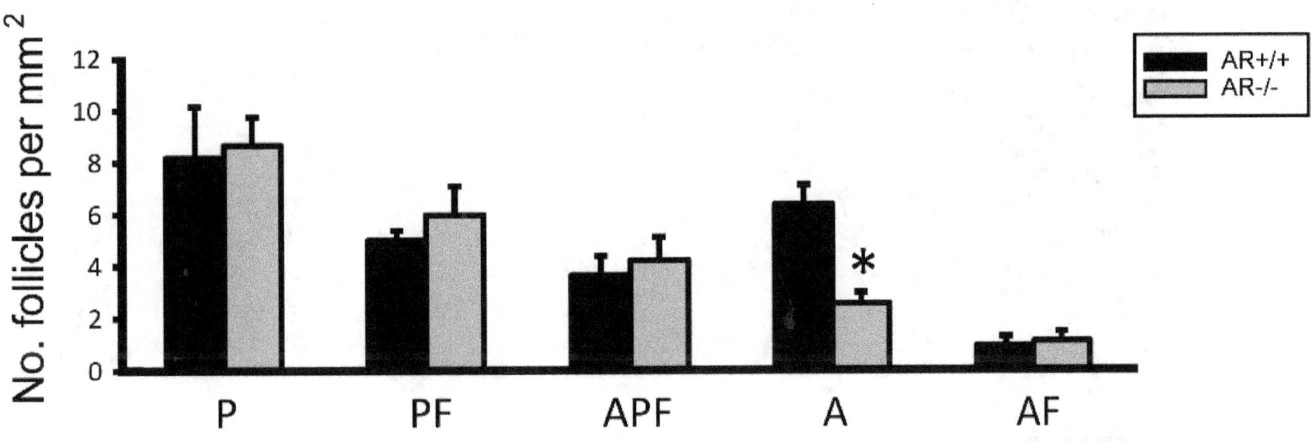

(B)

Figure 3. Morphological changes of ovarian follicle in $AR^{-/-}$ mice. (**A**) Comparison of morphology in ovaries of 4.5-week-old $AR^{+/+}$ and $AR^{-/-}$ female mice after PMSG-priming for 48 h ($n = 4$ mice per genotype). Representative hematoxylin and eosin-stained ovarian sections (**a**: $AR^{+/+}$, **b**,**c**: $AR^{+/+}$, **d**: $AR^{-/-}$ at 100× magnification; **e**,**f**: $AR^{-/-}$ at 400× magnification). The largest follicles had reached to Pedersen class 6 (incipient antral, white asterisk) or Pedersen class 7 (early antral, black asterisk) stages in $AR^{+/+}$ ovaries (**a**–**c**), whereas in $AR^{-/-}$ ovaries, the majority of follicles did not show antrum formation in this time period (**d**–**f**). Scale bar, 50 μm; (**B**) Statistical analysis of the number of the follicular compartments in $AR^{+/+}$ and $AR^{-/-}$ ovaries. The $AR^{-/-}$ female mice have few antral follicles relative to $AR^{+/+}$ mice. P, Primordial and primary follicle; PF, Preantral follicle; APF, Atretic primordial, primary, and preantral follicle; A, Antral follicle; AF, Atretic antral follicle. * $p < 0.05$, by Student's *t* test.

2.4. Alteration of Mitochondrial Morphology and Ultrastructure in Granulosa Cells from $AR^{-/-}$ Mice

To reveal the molecular mechanisms of ovarian follicles morphology changes, and impact on oocyte quality, we assayed the damage on the mitochondria of granulosa cells since the mitochondria are important for steroidogenisis and energy production [22,23]. We first examined the mitochondrial ultrastructure in granulosa cells using TEM on the ovaries of both $AR^{+/+}$ and $AR^{-/-}$ mice and found that most mitochondria of $AR^{+/+}$ granulosa cells contained highly folded inner membrane forming mitochondrial cristae, which were enveloped by an intact outer membrane. In contrast, the mitochondria of $AR^{-/-}$ granulosa cells showed increase electron density of the matrix, generally more round in appearance with fewer, disarrayed cristae, and contained vacuoles (white arrowhead in Figure 4A). Immunofluorescent staining of mitochondria exhibited a normal perinuclear distribution in $AR^{+/+}$ granulosa cells. In contrast, the mitochondria in $AR^{-/-}$ granulosa cells displayed aggregating distribution pattern and small "donut-like" vesicles appearance (Figure 4B).

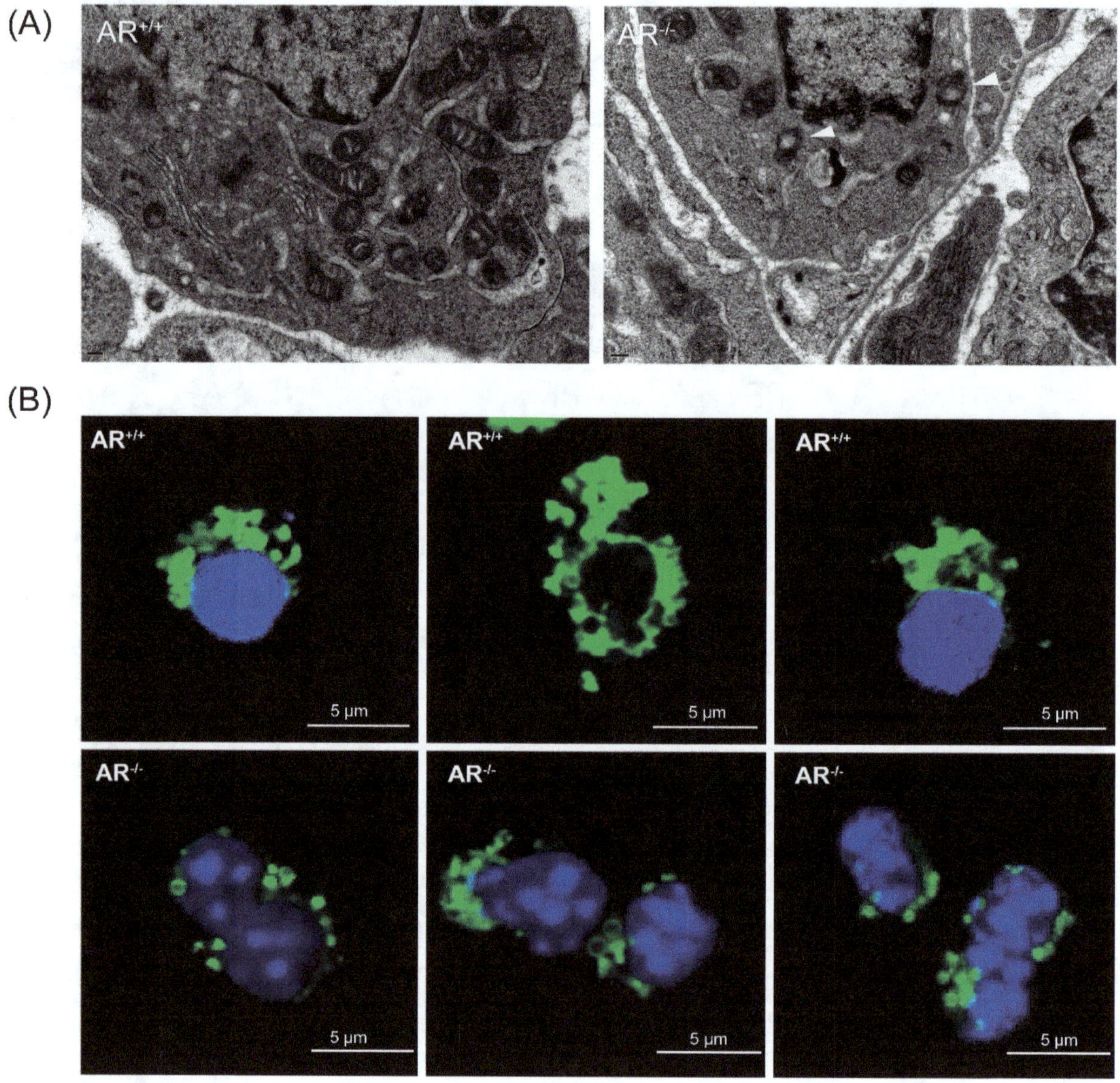

Figure 4. Alteration of mitochondrial morphology in granulosa cells of $AR^{-/-}$ mice. (**A**) Mitochondrial morphology in granulosa cells of $AR^{-/-}$ mice was visualized by transmission electron microscopy (TEM). Representative electron micrograph of mitochondria from control ($AR^{+/+}$) granulosa cells, showing bean-shaped structures with numerous transversely orientated cristae enveloped by an intact outer membrane; $AR^{-/-}$ granulosa cells displayed small spherical structures with fewer and disarrayed cristae, and the presence of large vacuoles in mitochondria (white arrowhead); (**B**) Mitochondrial morphology in granulosa cells of $AR^{-/-}$ mice was visualized by immunofluorescence. Granulosa cells collected from $AR^{+/+}$ and $AR^{-/-}$ mice were labeled with MitoTracker Green to visualize mitochondrial localization and co-stained with 4',6'-diamidino-2-phenylindole (DAPI) to visualize nuclei. Representative confocal sections of granulosa cells were shown. Mitochondria in granulosa cells from $AR^{-/-}$ mice displayed aggregating distribution pattern and small spherical structures.

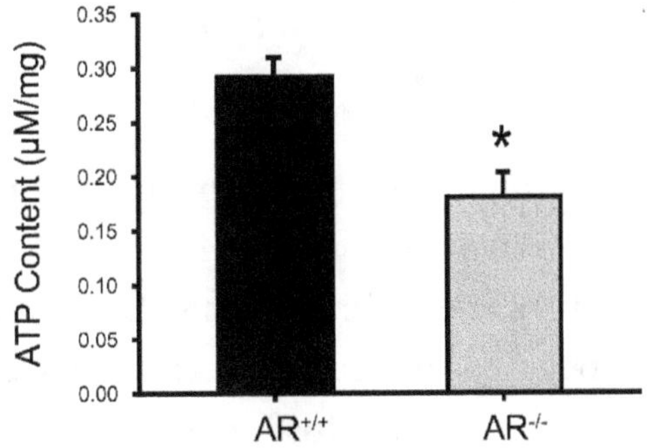

(A)

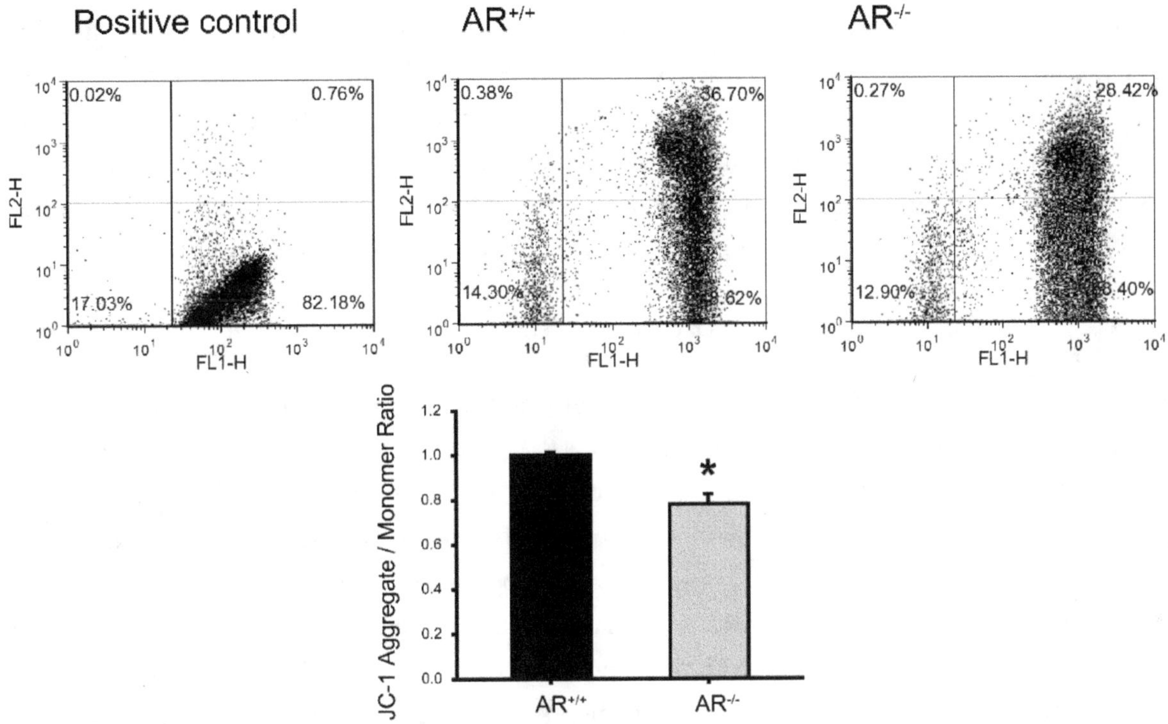

(B)

Figure 5. Metabolic dysfunction of mitochondria and reduced mitochondrial membrane potential in granulosa cells of $AR^{-/-}$ mice. (**A**) Reduced ATP content in granulosa cells of $AR^{-/-}$ mice. Granulosa cells removed from $AR^{+/+}$ and $AR^{-/-}$ mice were collected to determine the levels of ATP. Values are expressed as μM per mg protein. Histogram shows the average ATP content in granulosa cells from $AR^{+/+}$ and $AR^{-/-}$ mice. * $p < 0.05$, $n = 3$ mice per genotype; (**B**) Detection of mitochondrial membrane potential by flow cytometry analysis of JC-1 staining. Granulosa cells were collected from 4-week-old female mice ($AR^{+/+}$ and $AR^{-/-}$) injected with 48 h of 7.5 IU pregnant mare's serum gonadotropin (PMSG) treatment, and then stained with 5,5',6,6'-tetrachloro-1,1',3,3'-tetraethylbenzimidazolylcarbocyanine iodide (JC-1). JC-1 exists as a monomer in the cytosol (green fluorescent) and also accumulates as aggregates in the mitochondria (red fluorescent). In apoptotic and necrotic cells, JC-1 exists in monomeric form and stains the cytosol green. Results are expressed as the ratio of the aggregate to monomeric form of JC-1. The percentage is expressed as the mean ± SD ($n = 3$ mice per genotype). * $p < 0.05$.

2.5. Metabolic Dysfunction of Mitochondria and Reduced Mitochondrial Membrane Potential in Granulosa Cells of $AR^{-/-}$ Mice

Based on the above findings, we determine whether mitochondrial metabolism and mitochondria membrane potential were disturbed in granulosa cells of $AR^{-/-}$ mice. ATP synthesis by oxidative phosphorylation is the primary function associated with mitochondrial function [24]. Thus, the granulosa cells of $AR^{+/+}$ and $AR^{-/-}$ mice were processed to measure ATP content. As shown in Figure 5A, the ATP content were markedly decreased in the granulosa cells of $AR^{-/-}$ mice as compared to $AR^{+/+}$ mice (0.18 ± 0.02 μM/mg protein *vs.* 0.29 ± 0.02 μM/mg protein; $p < 0.05$). To detect the mitochondria membrane potential, we used flow cytometry analysis of JC-1 staining. JC-1 exists as a monomer in the cytosol (green fluorescent) and also accumulates as aggregates in the mitochondria (red fluorescent). In apoptotic and necrotic cells, JC-1 exists in monomeric form and stains the cytosol green. Results are expressed as the ratio of the aggregate to monomeric form of JC-1. As shown in Figure 5B, the mitochondria membrane potential were markedly reduced in the granulosa cells of $AR^{-/-}$ mice as compared to $AR^{+/+}$ mice, which suggesting a decline of mitochondrial function.

2.6. Decreased Mitochondrial Biogenesis in Granulosa Cells of $AR^{-/-}$ Mice

To investigate the potential effects of loss of AR in granulosa cells on mitochondrial biogenesis, we evaluated mtDNA content in granulosa cells from $AR^{+/+}$ and $AR^{-/-}$ mice by quantitative real-time PCR. Data are expressed as the ratio of mtDNA to nuclear DNA, as shown in Figure 6A (lower panel). We found that mtDNA content in granulosa cells were significantly lower in $AR^{-/-}$ mice than in $AR^{+/+}$ mice (0.2 ± 0.12 *vs.* 1.1 ± 0.2; $p < 0.05$). Figure 6A (upper panel) shows the mtDNA content by RT-PCR. Furthermore, we measured the mRNA levels of genes implicated in mitochondrial biogenesis, such as PGC-1α, PGC-1β, NRF1 and TFAM [25]. We found the granulosa cells of $AR^{-/-}$ mice exhibited decreased expression of PGC-1β, TFAM and NRF1 mRNA compared with $AR^{+/+}$ mice (Figure 6B). No significant difference was detected for PGC-1α transcripts (Figure 6B). Together these results revealed a deterioration of mitochondrial biogenic response in granulosa cells of $AR^{-/-}$ mice.

(A)

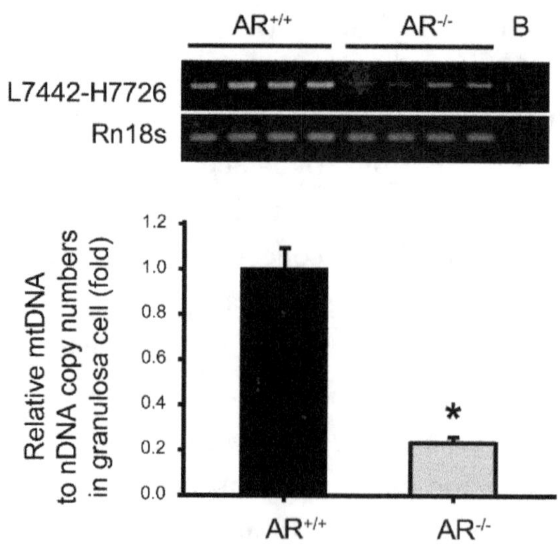

Figure 6. *Cont.*

(B)

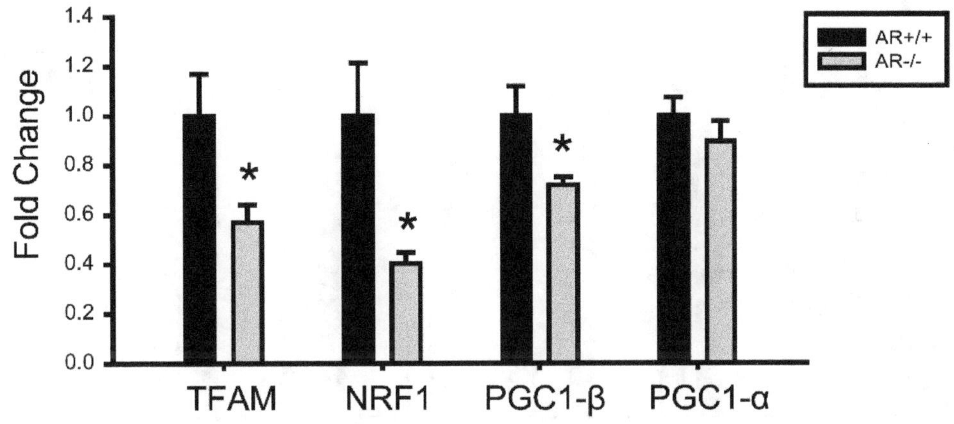

Figure 6. Decreased mitochondrial biogenesis in granulosa cells of $AR^{-/-}$ mice. (**A**) Decreased mitochondrial DNA (mtDNA) copy number in granulosa cells of $AR^{-/-}$ mice. mtDNA copy number was calculated using reverse transcription-PCR and quantitative real-time PCR by measuring the ratio of COX 2 (mtDNA) to Rn18s (nuclear DNA) DNA levels in granulosa cells of $AR^{+/+}$ and $AR^{-/-}$ mice. * $p < 0.05$, $n = 4$ mice per genotype; the lane marked B indicates blank control; (**B**) Decreased mitochondrial biogenesis in granulosa cells of $AR^{-/-}$ mice. The mRNA levels of genes implicated in mitochondrial biogenesis determined by real-time RT-PCR in granulosa cells of $AR^{+/+}$ and $AR^{-/-}$ mice. At least three experiments were performed and data are presented as the mean ± SEM. * $p < 0.05$, $n = 4$ mice per genotype. PGC-1: peroxisome proliferator-activated receptor γ coactivator-1. NRF1: nuclear respiratory factor 1. TFAM: mitochondrial transcription factor A.

2.7. Molecular Changes in Granulosa Cells of $AR^{-/-}$ Mice and the Effect of AR Deficiency on Serum E2 Levels

To understand the defects in granulosa cells during gonadotropin stimulation, we examined expression of genes that were important granulosa cell proliferation and differentiation markers during folliculogenesis. Granulosa cell differentiation is manifested during the process allowing the progression of preantral follicle to preovulatory follicle, which is dependent on sufficient FSH stimulation [26–28] and is marked by the acquisition of increased Cyp19a1 activity and luteinizing hormone (LH) receptor (LHCGR). Meanwhile, since androgen is known to augment the actions of FSH [3,4,7], we are specifically interested in evaluating the expression pattern of the genes encoding FSHR, LHCGR, and Cyp19a1 in $AR^{+/+}$ and $AR^{-/-}$ ovaries. A number of genes showed significant changes. The FSHR, LHCGR, Cyp11a1, Cyp19a1 and progesterone receptor (PR) expression were significantly decreased in granulosa cells of $AR^{-/-}$ mice (Figure 7A). These data indicate that preovulatory granulosa cells do not properly differentiate and develop sufficient aromatase activity in response to PMSG. Concomitantly, the level of serum estradiol (E2) was significantly decreased in $AR^{-/-}$ female mice as compared with $AR^{+/+}$ littermates (Figure 7B).

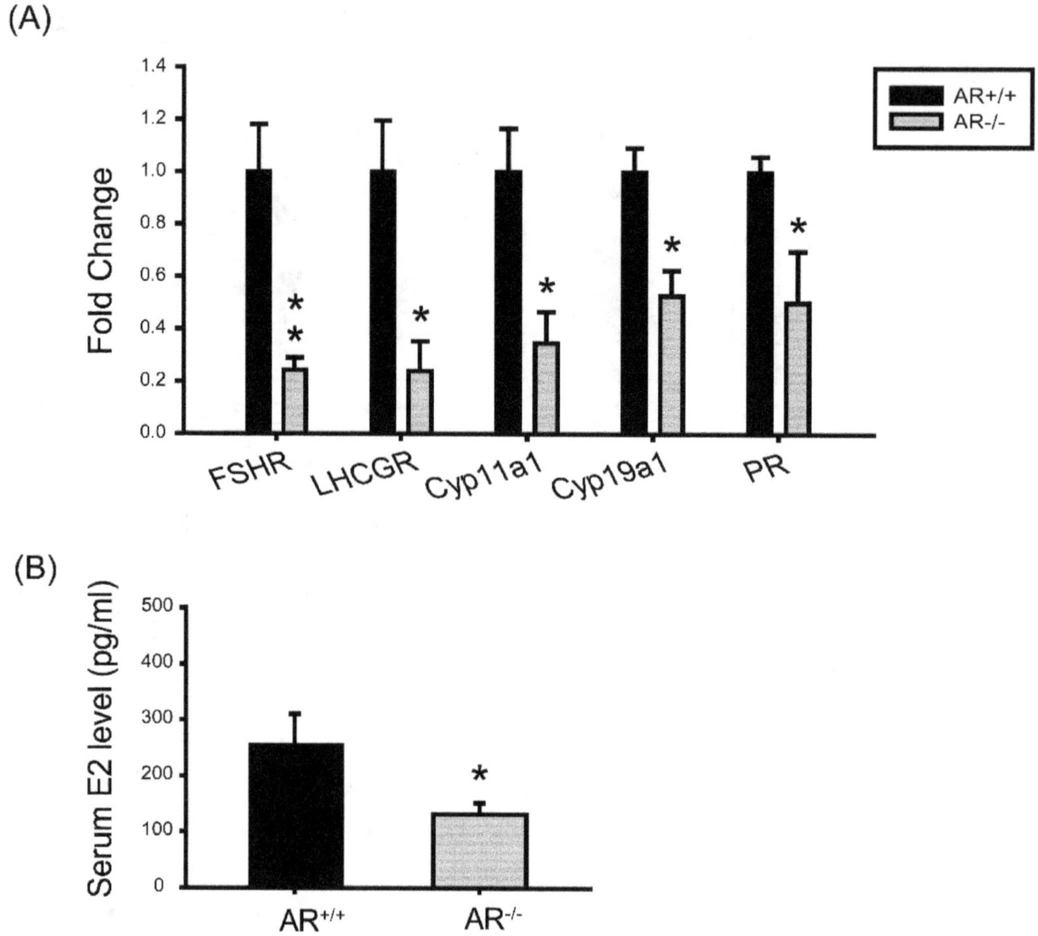

Figure 7. Molecular changes in granulosa cells of $AR^{-/-}$ mice and the effect of AR deficiency on serum estradiol (E2) levels. (**A**) The mRNA levels of genes that are important in granulosa cell differentiation during folliculogenesis and luteinization. At least three experiments were performed and data are presented as the mean ± SEM ($n = 4$ mice per genotype) of the fold changes; (**B**) Serum E2 level was measured by ELISA. ($n = 4$ mice per genotype). Data was expressed as the mean ± SEM. * $p < 0.05$, ** $p < 0.01$. FSHR: FSH receptor; LHCGR: Luteinizing hormone receptor; Cyp11a1: Cytochrome P450 side-chain-cleavage; Cyp19a1: Cytochrome P450 aromatase; PR: Progesterone receptor.

3. Discussion

The primary cause of reduced fecundity in $AR^{-/-}$ female mice is infrequent and inefficient ovulation [12–16]. In this study, we found a high frequency of morphological anomalies in the mitochondria of $AR^{-/-}$ granulosa cells. Mitochondria displayed an increased electron density of the matrix, small spherical structures with fewer, disarrayed cristae, and containing vacuoles under TEM (Figure 4A). It also displayed an aggregated distribution and a small "donut-like" vesicles appearance under immunofluorescence staining (Figure 4B) in the granulosa cells of $AR^{-/-}$ mice. These morphological changes are frequently referred to as "fragmented mitochondria" [29]. Evidences have suggested that fusion and fission of mitochondria affects the ability of cells to distribute their mitochondria to specific subcellular locations. It has been stated that trafficking of mitochondria are important for cellular function by placing them in appropriate locations relative to energy requiring processes [30]. Therefore, those fragmented and misplaced mitochondria in the granulosa cells of $AR^{-/-}$ mice might result in sub-cellular energy deficiency and even cell death [31,32]. Mitochondrial biogenesis requires the symphonious expression of mtDNA and nuclear genes that encode both mitochondrial proteins and their regulatory factors [33]. It has been proved that the PGC-1 family regulates mitochondrial biogenesis by serving as a coactivator of multiple transcription factors, such as NFR1 and TFAM [34,35]. Our quantitative real-time PCR analysis revealed that mtDNA copy number was significantly decreased in granulosa cells from $AR^{-/-}$ mice when compared to $AR^{+/+}$ mice. In agreement, PGC-1β, NRF1 and TFAM mRNAs were down-regulated in the $AR^{-/-}$ granulosa cells (Figure 6B). In the mean time, the average ATP contents and mitochondrial membrane potential were markedly reduced in the granulosa cells of $AR^{-/-}$ mice as compared to $AR^{+/+}$ mice (Figure 5A,B), suggesting a decline of mitochondrial function.

Within the follicle, cumulus cells surround the oocyte and directly contact with the oocyte via gap junctions [36]. The paracrine interactions between the oocyte and granulosa cells are bi-directional and critical for the development of granulosa cells, optimal oocyte quality and embryonic development [37,38]. Furthermore, earlier study has shown that oocytes enclosed by cumulus cells have higher ATP levels than those lacking cumulus cells [39], suggesting that cumulus cells provide ATP for oocyte development. Therefore, mitochondrial function is critical for the energy production in granulosa cells and then to participate as an energy source for oocyte maturation. The role of mitochondria in oocyte, in particular in the immature oocyte has not yet been identified. However, it is generally believed that mitochondria in the either oocyte or egg are in a transmitted state and undifferentiated. Therefore during oocyte maturation, the energy sources are believed from the surrounding cells [40]. Thus, in the current study, we demonstrated that the AR insufficiency that impact on mitochondria function will result in poor oocyte maturation rate. We found that 95% of $AR^{+/+}$ oocytes had reached to metaphase II, whereas a significantly lower maturation rate (60%) was observed in the $AR^{-/-}$ oocytes. This is consistence with earlier studies that such a declined ATP content in oocytes has been suggested to significantly affect oocyte quality, embryonic development and even the implantation process [41,42]. We speculated that the decreased ATP levels and reduced mitochondria membrane potential in the granulosa cells of $AR^{-/-}$ mice might contribute to the decreased oocyte competence to undergo final maturation, as observed in this study.

Our results also showed that PMSG-stimulated $AR^{-/-}$ granulosa cells exhibit reduced FSHR, Cyp11a1 and Cyp19a1 mRNA expressions (Figure 7A). It has been reported, in the both of hypophysectomized rat [43], as well as mutant mice that lacking either gonadotrophins [44], FSH [26], or FSHR [27,28], causes follicular development arrest at the preantral stage. Androgens, via the activation of AR, can increase the expression of FSHR and augment FSH-stimulated follicular differentiation [5–8]. Cyp11a1 and Cyp19a1 are stimulated by FSH in granulosa cells as part of the differentiation program induced by this hormone [45]. Earlier study by Wu [46] nicely showed that androgens have a direct stimulatory effect on Cyp19a1 and Cyp11a1 expression in rat ovarian granulosa cells. In agreement with those earlier studies, our results provide the clear evidence that AR is required for maximum FSH stimulation of Cyp11a1 and Cyp19a1 activity. Our results also showed that the level of serum E2 was significantly decreased in $AR^{-/-}$ mice as compared with $AR^{+/+}$ littermates (Figure 7B). Another important characteristic of preovulatory granulosa cells is the acquisition of LHCGR and PR, which provide the mechanism of follicle selection by which follicles can respond to the LH surge and ovulate [47].

Induction of LHCGR [48,49] and PR [50,51] expression in preovulatory granulosa cells is primarily activated by FSH. Herein, our data demonstrated that AR mediates this effect of androgen on FSHR expression further onwards regulation of LHCGR and PR levels in preovulatory granulosa cells. Granulosa cells of $AR^{-/-}$ mice exhibit a reduced LHCGR and PR mRNA levels relative to $AR^{+/+}$ controls (Figure 7A). We proposed that the reduced LHCGR levels in the granulosa cells of $AR^{-/-}$ mice might contribute to compromised response to LH surge and oocyte competence to undergo final maturation.

Combining our previous findings [12] with the results presented here, we conclude that $AR^{-/-}$ mice exhibit a reduced fertility with defective folliculogenesis and reduced corpus luteum formation might have been contributed by the following reasons. Mitochondrial dysfunction in granulosa cells may compromise the competence of oocytes in $AR^{-/-}$ ovaries. These data indicate that the intraovarian actions of AR and androgen are critical for enhancing the FSH action and the generation of fully differentiated follicles that exhibit the proper cellular organization (*i.e.*, antrum formation and cumulus-oocyte complex), optimal mitochondrial function, the necessary enzymatic activity, and the essential receptor signaling pathways. All of those factors converge to provide the follicle with the normal capacity to respond to the LH surge and expel a healthy oocyte that is fully competent for fertilization.

4. Materials and Methods

All the culture media and chemicals were purchased from Sigma-Aldrich Chemical Company, unless otherwise stated.

4.1. Generation of Female $AR^{-/-}$ Mice

All mouse studies were approved by the Animal Studies Committee at Taipei Medical University and conformed to the Guide for the Care and Use of Laboratory Animals published by the National Institutes of Health. $AR^{-/-}$ female mice and heterozygous female mice ($AR^{+/-}$) were generated by crossing AR-flox mice (exon 2 flanked by loxP sites) with transgenic β-actin-Cre (ACTB-Cre) mice, as previously described [52,53]. Briefly, male mice carrying floxed AR were mated with females genotyped with *AR/ar* ACTB-Cre ($AR^{+/-}$) to produce female $AR^{-/-}$ mice carrying the genotype *ar/ar* ACTB-Cre. Genotype of offspring was confirmed by PCR. Based on the sequence of the AR genomic DNA, three primers have been designed to distinguish the $AR^{+/+}$, $AR^{-/-}$, and floxed *AR* X chromosome on mice. The primer "select" is the 5' primer which is located in the intron 1 and its sequence is 5'-GTTGATACCTTAACCTCTGC-3'. The primer "2–9" is the 3' end primer which is located in intron 2 and its sequence is 5'-CTTACATGTACTGTGAGAGG-3'. The "2–3" primer is the 3' end primer which is located in the exon 2 and its sequence is 5'-CTTCAGCGGCTCTTTTGAAG-3'. If the mouse contains $AR^{+/+}$, the primer pairs (select and "2–9") will generate a product with 580 bp and the other pair of primers (select and "2–3") will generate a product with ~460 bp. If the mice carry $AR^{-/-}$, the primer pairs (select and "2–9") will generate a product with ~270 bp and the other pair of primers (select and "2–3") will not have a product. The expression of Cre and internal control interleukin 2 (IL-2) were confirmed by PCR during genotyping. The primer design and PCR conditions of Cre and IL-2 follow The Jackson Laboratory's suggestions as previously described [52,53].

4.2. Western Blot Analysis

To determine the expression levels of AR, the protein extracts of ovaries were subjected to sodium dodecyl sulfatepolyacrylamide denaturing gel electrophoresis. Proteins were transferred to hybond-P PVDF membranes (GE Healthcare, Little Chalfont, UK) and processed by routine procedures. Immunoreactive bands were visualized by blotting with primary antibodies against AR (1:1000, LS-C137965, LifeSpan Biosciences, Seattle, WA, USA), followed by incubation with horseradish peroxidase-conjugated secondary antibodies and detection with enhanced chemiluminescence (Thermo Scientific, Rockford, IL, USA).

4.3. Tissue Sampling, Oocyte in Vitro Maturation Assay, RNA Extraction and Analysis

To collect fully grown germinal vesicle (GV) oocytes and granulosa cells, a minimum of six mice per genotype (4.5 weeks old; $AR^{-/-}$ and $AR^{+/+}$) were treated with 7.5 IU PMSG by intraperitoneal injection. After 48 h, the ovaries were removed and placed in a dish containing M2 medium. For the oocyte collection, cumulus-oocyte complexes (COCs) and granulosa cells were harvested by manually puncturing of large antral follicles with a sterile needle. Oocytes were denuded with gentle pipetting with pulled-glass Pasteur pipettes as previously described [54]. Oocytes were then cultured in M2 media on the 37 °C heated stage for another 18 h in order to score maturation rates. For collection of granulosa cells, ovarian debris was removed by filtered cells through a 40 µm nylon mesh filter. Granulosa cells were collected by centrifugation and processed for total RNA extraction using an RNeasy kit (Qiagen, Chatsworth, CA, USA). The final concentration of granulosa cells is approximately about 1.5 million per mL. Granulosa cells represent a mixture of both mural and cumulus cells and are referred to as "granulosa cells" in the text. One µg of total RNA was reverse transcribed and subjected to real-time PCR using LightCycler 2.0 instrument (Roche Applied Science, Mannheim, Germany). In general, the real-time PCR was performed with SYBR Green PCR Master Mix (Roche Applied Science). Each sample was run in triplicate. Data were analyzed by a LightCycler Software 4.0 (Roche Applied Science). The mRNA levels of genes of interest in $AR^{-/-}$ mice were compared with those of their $AR^{+/+}$ littermates. Each gene expression pattern was confirmed using at least three pairs of $AR^{-/-}$ and $AR^{+/+}$ mice. Primer sequences used for studying gene expression were designed by Beacon Designer II software (Bio-Rad Lab., Irvine, CA, USA) and are listed below: PPARγ co-activator 1-α (PGC-1α)-forward (f): GACATAGAGTGTGCTGCTCTGGT; PGC-1α-reverse (r): GTTCGCAGGCTCATTGTTGT; PGC-1β-f: CGCTCCAGGAGACTGAATCCAG; PGC-1β-r: CTTGACTACTGTCTGTGAGGC; NRF1-f: TTACT CTGCTGTGGCTGATGG; NRF1-r: CCTCTGATGCTTGCGTCGTCT; TFAM-f: AGTTCATACCTTC GATTTTC; TFAM-r: TGACTTGGAGTTAGCTGC; FSHR-f: GAACGCCATTGAACTGAGATT; FSHR-r: CGGAGACTGGGAAGATTCTG; LHCGR-f: AGTCCATCACGCTGAAACTGT; LHCGR-r: GGCCTGCAATTTGGTGGAAG; Cytochrome P450 side-chain-cleavage (Cyp11a1)-f: GGTGGACA CGACCTCCATGA; Cyp11a1-r: TGCTGGCTTTGAGGAGTGGA; Cyp19a1-f: TTCGCTGAGAGACG TGGAGA; Cyp19a1-r: AGGATTGCTGCTTCGACCTC; Progesterone receptor (PR)-f: CTCAT GAGTCGGCCAGAGAT; PR-r: CACTGTCCTCTTCCACCTCC; β2-microglobulin (β2m)-f: ACC CTCATCAATGGCCTGTGGA; β2m-r: CATGGGCTTTGACCCTTGGG. Mouse β2m was used as a housekeeping gene.

4.4. Histology

Ovaries from 4.5-week-old $AR^{-/-}$ and $AR^{+/+}$ littermates (n = 4 per genotype) were obtained from PMSG treated animals. Ovaries were fixed in 4% paraformaldehyde at 4 °C and then paraffin embedded. Thereafter, 5-µm sections were taken at 30-µm intervals, mounted on slides, and subjected to hematoxylin and eosin staining for histological examination by light microscopy. All follicles with a visible oocyte nucleus were subsequently classified by stage of development. The follicle classification system was based on Pedersen's system [55]: (1) Primordial follicles—identified by an oocyte surrounded by a layer of flattened granulose progenitor cells; (2) Pedersen class 3 (primary)—identified by an oocyte surrounded by a layer of cuboid granulosa cells. There are 21 to 60 granulosa cells on the largest cross-section; (3) Pedersen classes 4–5 (secondary)—identified by an oocyte surrounded by two or more layers of granulosa cells. There are 61 to 400 granulosa cells on the largest cross-section; (4) Pedersen class 6 (incipient antral)—identified by a large oocyte with many layers of granulosa cells. The granulosa cells are separated by scattered areas of fluid. There are 401 to 600 granulosa cells on the largest cross-section; (5) Pedersen class 7 (early antral)—a follicle with a single cavity containing follicle fluid. There are more than 600 cells on the largest cross-section. The cumulus oophorus has formed; (6) Pedersen class 8 (Graafian follicle)—a large follicle with a single cavity with follicle fluid and a well-defined cumulus stalk. There are more than 600 cells on the largest cross-section. Follicles were considered as atretic based on any of the two morphometric criteria within a single cross section: A degenerated oocyte, three or more pyknotic nuclei or atretic bodies in granulosa cell layers or follicular antrum, disorganized granulosa cell layers, granulosa cells pulling away from basement membrane, and broken basement membrane as previously described [15,56].

4.5. Transmission Electron Microscopy (TEM)

For ultrastructural analysis of mitochondria, the ovaries from $AR^{-/-}$ and $AR^{+/+}$ mice were collected followed by 48 h PMSG treatment then fixed in a mixture of glutaraldehyde (1.5%) and paraformaldehyde (1.5%) in phosphate buffer at pH 7.3. They were then post-fixed in 1.0% osmium tetroxide, 1.5% potassium hexanoferrate, before being rinsed in cacodylate and 0.2 M sodium maleate buffers (pH 6.0) followed by block-staining with 1% uranyl acetate. Following dehydration, the ovaries were embedded in Epon and sectioned for TEM. Thin sections were cut, mounted on 200-mesh grids, stained with uranyl acetate and lead citrate, and examined using a H7100 Hitachi electron microscope (H7100, Hitachi High-Technologies Corporation, Tokyo, Japan). Digital images were captured using a MegaView III digital camera (OSIS Pro Software, Olympus Soft Imaging Solutions, Lakewood, CO, USA). The morphology of mitochondria in granulosa cells was examined.

4.6. Immunofluorescence

For mitochondrial localization, granulosa cells were collected and cultured overnight in µ-Slide 8 well plates (ibidi GmbH, Martinsried, Germany) to allow cell attachment. Granulosa cells were stained with 200 nM MitoTracker Green (Molecular Probes, Eugene, OR, USA) for 45 min at 37 °C and cells were then fixed with 4% paraformaldehyde for 20 min and treated with 0.5% Triton X-100 for 20 min, the liquid removed and blocked by 0.5% BSA in PBS 1 h. After empty the well, granulosa cells were counterstained with 4',6'-diamidino-2-phenylindole (DAPI; Sigma, St. Louis, MO, USA) for 10 min at 37 °C, samples were observed by *Delta Vision Elite* Microscope (Applied Precision, GE Healthcare, Olive Branch, MS, USA).

4.7. Determination of Mitochondrial DNA (mtDNA) Content in Granulosa Cells

Total DNA was extracted from granulosa cells using a DNeasy kit (Qiagen). mtDNA content was calculated using quantitative real-time PCR as described previously [57,58] by measuring the threshold cycle ratio (ΔC_t) of a mitochondria-encoded gene cytochrome c oxidase subunit 2 (COX2) (L7447, 5'-AATAGAACTTCCAATCCGTA-3', H7726, 5'-AAGGTTAACGCTCTTAGCTT-3') *vs.* a nuclear-encoded gene (18s-rRNA-f, 5'-CGGCTACCACATCCAAGGAA-3', 18s-rRNA-r, 5'-GCT GGAATTACCGCGGCT-3'). Data were expressed as mtDNA/nuclear DNA. All measurements were performed in triplicate.

4.8. Metabolite Analytic Assays

Granulosa cells were measured for ATP content by ATPlite™ Luminescence ATP Detection Assay System (Perkin Elmer Life and Analytical Sciences B.V., Groningen, The Netherlands). Granulosa cells were homogenized in 50 μL mammalian cell lysis buffer and then 50 μL substrates solution were added. Adapt the plate to darkness for 10 min and then measure the luminescence. Reactions were normalized to total protein, and metabolite contents were expressed as μM/mg protein.

4.9. Detection of Mitochondrial Membrane Potential by Flow Cytometry

Granulosa cells were collected from 4-week-old female mice ($AR^{+/+}$ and $AR^{-/-}$) injected with 7.5 IU PMSG 48 h later. To get positive control from cells treated with CCCP (carbonyl cyanide 3-chlorophenylhydrazone, Sigma Immunochemicals, St. Louis, MO, USA) to depolarized the mitochondrial membrane potential. Cells were incubated with 500 nM JC-1 (5,5',6,6'-tetrachloro-1,1',3,3'-tetraethylbenzimidazolylcarbocyanine iodide, Molecular Probes, Eugene, OR, USA) in the PBS for 15 min at 37 °C in the dark and then analyzed by flow cytometry (BD Biosciences, San Jose, CA, USA). The acquired data were further analyzed using FCS express V4 analysis software.

4.10. Assessment of Serum Hormone Levels

To determine the level of estradiol (E2) in the experimental animals, sera were collected from both $AR^{-/-}$ and $AR^{+/+}$ female mice that were previously treated with 7.5 IU PMSG for 48 h and stored at −20 °C before further analysis were carried out. Total E2 level was measured by ELISA kits following the manufacture's instruction (USCN Life Science, Wuhan, China). The volume of serum used for ELISA was 50 μL, the minimal detectable concentration was 4.38 pg/mL, and the intra- and interassay coefficients of variation were 8.7% and 9.8%, respectively.

4.11. Statistical Analysis

Data are presented as mean ± standard error mean (SEM), unless otherwise indicated. Statistical comparisons were made with Student's *t* test. $p \leq 0.05$ as considered to be significant.

Acknowledgments

We wish to thank the Laboratory of Reproductive Medicine in the department of Gynecology and Obstetrics at the Taipei Medical University Hospital for the laboratory procedures. This work was supported by National Science Council, Taiwan (NSC 96-2314-B-038-018-MY2) and by Taipei Medical University (TMU100-AE1-B02).

Author Contributions

Ruey-Sheng Wang and Heng-Yu Chang did the writing of paper and oocyte experiments as well as molecular biology experiments; Cheng-Heng Kao did the TEM experiments; Yi-Chen Wu and Shuyuan Yeh bred the knockout mice; Shu-Huei Kao, Chii-Ruey Tzeng and Chawnshang Chang supervised the project.

References

1. Chang, C.S.; Kokontis, J.; Liao, S.T. Molecular cloning of human and rat complementary DNA encoding androgen receptors. *Science* **1988**, *240*, 324–326.
2. Heinlein, C.A.; Chang, C. Androgen receptor in prostate cancer. *Endocr. Rev.* **2004**, *25*, 276–308.
3. Murray, A.A.; Gosden, R.G.; Allison, V.; Spears, N. Effect of androgens on the development of mouse follicles growing *in vitro*. *J. Reprod. Fertil.* **1998**, *113*, 27–33.
4. Vendola, K.A.; Zhou, J.; Adesanya, O.O.; Weil, S.J.; Bondy, C.A. Androgens stimulate early stages of follicular growth in the primate ovary. *J. Clin. Investig.* **1998**, *101*, 2622–2629.
5. Tetsuka, M.; Hillier, S.G. Differential regulation of aromatase and androgen receptor in granulosa cells. *J. Steroid Biochem. Mol. Biol.* **1997**, *61*, 233–239.
6. Weil, S.; Vendola, K.; Zhou, J.; Bondy, C.A. Androgen and follicle-stimulating hormone interactions in primate ovarian follicle development. *J. Clin. Endocrinol. Metab.* **1999**, *84*, 2951–2956.
7. Daniel, S.A.; Armstrong, D.T. Enhancement of follicle-stimulating hormone-induced aromatase activity by androgens in cultured rat granulosa cells. *Endocrinology* **1980**, *107*, 1027–1033.
8. Sen, A.; Prizant, H.; Light, A.; Biswas, A.; Hayes, E.; Lee, H.J.; Barad, D.; Gleicher, N.; Hammes, S.R. Androgens regulate ovarian follicular development by increasing follicle stimulating hormone receptor and microRNA-125b expression. *Proc. Natl. Acad. Sci. USA* **2014**, *111*, 3008–3013.
9. Hirai, M.; Hirata, S.; Osada, T.; Hagihara, K.; Kato, J. Androgen receptor mRNA in the rat ovary and uterus. *J. Steroid Biochem. Mol. Biol.* **1994**, *49*, 1–7.
10. Tetsuka, M.; Whitelaw, P.F.; Bremner, W.J.; Millar, M.R.; Smyth, C.D.; Hillier, S.G. Developmental regulation of androgen receptor in rat ovary. *J. Endocrinol.* **1995**, *145*, 535–543.
11. Gill, A.; Jamnongjit, M.; Hammes, S.R. Androgens promote maturation and signaling in mouse oocytes independent of transcription: A release of inhibition model for mammalian oocyte meiosis. *Mol. Endocrinol.* **2004**, *18*, 97–104.
12. Hu, Y.C.; Wang, P.H.; Yeh, S.; Wang, R.S.; Xie, C.; Xu, Q.; Zhou, X.; Chao, H.T.; Tsai, M.Y.; Chang, C. Subfertility and defective folliculogenesis in female mice lacking androgen receptor. *Proc. Natl. Acad. Sci. USA* **2004**, *101*, 11209–11214.

13. Shiina, H.; Matsumoto, T.; Sato, T.; Igarashi, K.; Miyamoto, J.; Takemasa, S.; Sakari, M.; Takada, I.; Nakamura, T.; Metzger, D.; *et al*. Premature ovarian failure in androgen receptor-deficient mice. *Proc. Natl. Acad. Sci. USA* **2006**, *103*, 224–229.
14. Walters, K.A.; Allan, C.M.; Jimenez, M.; Lim, P.R.; Davey, R.A.; Zajac, J.D.; Illingworth, P.; Handelsman, D.J. Female mice haploinsufficient for an inactivated androgen receptor (AR) exhibit age-dependent defects that resemble the AR null phenotype of dysfunctional late follicle development, ovulation, and fertility. *Endocrinology* **2007**, *148*, 674–684.
15. Sen, A.; Hammes, S.R. Granulosa cell-specific androgen receptors are critical regulators of ovarian development and function. *Mol. Endocrinol.* **2010**, *24*, 1393–1403.
16. Walters, K.A.; Middleton, L.J.; Joseph, S.R.; Hazra, R.; Jimenez, M.; Simanainen, U.; Allan, C.M.; Handelsman, D.J. Targeted loss of androgen receptor signaling in murine granulosa cells of preantral and antral follicles causes female subfertility. *Biol. Reprod.* **2012**, *87*, 151.
17. Spears, N.; Murray, A.A.; Allison, V.; Boland, N.I.; Gosden, R.G. Role of gonadotrophins and ovarian steroids in the development of mouse follicles *in vitro*. *J. Reprod. Fertil.* **1998**, *113*, 19–26.
18. Yang, M.Y.; Fortune, J.E. Testosterone stimulates the primary to secondary follicle transition in bovine follicles *in vitro*. *Biol. Reprod.* **2006**, *75*, 924–932.
19. Seifer, D.B.; DeJesus, V.; Hubbard, K. Mitochondrial deletions in luteinized granulosa cells as a function of age in women undergoing *in vitro* fertilization. *Fertil. Steril.* **2002**, *78*, 1046–1048.
20. Wang, Q.; Frolova, A.I.; Purcell, S.; Adastra, K.; Schoeller, E.; Chi, M.M.; Schedl, T.; Moley, K.H. Mitochondrial dysfunction and apoptosis in cumulus cells of type I diabetic mice. *PLoS ONE* **2010**, *5*, e15901.
21. Gannon, A.M.; Stampfli, M.R.; Foster, W.G. Cigarette smoke exposure elicits increased autophagy and dysregulation of mitochondrial dynamics in murine granulosa cells. *Biol. Reprod.* **2013**, *88*, 63.
22. Newmeyer, D.D.; Ferguson-Miller, S. Mitochondria: Releasing power for life and unleashing the machineries of death. *Cell* **2003**, *112*, 481–490.
23. Miller, W.L. Steroid hormone synthesis in mitochondria. *Mol. Cell. Endocrinol.* **2013**, *379*, 62–73.
24. Ramalho-Santos, J.; Varum, S.; Amaral, S.; Mota, P.C.; Sousa, A.P.; Amaral, A. Mitochondrial functionality in reproduction: From gonads and gametes to embryos and embryonic stem cells. *Hum. Reprod. Update* **2009**, *15*, 553–572.
25. Scarpulla, R.C. Transcriptional paradigms in mammalian mitochondrial biogenesis and function. *Physiol. Rev.* **2008**, *88*, 611–638.
26. Kumar, T.R.; Wang, Y.; Lu, N.; Matzuk, M.M. Follicle stimulating hormone is required for ovarian follicle maturation but not male fertility. *Nat. Genet.* **1997**, *15*, 201–204.
27. Dierich, A.; Sairam, M.R.; Monaco, L.; Fimia, G.M.; Gansmuller, A.; LeMeur, M.; Sassone-Corsi, P. Impairing follicle-stimulating hormone (FSH) signaling *in vivo*: Targeted disruption of the FSH receptor leads to aberrant gametogenesis and hormonal imbalance. *Proc. Natl. Acad. Sci. USA* **1998**, *95*, 13612–13617.
28. Danilovich, N.; Babu, P.S.; Xing, W.; Gerdes, M.; Krishnamurthy, H.; Sairam, M.R. Estrogen deficiency, obesity, and skeletal abnormalities in follicle-stimulating hormone receptor knockout (FORKO) female mice. *Endocrinology* **2000**, *141*, 4295–4308.
29. Detmer, S.A.; Chan, D.C. Functions and dysfunctions of mitochondrial dynamics. *Nat. Rev. Mol. Cell Biol.* **2007** *8*, 870–879.

30. Frazier, A.E.; Kiu, C.; Stojanovski, D.; Hoogenraad, N.J.; Ryan, M.T. Mitochondrial morphology and distribution in mammalian cells. *Biol. Chem.* **2006**, *387*, 1551–1558.
31. Martinou, J.C.; Youle, R.J. Mitochondria in apoptosis: Bcl-2 family members and mitochondrial dynamics. *Dev. Cell* **2011**, *21*, 92–101.
32. Chan, D.C. Mitochondria: Dynamic organelles in disease, aging, and development. *Cell* **2006**, *125*, 1241–1252.
33. Diaz, F.; Moraes, C.T. Mitochondrial biogenesis and turnover. *Cell Calcium* **2008**, *44*, 24–35.
34. Kelly, D.P.; Scarpulla, R.C. Transcriptional regulatory circuits controlling mitochondrial biogenesis and function. *Genes Dev.* **2004**, *18*, 357–368.
35. Lin, J.; Handschin, C.; Spiegelman, B.M. Metabolic control through the PGC-1 family of transcription coactivators. *Cell Metab.* **2005**, *1*, 361–370.
36. Gittens, J.E.; Barr, K.J.; Vanderhyden, B.C.; Kidder, G.M. Interplay between paracrine signaling and gap junctional communication in ovarian follicles. *J. Cell Sci.* **2005**, *118*, 113–122.
37. Eppig, J.J. Intercommunication between mammalian oocytes and companion somatic cells. *Bioessays* **1991**, *13*, 569–574.
38. Matzuk, M.M.; Burns, K.H.; Viveiros, M.M.; Eppig, J.J. Intercellular communication in the mammalian ovary: Oocytes carry the conversation. *Science* **2002**, *296*, 2178–2180.
39. Downs, S.M. The influence of glucose, cumulus cells, and metabolic coupling on ATP levels and meiotic control in the isolated mouse oocyte. *Dev. Biol.* **1995**, *167*, 502–512.
40. Sugiura, K.; Pendola, F.L.; Eppig, J.J. Oocyte control of metabolic cooperativity between oocytes and companion granulosa cells: Energy metabolism. *Dev. Biol.* **2005**, *279*, 20–30.
41. Quinn, P.; Wales, R.G. The relationships between the ATP content of preimplantation mouse embryos and their development *in vitro* during culture. *J. Reprod. Fertil.* **1973**, *35*, 301–309.
42. Van Blerkom, J.; Davis, P.W.; Lee, J. ATP content of human oocytes and developmental potential and outcome after *in vitro* fertilization and embryo transfer. *Hum. Reprod.* **1995**, *10*, 415–424.
43. Richards, J.S. Maturation of ovarian follicles: Actions and interactions of pituitary and ovarian hormones on follicular cell differentiation. *Physiol. Rev.* **1980**, *60*, 51–89.
44. Mason, A.J.; Hayflick, J.S.; Zoeller, R.T.; Young, W.S., 3rd.; Phillips, H.S.; Nikolics, K.; Seeburg, P.H. A deletion truncating the gonadotropin-releasing hormone gene is responsible for hypogonadism in the hpg mouse. *Science* **1986**, *234*, 1366–1371.
45. Hunzicker-Dunn, M.; Maizels, E.T. FSH signaling pathways in immature granulosa cells that regulate target gene expression: Branching out from protein kinase A. *Cell Signal.* **2006**, *18*, 1351–1359.
46. Wu, Y.G.; Bennett, J.; Talla, D.; Stocco, C. Testosterone, not 5α-dihydrotestosterone, stimulates LRH-1 leading to FSH-independent expression of Cyp19 and P450scc in granulosa cells. *Mol. Endocrinol.* **2011**, *25*, 656–668.
47. McGee, E.A.; Hsueh, A.J. Initial and cyclic recruitment of ovarian follicles. *Endocr. Rev.* **2000**, *21*, 200–214.
48. Segaloff, D.L.; Wang, H.Y.; Richards, J.S. Hormonal regulation of luteinizing hormone/chorionic gonadotropin receptor mRNA in rat ovarian cells during follicular development and luteinization. *Mol. Endocrinol.* **1990**, *4*, 1856–1865.

49. Kessel, B.; Liu, Y.X.; Jia, X.C.; Hsueh, A.J. Autocrine role of estrogens in the augmentation of luteinizing hormone receptor formation in cultured rat granulosa cells. *Biol. Reprod.* **1985**, *32*, 1038–1050.
50. Natraj, U.; Richards, J.S. Hormonal regulation, localization, and functional activity of the progesterone receptor in granulosa cells of rat preovulatory follicles. *Endocrinology* **1993**, *133*, 761–769.
51. Park-Sarge, O.K.; Mayo, K.E. Regulation of the progesterone receptor gene by gonadotropins and cyclic adenosine 3',5'-monophosphate in rat granulosa cells. *Endocrinology* **1994**, *134*, 709–718.
52. Yeh, S.; Tsai, M.Y.; Xu, Q.; Mu, X.M.; Lardy, H.; Huang, K.E.; Lin, H.; Yeh, S.D.; Altuwaijri, S.; Zhou, X.; *et al*. Generation and characterization of androgen receptor knockout (ARKO) mice: An *in vivo* model for the study of androgen functions in selective tissues. *Proc. Natl. Acad. Sci. USA* **2002**, *99*, 13498–13503.
53. Yeh, S.; Hu, Y.C.; Wang, P.H.; Xie, C.; Xu, Q.; Tsai, M.Y.; Dong, Z.; Wang, R.S.; Lee, T.H.; Chang, C. Abnormal mammary gland development and growth retardation in female mice and MCF7 breast cancer cells lacking androgen receptor. *J. Exp. Med.* **2003**, *198*, 1899–1908.
54. Chang, H.Y.; Minahan, K.; Merriman, J.A.; Jones, K.T. Calmodulin-dependent protein kinase γ 3 (CamKIIγ3) mediates the cell cycle resumption of metaphase II eggs in mouse. *Development* **2009**, *136*, 4077–4081.
55. Pedersen, T.; Peters, H. Proposal for a classification of oocytes and follicles in the mouse ovary. *J. Reprod. Fertil.* **1968**, *17*, 555–557.
56. Britt, K.L.; Drummond, A.E.; Cox, V.A.; Dyson, M.; Wreford, N.G.; Jones, M.E.; Simpson, E.R.; Findlay, J.K. An age-related ovarian phenotype in mice with targeted disruption of the *Cyp 19* (aromatase) gene. *Endocrinology* **2000**, *141*, 2614–2623.
57. Weng, S.W.; Lin, T.K.; Liou, C.W.; Chen, S.D.; Wei, Y.H.; Lee, H.C.; Chen, I.Y.; Hsieh, C.J.; Wang, P.W. Peripheral blood mitochondrial DNA content and dysregulation of glucose metabolism. *Diabetes Res. Clin. Pract.* **2009**, *83*, 94–99.
58. Bonnard, C.; Durand, A.; Peyrol, S.; Chanseaume, E.; Chauvin, M.A.; Morio, B.; Vidal, H.; Rieusset, J. Mitochondrial dysfunction results from oxidative stress in the skeletal muscle of diet-induced insulin-resistant mice. *J. Clin. Investig.* **2008**, *118*, 789–800.

7

Mitochondria: A Therapeutic Target for Parkinson's Disease?

Yu Luo [1,*], Alan Hoffer [1], Barry Hoffer [1] and Xin Qi [2,*]

[1] Department of Neurological Surgery, Case Western Reserve University, Cleveland, OH 44106, USA; E-Mails: Alan.Hoffer@uhhospitals.org (A.H.); bhoffer@intra.nida.nih.gov (B.H.)

[2] Department of Physiology, Case Western Reserve University, Cleveland, OH 44106, USA

* Authors to whom correspondence should be addressed; E-Mails: yxl710@case.edu (Y.L.); xxq38@case.edu (X.Q.)

Academic Editors: Jaime M. Ross and Giuseppe Coppotelli

Abstract: Parkinson's disease (PD) is one of the most common neurodegenerative disorders. The exact causes of neuronal damage are unknown, but mounting evidence indicates that mitochondrial-mediated pathways contribute to the underlying mechanisms of dopaminergic neuronal cell death both in PD patients and in PD animal models. Mitochondria are organized in a highly dynamic tubular network that is continuously reshaped by opposing processes of fusion and fission. Defects in either fusion or fission, leading to mitochondrial fragmentation, limit mitochondrial motility, decrease energy production and increase oxidative stress, thereby promoting cell dysfunction and death. Thus, the regulation of mitochondrial dynamics processes, such as fusion, fission and mitophagy, represents important mechanisms controlling neuronal cell fate. In this review, we summarize some of the recent evidence supporting that impairment of mitochondrial dynamics, mitophagy and mitochondrial import occurs in cellular and animal PD models and disruption of these processes is a contributing mechanism to cell death in dopaminergic neurons. We also summarize mitochondria-targeting therapeutics in models of PD, proposing that modulation of mitochondrial impairment might be beneficial for drug development toward treatment of PD.

Keywords: Parkinson's disease; mitochondrial dysfunction; mitochondrial dynamics

---

1. Introduction

Parkinson's disease (PD) is the second most common neurodegenerative disorder after Alzheimer's disease, affecting over 1% of the population older than 60 years of age. Clinically, it is diagnosed primarily based on motor abnormalities including bradykinesia, resting tremor, and cogwheel rigidity [1]. A key characteristic of pathology in PD is the degeneration of the nigrostriatal (NS) dopaminergic pathway which is one of the most important dopamine (DA) pathways in the brain and contains about 80% of the total brain DA. Despite a large number of studies on the pathogenesis of PD, there is still inconclusive evidence about why dopaminergic neurons are selectively degenerated. Currently, there is no effective restorative treatment available for PD, only symptomatic treatment is available.

Among a number of proposed mechanisms involved in PD pathogenesis, mitochondrial dysfunction has been repeatedly implicated as the cause of the death of DA neurons in PD [2–5]. Mitochondria are critical for many cellular functions, such as intermediary metabolism [6,7], redox signaling [8], calcium homeostasis [9–11], cell proliferation [12,13], development [14,15] and cell death [16–18]. Mitochondrial dysfunction is mainly characterized by the generation of reactive oxygen species (ROS), a defect in mitochondrial electron transport complex enzyme activities, ATP depletion, caspase 3 release and depletion of mitochondrial DNA. In this review, we summarize evidence on the critical involvement of mitochondria in both genetic mutation and environmental toxin-induced PD. We propose a causal role for mitochondrial dysfunction in the development of PD, because (1) neurotoxins causing parkinsonism, such as 1-methyl-4-phenyl-1,2,3,6-tetrahydropyridine (MPTP), rotenone, paraquat, induce dopaminergic neuronal death through direct inhibition of mitochondrial complex I activity; (2) mutant proteins from PD-related genes associate with mitochondria where they elicit diverse mitochondrial dysregulation and subsequently cause neuronal degeneration; (3) therapeutic agents that target mitochondrial protein or inhibit mitochondrial damage can reduce neuropathological phenotypes of PD in animal models and cells from PD patients.

2. Mitochondrial Dysfunction in Parkinson's Disease

Aberrant mitochondrial function is one of the major cytopathologies in PD and has been widely accepted as a central pathogenic mechanism underlying PD pathogenesis. Chronic systemic administration of rotenone, a specific complex I inhibitor and a pesticide, results in neuropathologic and behavioral changes in rats that are similar to human PD [19,20]. 1-methyl-4-phenyl-1,2,3,6-tetrahydropyridine (MPTP), a meperidine analog found to cause parkinsonism in humans, exerts its toxic effects through metabolism to 1-methyl-4-phenylpyridinium (MPP^+), another complex I inhibitor [19,21]. These compounds have long been used in animal models of PD. Furthermore a number of familial forms of PD are associated with mutations in genes encoding both mitochondrially targeted proteins and proteins involved in mitochondrial function and/or oxidative stress responses, including mutations in PINK-1, DJ-1, parkin, and leucine-rich repeat kinase 2 (LRRK2) [22]. Interestingly, one study reported that the size of mitochondria in dopaminergic neurons in the substantia nigra (SN; susceptible to PD degeneration) is smaller than in neighboring non-dopaminergic neurons or in dopaminergic neurons of the ventral tegmental area which is more resistant in PD, suggesting a basis for the increased vulnerability of SN neurons to subtle changes in mitochondrial maintenance and function [23].

In addition, increased oxidative stress due to mitochondrial compromise in PD model animals has been proposed to contribute to the degeneration of dopaminergic neurons [23].

Consistent with the evidence from basic science, clinical studies also showed that mitochondrial damage plays a predominant role in the development of PD in patients. Mild deficiency in mitochondrial respiratory electron transport chain NADH dehydrogenase (Complex I) activity has been reported in the substantia nigra [24] as well as platelets [25,26] and lymphocytes [27,28] in PD patients, suggesting a systemic inhibition of complex I activity in PD patients. Mitochondrial dysfunction could lead to increased oxidative stress. Indeed oxidative damage to lipids, proteins and DNA has been detected in brain tissue from PD patients [29,30]. A recent proteomic analysis of mitochondria-enriched fractions from post-mortem PD substantia nigra revealed differential expression of multiple mitochondrial proteins in PD brains as compared with control brains [23]. In further support for a "mitochondrial genetics" hypothesis for PD pathophysiology, Bender *et al.* [31] reported higher levels of mitochondrial DNA deletions in nigral neurons from PD patients. Moreover, both Bender *et al.* [31] and Kraytsberg *et al.* [32] reported higher levels of mitochondrial DNA deletions in nigral neurons of aged humans with sharp elevations starting shortly before age 70. This correlates with the known risk factor of age in PD. It is possible that there is an accumulation of mitochondrial dysfunction and of reactive oxygen species (ROS) damage during aging which needs to reach a critical threshold for cellular dysfunction and degeneration to be observed. How a systemic dysregulation in mitochondrial function or oxidative damage leads to tissue or cell type specific vulnerability still remains to be elucidated.

Taken together, although studies over many years on PD indicate an important role of mitochondria in PD-associated pathology, the process by which the mitochondria become dysfunctional in PD and whether correction of mitochondrial defects could provide neuroprotection in PD remain to be determined.

3. Environmental Toxins that Influence Mitochondrial Function

Some toxins used to model DA loss in PD, such as MPTP and rotenone, impair respiratory chain function by inhibiting complex I [33–36]. These complex I inhibitors replicate some of the key motor features of PD and lead to DA neuronal loss. Intravenous injection of the compound MPTP by drug addicts caused a condition that closely resembles the anatomic and clinical features of PD [37,38]. Multiple models have been developed in the laboratory in which the chronic infusion of the pesticide rotenone or combination of herbicides and pesticides [39] lead to a pattern of cell death and DA loss similar to that of PD. The precise role of environmental toxins in the cause of PD remains to be defined, but these data support the hypothesis that environmental toxins could introduce mitochondrial dysfunction and lead to parkinsonism in human. Indeed, epidemiological studies showed that the prevalence of sporadic PD is higher among farming communities [40]. Exposure to pesticides or herbicides elicited a three-to fourfold increased risk of developing PD [41]. All of these data suggest an environmental contribution to the etiology of sporadic PD.

4. Genetic Factors Associated with PD

In a twin study, Tanner *et al.* showed that for early onset cases, monozygotic concordance was twice that of dizygotic concordance, suggesting that genetic factors are important in the early onset PD [42].

Many examples of familial parkinsonism have also been reported (See review [43]). Mutations of several genes have been linked to familial PD and parkinsonian syndromes [44–46].

4.1. Mitochondrial DNA Mutations and Deletions in PD

Mitochondrial DNA (mtDNA) encodes 13 subunits of respiratory chain proteins, including seven complex I, one complex III, three complex IV, and two complex V subunits. Until now, there has been no PD-associated genetic mutations in mtDNA reported [47]. However, mtDNA deletions have been observed in individual dopaminergic neurons dissected from postmortem human substantia nigra tissue [31]. In addition, mutations in the gene encoding human mtDNA polymerase subunit γ (POLG) leads to clinical parkinsonism associated with multiple mtDNA deletions. [48,49]. Furthermore, proliferator-activated coactivator-1 α (PGC-1 α), one of the key regulators of mitochondrial biogenesis, was found to be decreased in PD patients through a genome-wide association study (GWAS) [50]. Directed deletion of transcriptional factor A (TFAM) in mouse DA neurons using the Cre-LoxP system, termed as MitoPark mice, causes marked deletion of mtDNA, severe impairment of oxidative phosphorylation and slowly progressive motor deficits in the DA system that mimic human parkinsonism as well as altered response to L-3,4-dihydroxyphenylalanine (L-DOPA) treatment [51–53]. The MitoPark mice of PD provided direct evidence that mitochondrial dysfunction in DA neurons can causes PD-related phenotypes. Consistently increased level of mtDNA deletions in the striatum of PD patients have been reported [54] and mtDNA deletions were significantly higher in neurons with impaired cytochrome oxidase activity [31,32]. These findings support a mitochondrial genetic contribution in PD.

4.2. Nuclear Gene Mutations Affecting Mitochondrial Function

Many of the PD susceptible genes identified are related to mitochondrial function (Table 1). These genes and their potential contribution to PD have previously been extensively reviewed [43]. PD linked genes that affect mitochondrial function include, but are not limited to, α-synuclein, Parkin, Phosphatase and tensin homologue (PTEN)-induced putative kinase 1 (PINK1), DJ-1 and LRRK2. In this review we will focus on their involvement and effects on dysregulation of mitochondrial dynamics, mitophagy and mitochondrial redox, and mitochondrial protein import.

Alpha-synuclein: α-synuclein is aggregation-prone protein which attains an increased propensity to aggregate because of the presence of its hydrophobic non-amyloid β component domain. Missense mutations in the alpha-synuclein gene are associated with autosomal dominant PD [55]. The fibrillar form of α-synucleinis a major component of Lewy bodies and has been demonstrated to trigger neurotoxicity in PD [56,57].

Parkin: Parkin is a RING finger containing ubiquitin E3 ligase. It is known to mediate poly-ubiquitination of its substrates for proteasomal degradation. Mutations in the parkin gene cause early onset juvenile autosomal recessive PD, and Parkin mutations are the most common cause of young onset PD. The loss of parkin E3 ligase activity results in accumulation of its substrates leading to neurotoxicity in autosomal recessive PD [58,59].

PINK1: Phosphatase and tensin homologue (PTEN)-induced putative kinase 1 (PINK1) is a serine/threonine kinase localized in mitochondria. Mutations in PINK1 are associated with a rare form of autosomal recessive PD. PINK mutations result in the loss of PINK1 function which leads to aberrant phosphorylation of its substrates to cause PD [60,61].

Table 1. Animal models of Parkinson's disease (PD)-related genes affect mitochondrial function.

Animal Models	Genetic Manipulation in Animals	Motor Phenotypes	PD Pathology and Mitochondrial Function	References
Alpha-Synuclein transgenic mice	hA53T alpha-Synuclein in mice; mPrP promoter	Severe leading to paralysis and premature death	Lewy body-like inclusion in older mice; mitochondrial dysfunction; no dopaminergic neuronal loss	[62–64]
	hA30P alpha-synuclein in mice; mThy-1 promoter	Severe leading to paralysis	Lewy body-like inclusion; sensorimotor neuronal loss in brain stem	[65]
	Alpha-synuclein overexpression in mice (Thy1 promoter)	Progressive declines in spontaneous and motor activity	No DA neuronal degeneration, mitochondrial dysregulation	[66,67]
	hA53T alpha-synuclein expressing in SN DA neurons of mice	Body weight loss; normal locomotion activity	Progressive DA neuronal loss; aberrant mitochondrial inclusion	[68]
LRRK2 transgenic or knock-in mice	LRRK2 R1441G mice; BAC promoter	Rearing activity decrease in older mice	DA neurite degeneration; Tau phosphorylation increase; no DA neuronal degeneration; mitochondrial dysfunction	[69,70]
	LRRK2 G2019S knock-in mice	Absent	Abnormal mitochondrial morphology; mitochondrial dysfunction; no DA neuronal loss	[71]
Parkin$^{-/-}$ mice	Parkin germline inactivation	Conflicting: either absent or subtle motor movement disturbance	Absent	[72,73]
DJ1$^{-/-}$ mice	DJ1 germline inactivation	Age-dependent declines in locomotor activity	Impaired DA update; no DA neuronal degeneration	[74]
PINK1$^{-/-}$ mice	PINK1 germline inactivation	Age-dependent declines in spontaneous activity	Impaired DA update; no DA neuronal degeneration; mitochondrial abnormalities	[75,76]
Mito-Park mice	DAT riven cre; loxed p TFAM	Begins at 3–4 months; declines in spontaneous and rearing activity	Abnormal mitochondrial aggregates; DA reduction in the striatum; progressive DA neuronal degeneration	[51]
Double-mutant mice	A53T alpha-synuclein overexpression in Parkin$^{-/-}$ mice	Absent	Altered mitochondrial structure and morphology; no DA neuronal loss	[77]

DJ-1: DJ-1 belongs to the peptidase C56 family of proteins. Wild-type DJ-1 can serve as a chaperone, protease, regulator of transcription and autophagy through redox regulation. DJ-1 seems to be cytoprotective specifically under conditions related to oxidative stress. Its protective action is the result of the modification of cysteine residues on DJ-1 to cysteine-sulfinic and cysteine-sulfonic acids under oxidative stress [78]. DJ-1 mutations associated with PD are rare and account for 1%–2% of autosomal recessive early-onset PD [79].

LRRK2: Leucine-rich repeat kinase 2 (LRRK2) is a protein encoded by the PARK8 locus. It has a conserved serine-threonine kinase mitogen-activated protein kinase kinase kinase (MAPKKK) domain, a member of the Roc (Ras of complex) GTPase family [80–82]. To date, there are over 50 variants identified in PD patients. The mutation G2019S (Gly2019 to Ser) that takes place in the MAPKKK domain has been recognized as the most common cause of dominant familial PD and accounts for up to 2% of sporadic PD cases [83]. The G2019S mutant augments the kinase activity of LRRK2, which is associated with increased toxicity in dopaminergic neurons [84].

4.3. Mitochondrial Dynamics Impairment in Parkinson's Disease

Mitochondria are organized in a highly dynamic tubular network that is continuously reshaped by opposing processes of fusion and fission [85]. Mitochondrial fission and fusion were first observed in yeast, and since have been observed in all mammalian cells [86]. A delicate balance is maintained between fusion and fission to ensure the normal function of mitochondria. Specifically the fusion process is important for mitochondrial interactions and communication, and fission facilitates the segregation of mitochondria into daughter cells and enhances mitochondrial renewal and distribution along cytoskeletal tracks. Fusion and fission events enable the proper exchanging and mixing of mitochondrial membranes and contents. This dynamic process controls not only mitochondrial morphology, but also the subcellular location and function of mitochondria. Defects in either fusion or fission limit mitochondrial motility, decrease energy production and increase oxidative stress, thereby promoting cell dysfunction and death [87,88]. The two opposing processes, fusion and fission, are controlled by evolutionarily conserved large GTPases that belong to the dynamin family of proteins. In mammalian cells, mitochondrial fusion is regulated by mitofusin-1 and -2 (MFN1/2) and optic atrophy 1 (OPA1), whereas mitochondrial fission is controlled by the dynamin-1-related protein, (Drp1) [89,90] and its mitochondrial adaptors such as Fis1, Mff and MiD49/51 [91–93]. Drp1 is primarily found in the cytosol, but it translocates from the cytosol to the mitochondrial surface in response to various cellular stimuli to regulate mitochondrial morphology [1]. At the mitochondrial surface, Drp1 is thought to wrap around the mitochondria to induce fission using its GTPase activity [94].

In terms of PD, no mutations in typical mitochondria fission and fusion genes have yet been identified in PD patients. However, increasing evidence from both toxin models and genetic mutations in PD animal models supports the hypothesis that mitochondrial dynamic regulation and dysfunction are involved in PD. Evidence from toxin-induced PD models support a role for mitochondrial fission/fusion in the pathogenesis of PD. The Parkinsonian neurotoxins, 6-hydroxy dopamine (6-OHDA), rotenone, and MPP^+, all induce mitochondrial fragmentation, leading to dopaminergic cell death in neuronal cultures [91,95,96]. Inhibition of pro-fission Drp1 or overexpression of pro-fusion protein mitofusin-1 (Mfn1) using genetic techniques prevents both neurotoxin-induced mitochondrial fission and neuronal cell death [97–99]. Loss of Mfn2 or conditional knockout of neuronal Mfn2 in mice has recently been reported to result in age-dependent motor deficits, followed by the loss of dopaminergic terminals in the striatum [100,101], suggesting a role for Mfn2 in parkinsonism. Also, Drp1 is required for synaptic formation [102] and lack of Drp1 leads to an impairment of brain development in mice [103,104]. Recently it has been reported that the Drp1 is critical for targeting mitochondria to the terminal synapses of dopaminergic neurons and deletion of Drp1 gene in dopaminergic neurons rapidly eliminates DA terminals in the caudate-putamen and causes cell bodies in the midbrain to degenerate and lose α-synuclein [105]. Taken together, all these support that molecular machinery which maintains the balance of fusion and fission dynamics in the cells might contribute to the pathogenesis of PD.

In addition, several genes, whose mutated forms are associated with familial PD, affect mitochondrial dynamics: these include PINK1, Parkin, LRRK2 and DJ-1 [45,94,96,106–110]. Among these genes, the role of PINK1/Parkin pathway in regulation of mitochondrial dynamics seems to be opposite from LRRK2 and DJ-1; mutations in PINK1/Parkin lead to mitochondrial fusion [44,111] whereas mutations in LRRK2 or DJ-1 promote mitochondrial fission [44,112]. Fibroblasts from PD patients carrying PINK1 or Parkin mutations exhibited a more fragmented mitochondrial network, showing mitochondrial dysfunction [113,114]. The mitochondrial network could also be reduced by the depletion of Drp1, or overexpression of OPA1 or Mfn2 [115–118]. Deficiency in DJ-1 in cell lines, cultured neurons and lymphoblasts derived from DJ-1-deficient patients displayed aberrant mitochondrial morphology [119]. Further, in double-mutant mice in which alpha-synuclein mutant is overexpressed and parkin is ablated, severe genotype-, age- and region-dependent mitochondrial morphological alterations were found in neuronal somata. The number of structurally altered mitochondria was significantly increased in the SN of these double-mutants mice [77]. These studies further support the involvement of vulnerable genes in mitochondrial dynamics regulation in PD.

Wild-type LRRK2 interacts and colocalizes with several key regulators of mitochondrial fusion/fission, suggesting that it might have multiple regulatory roles [120]. Furthermore, mutant *LRRK2* G2019S, the most common mutation in the population of familial PD patients, has been recently demonstrated to interact with Drp1 and to promote mitochondrial fragmentation, leading to mitochondrial dysfunction and neuronal abnormalities [121,122]. This fragmentation can be reduced by expression of the dominant-negative Drp1K38A or overexpression of the fusion protein Mfn2 [121,122]. iPS cells derived from PD patients carrying the G2019S mutation show excessive mitochondrial fission, aberrant autophagy and neuronal damage in DA neurons differentiated *in vitro* [123]. More importantly, treatment with P110, a selective peptide inhibitor of Drp1 recently developed in Qi's group [124,125], reduced mitochondrial fragmentation and damage, and corrected excessive autophagy. In this study it was also shown that G2019S mutated LRRK2 protein primarily phosphorylates Drp1 at T595 resulting in aberrant mitochondrial fragmentation [123].

Together, these findings suggest that impairment of mitochondrial dynamics might contribute to the pathogenesis and progression of PD.

5. Mitophagy/Autophagy Impairment in Parkinson's Disease

Autophagy is a process of cellular degradation in which cargos are degraded by autophagosomes fused with lysosomes [126]. In addition to non-selective cargos, autophagosomes can degrade protein aggregates or damaged mitochondria. Mitochondria-associated autophagy (mitophagy) is regulated by autophagy receptors that preferentially bind ubiquitylated mitochondria and subsequently recruit the autophagosome protein light chain 3 (LC3) through their LC3-interaction region (LIR) motif [127]. LC3 in mammals, also known as Atg8 in yeast, plays crucial roles in both autophagosome membrane biogenesis and cargo recognition [128,129]. In yeast, Atg32 functions as a receptor on mitochondria to initiate mitophagy through its interaction with Atg8 [130,131]. In mammals, FUNDC1 [132], p62 [133], BNIP3 [134], and AMBRA1 [135], have been recognized as receptors for mitophagy, all of which bind to LC3 via the LIR motif [136]. The implication of these receptors in mitophagy regulation, however, seems to be dependent of experimental conditions. Whether and how these receptors cooperatively regulate mitophagy remains to be determined.

A number of lines of evidence suggest a critical role for defective autophagy/mitophagy in neurodegeneration in PD. Cultured cells exposed to parkinsonian neurotoxins such as MPP^+, rotenone, or 6-OHDA showed an increased number of autophagosomes and associated neuronal cell death [125,137,138]. In recent years, PINK1-depenent activation of parkin has been recognized as a major pathway of mitophagy [139]. In mammalian cells, parkin is recruited to depolarized mitochondria, which are subsequently eliminated by autophagy. Such parkin recruitment to mitochondria depends on PINK1 accumulation on mitochondria and therefore PINK1 is a key molecule in the signal transduction of mitophagy [140]. The failure of PINK1/parkin-mediated mitophagic process leads to accumulation of damaged mitochondria, which results in an increase in ROS and cell death [141,142]. Because these studies were all conducted in cells with overexpression of PINK1 and Parkin and because Parkin at endogenous levels fail to mediate mitophagy in PD patient cells [143], the matter of whether these proteins, at the endogenous levels, cooperatively regulate mitophagy remains to be validated. Knock-down of DJ-1, another PD-related gene, resulted in decreased mitochondrial membrane potential, increased reactive oxygen species, excessive mitochondrial fragmentation and impaired autophagy [74,119,144]. Interestingly, overexpression of PINK1 and parkin can rescue mitochondrial fragmentation and dysfunction induced by the depletion of DJ-1 [145,146]. Given that wild-type and mutant DJ-1 can interact with PINK1 [147] and that Parkin, as an E3 ligase on mitochondria, catalyzes the ubiquitination of DJ-1 [148], PINK1, Parkin and DJ-1 may be operative in the pathway of PINK1/parkin-mediated mitophagy.

Pathogenic LRRK2-mediated autophagy has been observed in a variety of cell cultures [149,150], in neurons derived from patient-induced pluripotent stem cells [151] and in animal models in which a mutant form of LRRK2 is expressed [152]. Either knock-down of LRRK2 by siRNA or treatment with a LRRK2 kinase inhibitor caused an increase in autophagic fluxin-cultured cells [153,154]. Overexpression of LRRK2, especially mutant forms, seems to suppress autophagy [155]. In contrast, studies from other two groups show that mutant forms of LRRK2 may induce autophagy via an ERK1/2-dependent pathway [149,156]. Although the detailed mechanisms by which LRRK2 and its mutants mediate autophagy are not clear, LRRK2 may disrupt the balance of autophagy through damaging lysosome-related calcium storage and cargo degradation [157,158]. Further, our group recently reported excessive mitophagy in a variety of cells expressing the LRRK2 G2019S mutant, which was accompanied by mitochondrial depolarization, recruitment of p62 to the mitochondria, increased LC3II levels and lysosomal activity as well as death of dopaminergic neurons [123,159]. We showed that the LRRK2 G2019S mutant caused excessive mitophagyby phosphorylating its substrate including fission protein Drp1 and mitochondrial outer membrane protein Bcl_2 [123,159]. Thus, the pathway that LRRK2 mediation of mitophagy might be different from those occurred during autophagy.

Taken together, a large body of studies indicates that the different PD-related genes contribute to the pathogenesis of PD at the intersection of mitochondrial dysfunction and autophagy. The loss of mitochondrial membrane potential, which is associated with mitochondrial dysfunction, seems to be a common signal for mitochondria to be degraded via mitophagy. Moreover, occurrence of mitochondrial fragmentation due to impairment of mitochondrial fusion and fission often precedes the induction of mitophagy. Thus, it is possible that mitophagy may be a multistep process starting with the degradation of profusion/fission proteins, resulting in an imbalance of mitochondrial dynamics and the subsequent clearance of mitochondria [86]. However, the factors that regulate these processes involved in mitochondrial "quality control", including mitochondrial dynamics, mitochondria-associated degradation and mitophagy in PD, remain to be determined.

6. Mitochondrial Redox Signaling in Parkinson's Disease

Decreased Complex I activity in the SN of PD patients and animal models has been repeatedly observed. The defect in complex I results in impairment of electron transport and causes ROS accumulation in mitochondria which lead to neuronal degeneration. Neurotoxins causing parkinsonism, MPP^+ are selectively taken up into dopaminergic neurons in which it inhibits Complex I activity [160]. Rotenone also inhibits Complex I by impairing oxidative phosphorylation [161]. These studies demonstrate a contribution from ROS to the pathogenesis of dopamine neuronal loss in PD. The genetic PD-linked proteins play a significant role in this process. Alpha-synuclein mutant A53T can enter mitochondria where it binds to the Complex I subunit to inhibit Complex I activity, producing ROS [162,163]. Functional studies showed that alpha-synuclein associated with mitochondria induces cytochrome c release, increased calcium and ROS levels resulting in dopaminergic neuronal death [164]. In addition, expression of mutants of PD-related genes Parkin, PINK1, DJ-1 and LRRK2 in cultured cells all increased ROS. This evidence has been well-summarized in other reviews [29,165]. However, it is uncertain if such elevated ROS are directly caused by these mutants or through indirect cellular effects.

Besides ROS, reactive nitrogen species (RNS) mediating nitrosative stress is also implicated in SN neuronal loss in PD [166]. RNS are generated by the reaction of superoxide with nitric oxide (NO), which results in the production of peroxynitrite. NO inhibits several enzymes including complexes I and IV of the mitochondrial electron transport chain, which in turn lead to ROS generation [167,168]. Increased expression of iNOS and nNOS were observed in basal ganglia of postmortem brain of PD patients [169]. In the mouse MPTP model, there was a significant upregulation of iNOS associated with the gliosis in the SN [170]. Inhibition of nNOS has been reported to protect against neurotoxicity in MPTP-induced PD animal model [171,172]. These observations suggest that NO and its metabolite peroxynitriteare implicated in the pathogenesis of PD.

7. Mitochondrial Protein Import in Parkinson's Disease

Mitochondria possess their own DNA and translational machinery; however, there are only a small number of mitochondrial proteins encoded by mtDNA that are synthesized within the organelle. The majority of mitochondrial proteins are nuclear-encoded and have to be imported into the organelle. The translocase of the outer mitochondrial membrane (TOM complex) plays central roles in controlling protein entry to mitochondria [173]. The TOM complex is the main entry portal for most mitochondrial proteins that are synthesized in cytoplasm. The TOM complex contains seven subunits including TOM40, TOM22, two proprotein receptors, TOM20, and TOM70, and three smaller proteins, TOM5, TOM6 and TOM7. In general, imported proteins bind to one of these receptors. With the assistance of TOM22 and TOM5, the imported proteins pass the TOM 40 channel [174]. TOM complexes also need to cooperate with TIM (Translocase of the Inner Membrane) 23 complexes to import matrix-targeted proteins [175]. Thus, the TOM and the TIM23 complex direct the translocation of oxidative phosphorylation and metabolite transporter proteins to the inner membrane.

A number of studies have reported that PD-associated genes, alpha-synuclein, PINK1 and Parkin can interact with the TOM complex, disrupting the mitochondrial protein import. Alpha-synuclein can enter mitochondria where it mainly localizes at the inner membrane. The import of alpha-synuclein to mitochondria is through the TOM complex [176,177]. In postmortem brain samples of PD and in brain tissues of mice overexpressing alpha-synuclein, TOM 40 levels changed together with the levels of alpha-synuclein [177]. Expression with either wild-type alpha-synuclein or mutant A53T in cultured cells resulted in the loss of TOM40, whereas over-expression of TOM40 prevented the cellular damage caused by expression of either alpha-synuclein wild-type or its mutant A53T [177]. Thus, alpha-synuclein mutants may impair mitochondrial function by suppression of TOM40-dependent mitochondrial protein import pathways. PINK1 is also imported into the mitochondria via the TOM complex channel [178,179] and then is degraded in a membrane potential-dependent manner. Under mitochondrial depolarization PINK1, interacting with the TOM complex, recruits and activates Parkin, leading to degradation of mitochondrial outer membrane proteins and ultimately mitophagy [178]. Thus, pathogenic mutations in PINK1 and parkin may disrupt this pathway, resulting in the accumulation of dysfunctional mitochondria.

8. Potential Therapeutics Targeting Mitochondria for Treatment of PD

Given the importance of mitochondrial dysfunction in the pathogenesis of PD, therapeutics targeting mitochondria have been studied to prevent or treat PD. Although the cellular and pathological phenotypes from neurotoxin-induced and genetic mutant-associated PD models are different, the outcomes of mitochondrial dysfunction, including mitochondrial dynamic impairments, increased ROS, and impaired bioenergetics, seem to be common pathways. Thus, the mechanistic information on mitochondrial dysfunction in PD models provides potential targets for the development of therapeutic approaches for treatment of PD.

Here, we summarize reported therapeutic agents that reduce PD pathology in models of PD (see Table 2). We categorize these agents into (1) modulating PD-related genetic mutants; (2) modulating mitochondrial proteins; and (3) modulating the consequences of mitochondrial dysfunction. In Table 2, we list the neuroprotective effects of these agents especially in either animal models of PD or neuron derived patient-iPS cells.

Modulating PD-related genetic mutants: LRRK2 mutants are the most common genetic mutants in both familial and sporadic PD. Modulation of the LRRK2 kinase domain has been attractive for the development of therapeutics of PD. LRRK2 inhibitors, PF-06447475 and GW5074, have been shown to increase dopaminergic neuronal survival in primary neuronal cultures and neurons- derived from patient iPS cells [180,181]. Increasing glyoxalase activity of DJ-1 by supplying D-lactate and glycolate rescues the requirement for DJ-1 in maintenance of mitochondrial potential, increases cytocatalytic rate of DJ-1, and reduces neuronal death caused by paraquat and down-regulation of PINK1 [182]. However, thus far, these reagents are only tested in cultured cells so far. Whether treatment with these pharmacological agents protect neurons from PD in animal models remains to be determined.

Table 2. Therapeutic agents that target mitochondria for treatment of Parkinson's disease

Category	Agent	Molecular Action	PD Mode	Therapeutic Effects
Modulating PD-related genes	PF-06447475	LRRK2 kinase inhibitor	Transgenic rat with LRRK2 G2019S	Reduce behavioral and neuropathological phenotypes Reduce inflammation [183]
	GW5074	LRRK2 kinase inhibitor	DA neurons derived from PD patient iPS cells	Suppression of ROS Improve mitochondrial respiration Increase DA neuronal survival [180]
	FX2149	LRRK2 GTPase inhibitor	mouse inflammation model	Reduce neuroinflammation Inhibit microglial activity [184]
Modulating mitochondrial protein	Alda-1	ALDH2 activator	Rotenone- and MPTP-induced animal models	Improve mitochondrial membrane potential Inhibit mitochondrial ROS Reduce dopaminergic cell death [185]
	TRO40303	inhibitor of mitochondrial transition pore	Mice expressing alha-synuclein	Upregulate mitochondrial proteins Increase TH expression [186]
	P110	Peptide inhibitor of Drp1	DA neurons from LRRK2 G2019S PD patient iPS cells	Improve mitochondrial membrane potential Inhibit mitochondrial ROS Increase mitochondrial integrity Reduce autophagy Improve DA neuronal morphology and survival [123]
	Mdivi-1	Inhibitor of mitochondrial fragmentation	MPTP-induced mouse PD model $PINK1^{-/-}$ mouse model	Improve mitochondrial morphology Improve mouse behavioral outcome Reduce DA neuronal loss in SN Restore dopamine level [187]

Table 2. *Cont.*

Category	Agent	Molecular Action	PD Mode	Therapeutic Effects
Modulating mitochondrial dysfunction	Q1	8-OH-quinoline-based iron chelator	MPTP-induced mouse PD model	Reduce DA neuronal degeneration in SN Decrease mitochondrial iron pool [188]
	Rapamycin	mTOR inhibitor	6-OHDA-induced rat PD model	Inhibit oxidative stress Inhibit mitochondrial apoptosis [189]
	Edaravone	ROS scavenger	Rotenone-induced rat PD model	Inhibit mitochondrial apoptosis Reduce ROS [190]
	Melatonin	Antioxidant	Rotenone-induced rat PD model 6-OHDA-induced rat PD model	Suppress calcium level Inhibit mitochondrial ROS Enhance complex I activity [191,192]
	Quercetin	Bioflavonoid	Rotenone-induced rat PD model	Inhibit mitochondrial ROS generation Inhibit p53 level Inhibit nuclear translocation of NF-kappaB Inhibit mitochondrial apoptosis [193]
	CNB-001	Curcumin derivative	MPTP-induced mouse PD model	Improve mitochondrial morphology Inhibit mitochondrial apoptotic pathway Improve mitochondrial membrane potential [194]
	Alpha-Lipoic acid	Antioxidant	Rotenone-induced rat PD model	Increase mitochondrial complex I activity Inhibit ROS generation Increase mitochondrial biogenesis Increase glutathione [195]
	Lycopene	Chemical carotene	Rotenone-induced rat PD model	Inhibit mitochondrial apoptotic pathway Increase SOD activity Increase glutathione Inhibit lipid peroxidation [196]

Modulating mitochondrial proteins: Aldehyde dehydrogenase 2 (ALDH2), located in mitochondrial matrix, functions as a cellular protector against oxidative stress by detoxification of cytotoxic aldehydes. Alda-1 is a small molecule that enhances ALDH enzyme activity and protects against oxidative toxicity [197]. Treatment with Alda-1 can reduce rotenone-induced apoptosis in both SH-SY5Y cells and primary dopaminergic neurons. Moreover, intraperitoneal administration of Alda-1 can improve mitochondrial membrane potential, inhibit mitochondrial ROS and reduce death of tyrosine hydroxylase (TH)-positive dopaminergic neurons in rotenone- or MPTP-induced PD animal models [185]. Cholesterol oximes such as olesoxime and TRO40303 are small molecules that interact with the mitochondrial outer membrane protein VDAC and limit opening of the mitochondrial transition pore in response to oxidative stress [198]. Olesoxime can protect differentiated SHSY-5Y cells from cell death, and reduce neurite retraction and cytoplasmic shrinkage induced by alpha-synuclein overexpression [199]. Low dose TRO40303 upregulates a number of mitochondrial proteins including Drp1 and VDAC and enhances expression of tyrosine hydroxylase in mice overexpressing alpha-synuclein [186]. As mentioned above, our group has developed a peptide inhibitor P110 that selectively blocks the protein-protein interactions between Drp1 and its mitochondrial adaptor Fis1 [125]. Treatment with P110 significantly reduced mitochondrial fragmentation, decreased mitochondrial ROS and improved mitochondrial integrity in dopaminergic neurons exposed to MPP^+ in neurons expressing the LRRK2 G2019S mutation, and in dopaminergic neurons derived from LRRK2 G2019S patient-iPS cells [123,125]. Importantly, we showed that treatment with P110 had minor effects on mitochondrial dynamics and neuronal survival under physiological conditions [123,125]. In addition, Mdivi-1, an inhibitor of mitochondrial fragmentation, was reported to reduce behavioral and neuropathological phenotypes in an MPTP-induced PD mouse model, in addition to its protective effects in dopaminergic neurons exposed to neurotoxins [187].

Modulating the consequences of mitochondrial dysfunction: Modulation of downstream mitochondrial dysfunction may also provide therapeutic opportunities, in both sporadic and familial PD. In the past decades, a number of natural products and small molecules have been reported to protect against neuropathology associated with PD, at least in part through protecting mitochondria. Treatment with these agents has been shown to reduce mitochondrial ROS, inhibit mitochondrial apoptotic pathways, and increase mitochondrial complex I activity. As a consequence, treatment with these agents can reduce neuronal degeneration in PD in culture and in animals. These agents have been extensively reviewed [200–203]. Here, we only list those that have significant protection in animal models of PD (Table 2).

9. Concluding Remarks

Accumulating evidence supports the hypothesis that mitochondrial abnormalities and dysfunction could critically influence neuronal degeneration in both sporadic and faimilial PD. Dysregulation of mitochondrial dynamics and mitophagy have been centrally implicated in the neuropathology of PD. This evidence supports the idea that mitochondrial damage might be a primary cause initiating the progression of PD. Thus, targeting mitochondria may offer the opportunities for drug development to treat neurodegenerative diseases such as parkinsonism.

References

1. Duvoisin, R.C. Overview of Parkinson's disease. *Ann. N. Y. Acad. Sci.* **1992**, *648*, 187–193.
2. Dauer, W.; Przedborski, S. Parkinson's disease: Mechanisms and models. *Neuron* **2003**, *39*, 889–909.
3. Dawson, T.M.; Dawson, V.L. Molecular pathways of neurodegeneration in Parkinson's disease. *Science* **2003**, *302*, 819–822.
4. Yao, Z.; Wood, N.W. Cell death pathways in Parkinson's disease: Role of mitochondria. *Antioxid. Redox Signal.* **2009**, *11*, 2135–2149.
5. Ellis, C.E.; Murphy, E.J.; Mitchell, D.C.; Golovko, M.Y.; Scaglia, F.; Barcelo-Coblijn, G.C.; Nussbaum, R.L. Mitochondrial lipid abnormality and electron transport chain impairment in mice lacking alpha-synuclein. *Mol. Cell. Biol.* **2005**, *25*, 10190–10201.
6. Herrmann, J.M.; Longen, S.; Weckbecker, D.; Depuydt, M. Biogenesis of mitochondrial proteins. *Adv. Exp. Med. Biol.* **2012**, *748*, 41–64.
7. Sas, K.; Robotka, H.; Toldi, J.; Vecsei, L. Mitochondria, metabolic disturbances, oxidative stress and the kynurenine system, with focus on neurodegenerative disorders. *J. Neurol. Sci.* **2007**, *257*, 221–239.
8. Daiber, A. Redox signaling (cross-talk) from and to mitochondria involves mitochondrial pores and reactive oxygen species. *Biochim. Biophys. Acta* **2010**, *1797*, 897–906.
9. Bononi, A.; Missiroli, S.; Poletti, F.; Suski, J.M.; Agnoletto, C.; Bonora, M.; de Marchi, E.; Giorgi, C.; Marchi, S.; Patergnani, S.; *et al.* Mitochondria-associated membranes (MAMs) as hotspot Ca^{2+} signaling units. *Adv. Exp. Med. Biol.* **2012**, *740*, 411–437.
10. Rizzuto, R.; de Stefani, D.; Raffaello, A.; Mammucari, C. Mitochondria as sensors and regulators of calcium signalling. *Nat. Rev. Mol. Cell Biol.* **2012**, *13*, 566–578.
11. Cali, T.; Ottolini, D.; Brini, M. Mitochondrial Ca^{2+} as a key regulator of mitochondrial activities. *Adv. Exp. Med. Biol.* **2012**, *942*, 53–73.
12. Osteryoung, K.W.; Nunnari, J. The division of endosymbiotic organelles. *Science* **2003**, *302*, 1698–704.
13. Antico Arciuch, V.G.; Elguero, M.E.; Poderoso, J.J.; Carreras, M.C. Mitochondrial regulation of cell cycle and proliferation. *Antioxid. Redox Signal.* **2012**, *16*, 1150–1180.
14. Moyes, C.D.; Hood, D.A. Origins and consequences of mitochondrial variation in vertebrate muscle. *Annu. Rev. Physiol.* **2003**, *65*, 177–201.
15. Schatten, H.; Prather, R.S.; Sun, Q.Y. The significance of mitochondria for embryo development in cloned farm animals. *Mitochondrion* **2005**, *5*, 303–321.
16. Giorgi, C.; Baldassari, F.; Bononi, A.; Bonora, M.; de Marchi, E.; Marchi, S.; Missiroli, S.; Patergnani, S.; Rimessi, A.; Suski, J.M.; *et al.* Mitochondrial Ca^{2+} and apoptosis. *Cell Calcium* **2012**, *52*, 36–43.

17. Rasola, A.; Bernardi, P. Mitochondrial permeability transition in Ca^{2+}-dependent apoptosis and necrosis. *Cell Calcium* **2011**, *50*, 222–233.
18. Wang, C.; Youle, R.J. The role of mitochondria in apoptosis*. *Annu. Rev. Genet.* **2009**, *43*, 95–118.
19. Burbach, J.P.; Smits, S.; Smidt, M.P. Transcription factors in the development of midbrain dopamine neurons. *Ann. N. Y. Acad. Sci.* **2003**, *991*, 61–68.
20. Wallen, A.; Perlmann, T. Transcriptional control of dopamine neuron development. *Ann. N. Y. Acad. Sci.* **2003**, *991*, 48–60.
21. Dreyer, S.D.; Zhou, G.; Baldini, A.; Winterpacht, A.; Zabel, B.; Cole, W.; Johnson, R.L.; Lee, B. Mutations in LMX1B cause abnormal skeletal patterning and renal dysplasia in nail patella syndrome. *Nat. Genet.* **1998**, *19*, 47–50.
22. Smidt, M.P.; van Schaick, H.S.; Lanctot, C.; Tremblay, J.J.; Cox, J.J.; van der Kleij, A.A.; Wolterink, G.; Drouin, J.; Burbach, J.P. A homeodomain gene Ptx3 has highly restricted brain expression in mesencephalic dopaminergic neurons. *Proc. Natl Acad. Sci. USA* **1997**, *94*, 13305–13310.
23. Smidt, M.P.; Asbreuk, C.H.; Cox, J.J.; Chen, H.; Johnson, R.L.; Burbach, J.P. A second independent pathway for development of mesencephalic dopaminergic neurons requires Lmx1b. *Nat. Neurosci.* **2000**, *3*, 337–341.
24. Mann, V.M.; Cooper, J.M.; Daniel, S.E.; Srai, K.; Jenner, P.; Marsden, C.D.; Schapira, A.H. Complex I, iron, and ferritin in Parkinson's disease substantia nigra. *Ann. Neurol.* **1994**, *36*, 876–881.
25. Blandini, F.; Nappi, G.; Greenamyre, J.T. Quantitative study of mitochondrial complex I in platelets of parkinsonian patients. *Mov. Disord.* **1998**, *13*, 11–15.
26. Haas, R.H.; Nasirian, F.; Nakano, K.; Ward, D.; Pay, M.; Hill, R.; Shults, C.W. Low platelet mitochondrial complex I and complex II/III activity in early untreated Parkinson's disease. *Ann. Neurol.* **1995**, *37*, 714–722.
27. Barroso, N.; Campos, Y.; Huertas, R.; Esteban, J.; Molina, J.A.; Alonso, A.; Gutierrez-Rivas, E.; Arenas, J. Respiratory chain enzyme activities in lymphocytes from untreated patients with Parkinson disease. *Clin. Chem.* **1993**, *39*, 667–69.
28. Yoshino, H.; Nakagawa-Hattori, Y.; Kondo, T.; Mizuno, Y. Mitochondrial complex I and II activities of lymphocytes and platelets in Parkinson's disease. *J. Neural Transm Parkinsons Dis. Dement. Sect.* **1992**, *4*, 27–34.
29. Dias, V.; Junn, E.; Mouradian, M.M. The role of oxidative stress in Parkinson's disease. *J. Parkinsons Dis.* **2013**, *3*, 461–491.
30. Hwang, O. Role of oxidative stress in Parkinson's disease. *Exp. Neurobiol.* **2013**, *22*, 11–17.
31. Bender, A.; Krishnan, K.J.; Morris, C.M.; Taylor, G.A.; Reeve, A.K.; Perry, R.H.; Jaros, E.; Hersheson, J.S.; Betts, J.; Klopstock, T.; *et al.* High levels of mitochondrial DNA deletions in substantia nigra neurons in aging and Parkinson disease. *Nat. Genet.* **2006**, *38*, 515–517.
32. Kraytsberg, Y.; Kudryavtseva, E.; McKee, A.C.; Geula, C.; Kowall, N.W.; Khrapko, K. Mitochondrial DNA deletions are abundant and cause functional impairment in aged human substantia nigra neurons. *Nat. Genet.* **2006**, *38*, 518–520.
33. Langston, J.W.; Ballard, P.; Tetrud, J.W.; Irwin, I. Chronic Parkinsonism in humans due to a product of meperidine-analog synthesis. *Science* **1983**, *219*, 979–980.

34. Betarbet, R.; Sherer, T.B.; MacKenzie, G.; Garcia-Osuna, M.; Panov, A.V.; Greenamyre, J.T. Chronic systemic pesticide exposure reproduces features of Parkinson's disease. *Nat. Neurosci.* **2000**, *3*, 1301–1306.
35. Testa, C.M.; Sherer, T.B.; Greenamyre, J.T. Rotenone induces oxidative stress and dopaminergic neuron damage in organotypic substantia nigra cultures. *Brain Res. Mol. Brain Res.* **2005**, *134*, 109–118.
36. Smeyne, R.J.; Jackson-Lewis, V. The MPTP model of Parkinson's disease. *Brain Res. Mol. Brain Res.* **2005**, *134*, 57–66.
37. Langston, J.W.; Ballard, P.A., Jr. Parkinson's disease in a chemist working with 1-methyl-4-phenyl-1,2,5,6-tetrahydropyridine. *N. Engl. J. Med.* **1983**, *309*, 310.
38. Kopin, I.J. Toxins and Parkinson's disease: MPTP parkinsonism in humans and animals. *Adv. Neurol.* **1987**, *45*, 137–144.
39. Barbeau, A. Parkinson's disease: Clinical features and etiopahthology. In *Amsterdam Handbook of Clinical Neurology*; Viken, P.J., Bruyn, G.W., Klawans, H.L., Eds.; Elsevier Science Publishers: Amsterdam, The Netherlands, 1986; pp. 87–108.
40. Jenner, P.; Olanow, C.W. Understanding cell death in Parkinson's disease. *Ann. Neurol.* **1998**, *44* (Suppl. 1), S72–S84.
41. Nunes, I.; Tovmasian, L.T.; Silva, R.M.; Burke, R.E.; Goff, S.P. Pitx3 is required for development of substantia nigra dopaminergic neurons. *Proc. Natl. Acad. Sci. USA* **2003**, *100*, 4245–4250.
42. Van den Munckhof, P.; Luk, K.C.; Ste-Marie, L.; Montgomery, J.; Blanchet, P.J.; Sadikot, A.F.; Drouin, J. Pitx3 is required for motor activity and for survival of a subset of midbrain dopaminergic neurons. *Development* **2003**, *130*, 2535–2542.
43. Klein, C.; Westenberger, A. Genetics of Parkinson's disease. *Cold Spring Harb. Perspect. Med.* **2012**, *2*, a008888.
44. Zetterstrom, R.H.; Solomin, L.; Jansson, L.; Hoffer, B.J.; Olson, L.; Perlmann, T. Dopamine neuron agenesis in Nurr1-deficient mice. *Science* **1997**, *276*, 248–250.
45. Wallen, A.; Zetterstrom, R.H.; Solomin, L.; Arvidsson, M.; Olson, L.; Perlmann, T. Fate of mesencephalic AHD2-expressing dopamine progenitor cells in NURR1 mutant mice. *Exp. Cell Res.* **1999**, *253*, 737–46.
46. Smidt, M.P.; Smits, S.M.; Burbach, J.P. Molecular mechanisms underlying midbrain dopamine neuron development and function. *Eur. J. Pharmacol.* **2003**, *480*, 75–88.
47. Franco-Iborra, S.; Vila, M.; Perier, C. The Parkinson Disease Mitochondrial Hypothesis: Where Are We at? *Neuroscientist* **2015**, doi:10.1177/1073858415574600.
48. Luoma, P.; Melberg, A.; Rinne, J.O.; Kaukonen, J.A.; Nupponen, N.N.; Chalmers, R.M.; Oldfors, A.; Rautakorpi, I.; Peltonen, L.; Majamaa, K.; *et al.* Parkinsonism, premature menopause, and mitochondrial DNA polymerase gamma mutations: Clinical and molecular genetic study. *Lancet* **2004**, *364*, 875–882.
49. Henchcliffe, C.; Beal, M.F. Mitochondrial biology and oxidative stress in Parkinson disease pathogenesis. *Nat. Clin. Pract. Neurol.* **2008**, *4*, 600–609.

50. Zheng, B.; Liao, Z.; Locascio, J.J.; Lesniak, K.A.; Roderick, S.S.; Watt, M.L.; Eklund, A.C.; Zhang-James, Y.; Kim, P.D.; Hauser, M.A.; *et al.* PGC-1α, a potential therapeutic target for early intervention in Parkinson's disease. *Sci. Transl Med.* **2010**, *2*, 52ra73.
51. Ekstrand, M.I.; Terzioglu, M.; Galter, D.; Zhu, S.; Hofstetter, C.; Lindqvist, E.; Thams, S.; Bergstrand, A.; Hansson, F.S.; Trifunovic, A.; *et al.* Progressive parkinsonism in mice with respiratory-chain-deficient dopamine neurons. *Proc. Natl. Acad. Sci. USA* **2007**, *104*, 1325–1330.
52. Ekstrand, M.I.; Galter, D. The MitoPark Mouse—An animal model of Parkinson's disease with impaired respiratory chain function in dopamine neurons. *Parkinsonism Relat. Disord.* **2009**, *15* (Suppl. 3), S185–S188.
53. Gellhaar, S.; Marcellino, D.; Abrams, M.B.; Galter, D. Chronic L-DOPA induces hyperactivity, normalization of gait and dyskinetic behavior in MitoPark mice. *Genes Brain Behav.* **2015**, *14*, 260–270.
54. Ikebe, S.; Tanaka, M.; Ohno, K.; Sato, W.; Hattori, K.; Kondo, T.; Mizuno, Y.; Ozawa, T. Increase of deleted mitochondrial DNA in the striatum in Parkinson's disease and senescence. *Biochem. Biophys. Res. Commun.* **1990**, *170*, 1044–1048.
55. Lee, V.M.; Trojanowski, J.Q. Mechanisms of Parkinson's disease linked to pathological alpha-synuclein: New targets for drug discovery. *Neuron* **2006**, *52*, 33–38.
56. Peelaerts, W.; Bousset, L.; van der Perren, A.; Moskalyuk, A.; Pulizzi, R.; Giugliano, M.; van den Haute, C.; Melki, R.; Baekelandt, V. Synuclein strains cause distinct synucleinopathies after local and systemic administration. *Nature* **2015**, *522*, 340–344.
57. Selkoe, D.; Dettmer, U.; Luth, E.; Kim, N.; Newman, A.; Bartels, T. Defining the native state of alpha-synuclein. *Neurodegener. Dis.* **2014**, *13*, 114–117.
58. Riess, O.; Jakes, R.; Kruger, R. Genetic dissection of familial Parkinson's disease. *Mol. Med. Today* **1998**, *4*, 438–444.
59. De Silva, H.R.; Khan, N.L.; Wood, N.W. The genetics of Parkinson's disease. *Curr. Opin. Genet. Dev.* **2000**, *10*, 292–298.
60. Tan, J.M.; Dawson, T.M. Parkin blushed by PINK1. *Neuron* **2006**, *50*, 527–529.
61. Kubo, S.; Hattori, N.; Mizuno, Y. Recessive Parkinson's disease. *Mov. Disord.* **2006**, *21*, 885–893.
62. Giasson, B.I.; Duda, J.E.; Quinn, S.M.; Zhang, B.; Trojanowski, J.Q.; Lee, V.M. Neuronal alpha-synucleinopathy with severe movement disorder in mice expressing A53T human alpha-synuclein. *Neuron* **2002**, *34*, 521–533.
63. Lee, M.K.; Stirling, W.; Xu, Y.; Xu, X.; Qui, D.; Mandir, A.S.; Dawson, T.M.; Copeland, N.G.; Jenkins, N.A.; Price, D.L. Human alpha-synuclein-harboring familial Parkinson's disease-linked Ala-53 —> Thr mutation causes neurodegenerative disease with alpha-synuclein aggregation in transgenic mice. *Proc. Natl. Acad. Sci. USA* **2002**, *99*, 8968–8973.
64. Gispert, S.; Del Turco, D.; Garrett, L.; Chen, A.; Bernard, D.J.; Hamm-Clement, J.; Korf, H.W.; Deller, T.; Braak, H.; Auburger, G.; *et al.* Transgenic mice expressing mutant A53T human alpha-synuclein show neuronal dysfunction in the absence of aggregate formation. *Mol. Cell. Neurosci.* **2003**, *24*, 419–429.

65. Neumann, M.; Kahle, P.J.; Giasson, B.I.; Ozmen, L.; Borroni, E.; Spooren, W.; Muller, V.; Odoy, S.; Fujiwara, H.; Hasegawa, M.; *et al.* Misfolded proteinase K-resistant hyperphosphorylated alpha-synuclein in aged transgenic mice with locomotor deterioration and in human alpha-synucleinopathies. *J. Clin. Investig.* **2002**, *110*, 1429–11439.
66. Rockenstein, E.; Mallory, M.; Hashimoto, M.; Song, D.; Shults, C.W.; Lang, I.; Masliah, E. Differential neuropathological alterations in transgenic mice expressing alpha-synuclein from the platelet-derived growth factor and Thy-1 promoters. *J. Neurosci. Res.* **2002**, *68*, 568–578.
67. Fleming, S.M.; Salcedo, J.; Fernagut, P.O.; Rockenstein, E.; Masliah, E.; Levine, M.S.; Chesselet, M.F. Early and progressive sensorimotor anomalies in mice overexpressing wild-type human alpha-synuclein. *J. Neurosci.* **2004**, *24*, 9434–9440.
68. Chen, L.; Xie, Z.; Turkson, S.; Zhuang, X. A53T human alpha-synuclein overexpression in transgenic mice induces pervasive mitochondria macroautophagy defects preceding dopamine neuron degeneration. *J. Neurosci.* **2015**, *35*, 890–905.
69. Li, Y.; Liu, W.; Oo, T.F.; Wang, L.; Tang, Y.; Jackson-Lewis, V.; Zhou, C.; Geghman, K.; Bogdanov, M.; Przedborski, S.; *et al.* Mutant LRRK2(R1441G) BAC transgenic mice recapitulate cardinal features of Parkinson's disease. *Nat. Neurosci.* **2009**, *12*, 826–828.
70. Dranka, B.P.; Gifford, A.; McAllister, D.; Zielonka, J.; Joseph, J.; O'Hara, C.L.; Stucky, C.L.; Kanthasamy, A.G.; Kalyanaraman, B. A novel mitochondrially-targeted apocynin derivative prevents hyposmia and loss of motor function in the leucine-rich repeat kinase 2 (LRRK2(R1441G)) transgenic mouse model of Parkinson's disease. *Neurosci. Lett.* **2014**, *583*, 159–164.
71. Yue, M.; Hinkle, K.M.; Davies, P.; Trushina, E.; Fiesel, F.C.; Christenson, T.A.; Schroeder, A.S.; Zhang, L.; Bowles, E.; Behrouz, B.; *et al.* Progressive dopaminergic alterations and mitochondrial abnormalities in LRRK2 G2019S knock-in mice. *Neurobiol. Dis.* **2015**, *78*, 172–195.
72. Goldberg, M.S.; Fleming, S.M.; Palacino, J.J.; Cepeda, C.; Lam, H.A.; Bhatnagar, A.; Meloni, E.G.; Wu, N.; Ackerson, L.C.; Klapstein, G.J.; *et al.* Parkin-deficient mice exhibit nigrostriatal deficits but not loss of dopaminergic neurons. *J. Biol. Chem.* **2003**, *278*, 43628–43635.
73. Perez, F.A.; Palmiter, R.D. Parkin-deficient mice are not a robust model of parkinsonism. *Proc. Natl. Acad. Sci. USA* **2005**, *102*, 2174–2179.
74. Lopert, P.; Patel, M. Brain mitochondria from DJ-1 knockout mice show increased respiration-dependent hydrogen peroxide consumption. *Redox Biol.* **2014**, *2*, 667–672.
75. Akundi, R.S.; Huang, Z.; Eason, J.; Pandya, J.D.; Zhi, L.; Cass, W.A.; Sullivan, P.G.; Bueler, H. Increased mitochondrial calcium sensitivity and abnormal expression of innate immunity genes precede dopaminergic defects in Pink1-deficient mice. *PLoS ONE* **2011**, *6*, e16038.
76. Gispert, S.; Ricciardi, F.; Kurz, A.; Azizov, M.; Hoepken, H.H.; Becker, D.; Voos, W.; Leuner, K.; Muller, W.E.; Kudin, A.P.; *et al.* Parkinson phenotype in aged PINK1-deficient mice is accompanied by progressive mitochondrial dysfunction in absence of neurodegeneration. *PLoS ONE* **2009**, *4*, e5777.
77. Stichel, C.C.; Zhu, X.R.; Bader, V.; Linnartz, B.; Schmidt, S.; Lubbert, H. Mono- and double-mutant mouse models of Parkinson's disease display severe mitochondrial damage. *Hum. Mol. Genet.* **2007**, *16*, 2377–2393.

78. Bonifati, V.; Oostra, B.A.; Heutink, P. Linking DJ-1 to neurodegeneration offers novel insights for understanding the pathogenesis of Parkinson's disease. *J. Mol. Med. (Berl.)* **2004**, *82*, 163–174.
79. Abou-Sleiman, P.M.; Healy, D.G.; Wood, N.W. Causes of Parkinson's disease: Genetics of DJ-1. *Cell Tissue Res.* **2004**, *318*, 185–188.
80. Zimprich, A.; Biskup, S.; Leitner, P.; Lichtner, P.; Farrer, M.; Lincoln, S.; Kachergus, J.; Hulihan, M.; Uitti, R.J.; Calne, D.B.; *et al.* Mutations in LRRK2 cause autosomal-dominant parkinsonism with pleomorphic pathology. *Neuron* **2004**, *44*, 601–607.
81. Paisan-Ruiz, C.; Jain, S.; Evans, E.W.; Gilks, W.P.; Simon, J.; van der Brug, M.; de Munain, A.L.; Aparicio, S.; Gil, A.M.; Khan, N.; *et al.* Cloning of the gene containing mutations that cause PARK8-linked Parkinson's disease. *Neuron* **2004**, *44*, 595–600.
82. West, A.B.; Moore, D.J.; Choi, C.; Andrabi, S.A.; Li, X.; Dikeman, D.; Biskup, S.; Zhang, Z.; Lim, K.L.; Dawson, V.L.; *et al.* Parkinson's disease-associated mutations in LRRK2 link enhanced GTP-binding and kinase activities to neuronal toxicity. *Hum. Mol. Genet.* **2007**, *16*, 223–232.
83. Cookson, M.R. The role of leucine-rich repeat kinase 2 (LRRK2) in Parkinson's disease. *Nat. Rev. Neurosci.* **2010**, *11*, 791–797.
84. West, A.B.; Moore, D.J.; Biskup, S.; Bugayenko, A.; Smith, W.W.; Ross, C.A.; Dawson, V.L.; Dawson, T.M. Parkinson's disease-associated mutations in leucine-rich repeat kinase 2 augment kinase activity. *Proc. Natl. Acad. Sci. USA* **2005**, *102*, 16842–16847.
85. Simon, H.H.; Saueressig, H.; Wurst, W.; Goulding, M.D.; O'Leary, D.D. Fate of midbrain dopaminergic neurons controlled by the engrailed genes. *J. Neurosci.* **2001**, *21*, 3126–3134.
86. Yu-chin Su, X.Q.; Impairment of mitochondrial dynamics: A target for the treatment of neurological disorders? *Future Med.* **2013**, *8*, 333–346.
87. Tanner, C.M.; Goldman, S.M. Epidemiology of Parkinson's disease. *Neurol. Clin.* **1996**, *14*, 317–35.
88. Caradoc-Davies, T.H.; Weatherall, M.; Dixon, G.S.; Caradoc-Davies, G.; Hantz, P. Is the prevalence of Parkinson's disease in New Zealand really changing? *Acta Neurol. Scand.* **1992**, *86*, 40–44.
89. Morgante, L.; Rocca, W.A.; di Rosa, A.E.; de Domenico, P.; Grigoletto, F.; Meneghini, F.; Reggio, A.; Savettieri, G.; Castiglione, M.G.; Patti, F.; *et al.* Prevalence of Parkinson's disease and other types of parkinsonism: A door-to-door survey in three Sicilian municipalities. The Sicilian Neuro-Epidemiologic Study (SNES) Group. *Neurology* **1992**, *42*, 1901–1907.
90. Mutch, W.J.; Dingwall-Fordyce, I.; Downie, A.W.; Paterson, J.G.; Roy, S.K. Parkinson's disease in a Scottish city. *Br. Med. J. (Clin. Res. Ed.)* **1986**, *292*, 534–536.
91. Polymeropoulos, M.H.; Higgins, J.J.; Golbe, L.I.; Johnson, W.G.; Ide, S.E.; di Iorio, G.; Sanges, G.; Stenroos, E.S.; Pho, L.T.; Schaffer, A.A.; *et al.* Mapping of a gene for Parkinson's disease to chromosome 4q21-q23. *Science* **1996**, *274*, 1197–1199.
92. Dekker, M.C.; Bonifati, V.; van Duijn, C.M. Parkinson's disease: Piecing together a genetic jigsaw. *Brain* **2003**, *126*, 1722–1733.
93. Langston, J.W. Epidemiology *versus* genetics in Parkinson's disease: Progress in resolving an age-old debate. *Ann. Neurol.* **1998**, *44* (Suppl. 1), S45–S52.

94. Bowers, W.J.; Howard, D.F.; Federoff, H.J. Gene therapeutic strategies for neuroprotection: Implications for Parkinson's disease. *Exp. Neurol.* **1997**, *144*, 58–68.
95. Kitada, T.; Asakawa, S.; Hattori, N.; Matsumine, H.; Yamamura, Y.; Minoshima, S.; Yokochi, M.; Mizuno, Y.; Shimizu, N. Mutations in the parkin gene cause autosomal recessive juvenile parkinsonism. *Nature* **1998**, *392*, 605–608.
96. Lincoln, S.; Vaughan, J.; Wood, N.; Baker, M.; Adamson, J.; Gwinn-Hardy, K.; Lynch, T.; Hardy, J.; Farrer, M. Low frequency of pathogenic mutations in the ubiquitin carboxy-terminal hydrolase gene in familial Parkinson's disease. *Neuroreport* **1999**, *10*, 427–429.
97. Golbe, L.I.; Di Iorio, G.; Bonavita, V.; Miller, D.C.; Duvoisin, R.C. A large kindred with autosomal dominant Parkinson's disease. *Ann. Neurol.* **1990**, *27*, 276–282.
98. Kruger, R.; Kuhn, W.; Muller, T.; Woitalla, D.; Graeber, M.; Kosel, S.; Przuntek, H.; Epplen, J.T.; Schols, L.; Riess, O. Ala30Pro mutation in the gene encoding alpha-synuclein in Parkinson's disease. *Nat. Genet.* **1998**, *18*, 106–108.
99. Polymeropoulos, M.H.; Lavedan, C.; Leroy, E.; Ide, S.E.; Dehejia, A.; Dutra, A.; Pike, B.; Root, H.; Rubenstein, J.; Boyer, R.; *et al.* Mutation in the alpha-synuclein gene identified in families with Parkinson's disease. *Science* **1997**, *276*, 2045–2047.
100. Pham, A.H.; Meng, S.; Chu, Q.N.; Chan, D.C. Loss of Mfn2 results in progressive, retrograde degeneration of dopaminergic neurons in the nigrostriatal circuit. *Hum. Mol. Genet.* **2012**, *21*, 4817–4826.
101. Lee, S.; Sterky, F.H.; Mourier, A.; Terzioglu, M.; Cullheim, S.; Olson, L.; Larsson, N.G. Mitofusin 2 is necessary for striatal axonal projections of midbrain dopamine neurons. *Hum. Mol. Genet.* **2012**, *21*, 4827–35.
102. Li, H.; Chen, Y.; Jones, A.F.; Sanger, R.H.; Collis, L.P.; Flannery, R.; McNay, E.C.; Yu, T.; Schwarzenbacher, R.; Bossy, B.; *et al.* Bcl-xL induces Drp1-dependent synapse formation in cultured hippocampal neurons. *Proc. Natl. Acad. Sci. USA* **2008**, *105*, 2169–2174.
103. Ishihara, N.; Nomura, M.; Jofuku, A.; Kato, H.; Suzuki, S.O.; Masuda, K.; Otera, H.; Nakanishi, Y.; Nonaka, I.; Goto, Y.; *et al.* Mitochondrial fission factor Drp1 is essential for embryonic development and synapse formation in mice. *Nat. Cell. Biol.* **2009**, *11*, 958–966.
104. Wakabayashi, J.; Zhang, Z.; Wakabayashi, N.; Tamura, Y.; Fukaya, M.; Kensler, T.W.; Iijima, M.; Sesaki, H. The dynamin-related GTPase Drp1 is required for embryonic and brain development in mice. *J. Cell Biol.* **2009**, *186*, 805–816.
105. Berthet, A.; Margolis, E.B.; Zhang, J.; Hsieh, I.; Zhang, J.; Hnasko, T.S.; Ahmad, J.; Edwards, R.H.; Sesaki, H.; Huang, E.J.; *et al.* Loss of mitochondrial fission depletes axonal mitochondria in midbrain dopamine neurons. *J. Neurosci.* **2014**, *34*, 14304–14317.
106. Zarranz, J.J.; Alegre, J.; Gomez-Esteban, J.C.; Lezcano, E.; Ros, R.; Ampuero, I.; Vidal, L.; Hoenicka, J.; Rodriguez, O.; Atares, B.; *et al.* The new mutation, E46K, of alpha-synuclein causes Parkinson and Lewy body dementia. *Ann. Neurol.* **2004**, *55*, 164–173.
107. Bonifati, V.; Rizzu, P.; Squitieri, F.; Krieger, E.; Vanacore, N.; van Swieten, J.C.; Brice, A.; van Duijn, C.M.; Oostra, B.; Meco, G.; *et al.* DJ-1(PARK7), a novel gene for autosomal recessive, early onset parkinsonism. *Neurol. Sci.* **2003**, *24*, 159–160.
108. Iwatsubo, T.; Ito, G.; Takatori, S.; Hannno, Y.; Kuwahara, T. Pathogenesis of Parkinson's disease: Implications from familial Parkinson's disease. *Rinsho Shinkeigaku* **2005**, *45*, 899–901.

109. Bialecka, M.; Hui, S.; Klodowska-Duda, G.; Opala, G.; Tan, E.K.; Drozdzik, M. Analysis of LRRK 2 G 2019 S and I 2020 T mutations in Parkinson's disease. *Neurosci. Lett.* **2005**, *390*, 1–3.
110. Smith, R.G. The aging process: Where are the drug opportunities? *Curr. Opin. Chem. Biol.* **2000**, *4*, 371–376.
111. Le, W.; Conneely, O.M.; He, Y.; Jankovic, J.; Appel, S.H. Reduced Nurr1 expression increases the vulnerability of mesencephalic dopamine neurons to MPTP-induced injury. *J. Neurochem.* **1999**, *73*, 2218–2221.
112. Xiao, Q.; Castillo, S.O.; Nikodem, V.M. Distribution of messenger RNAs for the orphan nuclear receptors Nurr1 and Nur77 (NGFI-B) in adult rat brain using in situ hybridization. *Neuroscience* **1996**, *75*, 221–230.
113. Rakovic, A.; Grunewald, A.; Seibler, P.; Ramirez, A.; Kock, N.; Orolicki, S.; Lohmann, K.; Klein, C. Effect of endogenous mutant and wild-type PINK1 on Parkin in fibroblasts from Parkinson disease patients. *Hum. Mol. Genet.* **2010**, *19*, 3124–3137.
114. Rakovic, A.; Grunewald, A.; Kottwitz, J.; Bruggemann, N.; Pramstaller, P.P.; Lohmann, K.; Klein, C. Mutations in PINK1 and Parkin Impair Ubiquitination of Mitofusins in Human Fibroblasts. *PLoS ONE* **2011**, *6*, e16746.
115. Jin, H.J.; Li, C.G. Tanshinone IIA and Cryptotanshinone Prevent Mitochondrial Dysfunction in Hypoxia-Induced H9c2 Cells: Association to Mitochondrial ROS, Intracellular Nitric Oxide, and Calcium Levels. *Evid. Based Complement. Alternat. Med.* **2013**, *2013*, 610694.
116. Poole, A.C.; Thomas, R.E.; Andrews, L.A.; McBride, H.M.; Whitworth, A.J.; Pallanck, L.J. The PINK1/Parkin pathway regulates mitochondrial morphology. *Proc. Natl. Acad. Sci. USA* **2008**, *105*, 1638–1643.
117. Yang, Y.; Ouyang, Y.; Yang, L.; Beal, M.F.; McQuibban, A.; Vogel, H.; Lu, B. Pink1 regulates mitochondrial dynamics through interaction with the fission/fusion machinery. *Proc. Natl. Acad. Sci. USA* **2008**, *105*, 7070–7075.
118. Yu, W.; Sun, Y.; Guo, S.; Lu, B. The PINK1/Parkin pathway regulates mitochondrial dynamics and function in mammalian hippocampal and dopaminergic neurons. *Hum. Mol. Genet.* **2011**, *20*, 3227–3240.
119. Wang, X.; Petrie, T.G.; Liu, Y.; Liu, J.; Fujioka, H.; Zhu, X. Parkinson's disease-associated DJ-1 mutations impair mitochondrial dynamics and cause mitochondrial dysfunction. *J. Neurochem.* **2012**, *121*, 830–839.
120. Ryan, B.J.; Hoek, S.; Fon, E.A.; Wade-Martins, R. Mitochondrial dysfunction and mitophagy in Parkinson's: From familial to sporadic disease. *Trends Biochem. Sci.* **2015**, *40*, 200–210.
121. Niu, J.; Yu, M.; Wang, C.; Xu, Z. Leucine-rich repeat kinase 2 disturbs mitochondrial dynamics via Dynamin-like protein. *J. Neurochem.* **2012**, *122*, 650–658.
122. Wang, X.; Yan, M.H.; Fujioka, H.; Liu, J.; Wilson-Delfosse, A.; Chen, S.G.; Perry, G.; Casadesus, G.; Zhu, X. LRRK2 regulates mitochondrial dynamics and function through direct interaction with DLP1. *Hum. Mol. Genet.* **2012**, *21*, 1931–1944.
123. Su, Y.C.; Qi, X. Inhibition of excessive mitochondrial fission reduced aberrant autophagy and neuronal damage caused by LRRK2 G2019S mutation. *Hum. Mol. Genet.* **2013**, *22*, 4545–4561.

124. Guo, X.; Disatnik, M.H.; Monbureau, M.; Shamloo, M.; Mochly-Rosen, D.; Qi, X. Inhibition of mitochondrial fragmentation diminishes Huntington's disease-associated neurodegeneration. *J. Clin. Investig.* **2013**, *123*, 5371–5388.
125. Qi, X.; Qvit, N.; Su, Y.C.; Mochly-Rosen, D. A novel Drp1 inhibitor diminishes aberrant mitochondrial fission and neurotoxicity. *J. Cell Sci.* **2013**, *126*, 789–802.
126. Yamamoto, A.; Yue, Z. Autophagy and its normal and pathogenic states in the brain. *Annu. Rev. Neurosci.* **2014**, *37*, 55–78.
127. Redmann, M.; Dodson, M.; Boyer-Guittaut, M.; Darley-Usmar, V.; Zhang, J. Mitophagy mechanisms and role in human diseases. *Int. J. Biochem. Cell Biol.* **2014**, *53*, 127–33.
128. Fass, E.; Amar, N.; Elazar, Z. Identification of essential residues for the C-terminal cleavage of the mammalian LC3: A lesson from yeast Atg8. *Autophagy* **2007**, *3*, 48–50.
129. Kabeya, Y.; Mizushima, N.; Ueno, T.; Yamamoto, A.; Kirisako, T.; Noda, T.; Kominami, E.; Ohsumi, Y.; Yoshimori, T. LC3, a mammalian homologue of yeast Apg8p, is localized in autophagosome membranes after processing. *EMBO J.* **2000**, *19*, 5720–5728.
130. Kanki, T.; Wang, K.; Cao, Y.; Baba, M.; Klionsky, D.J. Atg32 is a mitochondrial protein that confers selectivity during mitophagy. *Dev. Cell* **2009**, *17*, 98–109.
131. Okamoto, K.; Kondo-Okamoto, N.; Ohsumi, Y. Mitochondria-anchored receptor Atg32 mediates degradation of mitochondria via selective autophagy. *Dev. Cell* **2009**, *17*, 87–97.
132. Liu, L.; Feng, D.; Chen, G.; Chen, M.; Zheng, Q.; Song, P.; Ma, Q.; Zhu, C.; Wang, R.; Qi, W.; *et al.* Mitochondrial outer-membrane protein FUNDC1 mediates hypoxia-induced mitophagy in mammalian cells. *Nat. Cell Biol.* **2012**, *14*, 177–185.
133. Narendra, D.; Kane, L.A.; Hauser, D.N.; Fearnley, I.M.; Youle, R.J. p62/SQSTM1 is required for Parkin-induced mitochondrial clustering but not mitophagy; VDAC1 is dispensable for both. *Autophagy* **2010**, *6*, 1090–1106.
134. Shi, R.Y.; Zhu, S.H.; Li, V.; Gibson, S.B.; Xu, X.S.; Kong, J.M. BNIP3 interacting with LC3 triggers excessive mitophagy in delayed neuronal death in stroke. *CNS Neurosci. Ther.* **2014**, *20*, 1045–1055.
135. Strappazzon, F.; Nazio, F.; Corrado, M.; Cianfanelli, V.; Romagnoli, A.; Fimia, G.M.; Campello, S.; Nardacci, R.; Piacentini, M.; Campanella, M.; *et al.* AMBRA1 is able to induce mitophagy via LC3 binding, regardless of PARKIN and p62/SQSTM1. *Cell. Death Differ.* **2015**, *22*, 419–432.
136. Johansen, T.; Lamark, T. Selective autophagy mediated by autophagic adapter proteins. *Autophagy* **2011**, *7*, 279–296.
137. Wu, F.; Xu, H.D.; Guan, J.J.; Hou, Y.S.; Gu, J.H.; Zhen, X.C., Qin, Z.H. Rotenone impairs autophagic flux and lysosomal functions in Parkinson's disease. *Neuroscience* **2015**, *284*, 900–911.
138. Arsikin, K.; Kravic-Stevovic, T.; Jovanovic, M.; Ristic, B.; Tovilovic, G.; Zogovic, N.; Bumbasirevic, V.; Trajkovic, V.; Harhaji-Trajkovic, L. Autophagy-dependent and -independent involvement of AMP-activated protein kinase in 6-hydroxydopamine toxicity to SH-SY5Y neuroblastoma cells. *Biochim. Biophys. Acta* **2012**, *1822*, 1826–1836.
139. Youle, R.J.; Narendra, D.P. Mechanisms of mitophagy. *Nat. Rev. Mol. Cell Biol.* **2011**, *12*, 9–14.
140. Narendra, D.P.; Youle, R.J. Targeting mitochondrial dysfunction: Role for PINK1 and Parkin in mitochondrial quality control. *Antioxid. Redox Signal.* **2011**, *14*, 1929–1938.

141. Ashrafi, G.; Schlehe, J.S.; LaVoie, M.J.; Schwarz, T.L. Mitophagy of damaged mitochondria occurs locally in distal neuronal axons and requires PINK1 and Parkin. *J. Cell Biol.* **2014**, *206*, 655–670.
142. Geisler, S.; Holmstrom, K.M.; Skujat, D.; Fiesel, F.C.; Rothfuss, O.C.; Kahle, P.J.; Springer, W. PINK1/Parkin-mediated mitophagy is dependent on VDAC1 and p62/SQSTM1. *Nat. Cell Biol.* **2010**, *12*, 119–131.
143. Rakovic, A.; Shurkewitsch, K.; Seibler, P.; Grunewald, A.; Zanon, A.; Hagenah, J.; Krainc, D.; Klein, C. Phosphatase and tensin homolog (PTEN)-induced putative kinase 1 (PINK1)-dependent ubiquitination of endogenous Parkin attenuates mitophagy: Study in human primary fibroblasts and induced pluripotent stem cell-derived neurons. *J. Biol. Chem.* **2013**, *288*, 2223–2237.
144. McCoy, M.K.; Cookson, M.R. DJ-1 regulation of mitochondrial function and autophagy through oxidative stress. *Autophagy* **2011**, *7*, 531–532.
145. Thomas, K.J.; McCoy, M.K.; Blackinton, J.; Beilina, A.; van der Brug, M.; Sandebring, A.; Miller, D.; Maric, D.; Cedazo-Minguez, A.; Cookson, M.R. DJ-1 acts in parallel to the PINK1/parkin pathway to control mitochondrial function and autophagy. *Hum. Mol. Genet.* **2011**, *20*, 40–50.
146. Irrcher, I.; Aleyasin, H.; Seifert, E.L.; Hewitt, S.J.; Chhabra, S.; Phillips, M.; Lutz, A.K.; Rousseaux, M.W.; Bevilacqua, L.; Jahani-Asl, A.; *et al.* Loss of the Parkinson's disease-linked gene DJ-1 perturbs mitochondrial dynamics. *Hum. Mol. Genet.* **2010**, *19*, 3734–46.
147. Tang, B.; Xiong, H.; Sun, P.; Zhang, Y.; Wang, D.; Hu, Z.; Zhu, Z.; Ma, H.; Pan, Q.; Xia, J.H.; *et al.* Association of PINK1 and DJ-1 confers digenic inheritance of early-onset Parkinson's disease. *Hum. Mol. Genet.* **2006**, *15*, 1816–1825.
148. Xiong, H.; Wang, D.; Chen, L.; Choo, Y.S.; Ma, H.; Tang, C.; Xia, K.; Jiang, W.; Ronai, Z.; Zhuang, X.; *et al.* Parkin, PINK1, and DJ-1 form a ubiquitin E3 ligase complex promoting unfolded protein degradation. *J. Clin. Investig.* **2009**, *119*, 650–660.
149. Plowey, E.D.; Cherra, S.J., 3rd; Liu, Y.J.; Chu, C.T. Role of autophagy in G2019S-LRRK2-associated neurite shortening in differentiated SH-SY5Y cells. *J. Neurochem.* **2008**, *105*, 1048–1056.
150. Yakhine-Diop, S.M.; Bravo-San Pedro, J.M.; Gomez-Sanchez, R.; Pizarro-Estrella, E.; Rodriguez-Arribas, M.; Climent, V.; Aiastui, A.; Lopez de Munain, A.; Fuentes, J.M.; Gonzalez-Polo, R.A. G2019S LRRK2 mutant fibroblasts from Parkinson's disease patients show increased sensitivity to neurotoxin 1-methyl-4-phenylpyridinium dependent of autophagy. *Toxicology* **2014**, *324*, 1–9.
151. Sanchez-Danes, A.; Richaud-Patin, Y.; Carballo-Carbajal, I.; Jimenez-Delgado, S.; Caig, C.; Mora, S.; di Guglielmo, C.; Ezquerra, M.; Patel, B.; Giralt, A.; *et al.* Disease-specific phenotypes in dopamine neurons from human iPS-based models of genetic and sporadic Parkinson's disease. *EMBO Mol. Med.* **2012**, *4*, 380–395.
152. Lachenmayer, M.L.; Yue, Z. Genetic animal models for evaluating the role of autophagy in etiopathogenesis of Parkinson disease. *Autophagy* **2012**, *8*, 1837–1838.
153. Alegre-Abarrategui, J.; Christian, H.; Lufino, M.M.; Mutihac, R.; Venda, L.L.; Ansorge, O.; Wade-Martins, R. LRRK2 regulates autophagic activity and localizes to specific membrane microdomains in a novel human genomic reporter cellular model. *Hum. Mol. Genet.* **2009**, *18*, 4022–4034.

154. Saez-Atienzar, S.; Bonet-Ponce, L.; Blesa, J.R.; Romero, F.J.; Murphy, M.P.; Jordan, J.; Galindo, M.F. The LRRK2 inhibitor GSK2578215A induces protective autophagy in SH-SY5Y cells: Involvement of Drp-1-mediated mitochondrial fission and mitochondrial-derived ROS signaling. *Cell Death Dis.* **2014**, *5*, e1368.

155. Gomez-Suaga, P.; Luzon-Toro, B.; Churamani, D.; Zhang, L.; Bloor-Young, D.; Patel, S.; Woodman, P.G.; Churchill, G.C.; Hilfiker, S. Leucine-rich repeat kinase 2 regulates autophagy through a calcium-dependent pathway involving NAADP. *Hum. Mol. Genet.* **2012**, *21*, 511–525.

156. Bravo-San Pedro, J.M.; Niso-Santano, M.; Gomez-Sanchez, R.; Pizarro-Estrella, E.; Aiastui-Pujana, A.; Gorostidi, A.; Climent, V.; Lopez de Maturana, R.; Sanchez-Pernaute, R.; Lopez de Munain, A.; *et al.* The LRRK2 G2019S mutant exacerbates basal autophagy through activation of the MEK/ERK pathway. *Cell. Mol. Life Sci.* **2013**, *70*, 121–136.

157. Gomez-Suaga, P.; Hilfiker, S. LRRK2 as a modulator of lysosomal calcium homeostasis with downstream effects on autophagy. *Autophagy* **2012**, *8*, 692–693.

158. Orenstein, S.J.; Kuo, S.H.; Tasset, I.; Arias, E.; Koga, H.; Fernandez-Carasa, I.; Cortes, E.; Honig, L.S.; Dauer, W.; Consiglio, A.; *et al.* Interplay of LRRK2 with chaperone-mediated autophagy. *Nat. Neurosci.* **2013**, *16*, 394–406.

159. Su, Y.C.; Guo, X.;Qi, X. Threonine 56 phosphorylation of Bcl-2 is required for LRRK2 G2019S-induced mitochondrial depolarization and autophagy. *Biochim. Biophys. Acta* **2015**, *1852*, 12–21.

160. Ramsay, R.R.; Salach, J.I.; Singer, T.P. Uptake of the neurotoxin 1-methyl-4-phenylpyridine (MPP+) by mitochondria and its relation to the inhibition of the mitochondrial oxidation of NAD^+-linked substrates by MPP+. *Biochem. Biophys. Res. Commun.* **1986**, *134*, 743–748.

161. Marey-Semper, I.; Gelman, M.; Levi-Strauss, M. The high sensitivity to rotenone of striatal dopamine uptake suggests the existence of a constitutive metabolic deficiency in dopaminergic neurons from the substantia nigra. *Eur. J. Neurosci.* **1993**, *5*, 1029–1034.

162. Chinta, S.J.; Mallajosyula, J.K.; Rane, A.; Andersen, J.K. Mitochondrial alpha-synuclein accumulation impairs complex I function in dopaminergic neurons and results in increased mitophagy *in vivo*. *Neurosci. Lett.* **2010**, *486*, 235–239.

163. Devi, L.; Raghavendran, V.; Prabhu, B.M.; Avadhani, N.G.; Anandatheerthavarada, H.K. Mitochondrial import and accumulation of alpha-synuclein impair complex I in human dopaminergic neuronal cultures and Parkinson disease brain. *J. Biol. Chem.* **2008**, *283*, 9089–9100.

164. Martin, L.J.; Pan, Y.; Price, A.C.; Sterling, W.; Copeland, N.G.; Jenkins, N.A.; Price, D.L.; Lee, M.K. Parkinson's disease alpha-synuclein transgenic mice develop neuronal mitochondrial degeneration and cell death. *J. Neurosci.* **2006**, *26*, 41–50.

165. Zuo, L.; Motherwell, M.S. The impact of reactive oxygen species and genetic mitochondrial mutations in Parkinson's disease. *Gene* **2013**, *532*, 18–23.

166. Nakamura, T.; Prikhodko, O.A.; Pirie, E.; Nagar, S.; Akhtar, M.W.; Oh, C.K.; McKercher, S.R.; Ambasudhan, R.; Okamoto, S.I.; Lipton, S.A. Aberrant protein S-nitrosylation contributes to the pathophysiology of neurodegenerative diseases. *Neurobiol. Dis.* **2015**, doi:10.1016/j.nbd.2015.03.017.

167. Van Muiswinkel, F.L.; Steinbusch, H.W.; Drukarch, B.; de Vente, J. Identification of NO-producing and -receptive cells in mesencephalic transplants in a rat model of Parkinson's disease: A study using NADPH-d enzyme- and NOSc/cGMP immunocytochemistry. *Ann. N. Y. Acad. Sci.* **1994**, *738*, 289–304.

168. Gu, Z.; Nakamura, T.; Lipton, S.A. Redox reactions induced by nitrosative stress mediate protein misfolding and mitochondrial dysfunction in neurodegenerative diseases. *Mol. Neurobiol.* **2010**, *41*, 55–72.

169. Levecque, C.; Elbaz, A.; Clavel, J.; Richard, F.; Vidal, J.S.; Amouyel, P.; Tzourio, C.; Alperovitch, A.; Chartier-Harlin, M.C. Association between Parkinson's disease and polymorphisms in the nNOS and iNOS genes in a community-based case-control study. *Hum. Mol. Genet.* **2003**, *12*, 79–86.

170. Joniec, I.; Ciesielska, A.; Kurkowska-Jastrzebska, I.; Przybylkowski, A.; Czlonkowska, A.; Czlonkowski, A. Age- and sex-differences in the nitric oxide synthase expression and dopamine concentration in the murine model of Parkinson's disease induced by 1-methyl-4-phenyl-1,2,3,6-tetrahydropyridine. *Brain Res.* **2009**, *1261*, 7–19.

171. Watanabe, Y.; Kato, H.; Araki, T. Protective action of neuronal nitric oxide synthase inhibitor in the MPTP mouse model of Parkinson's disease. *Metab. Brain Dis.* **2008**, *23*, 51–69.

172. Castagnoli, K.; Palmer, S.; Castagnoli, N., Jr. Neuroprotection by (R)-deprenyl and 7-nitroindazole in the MPTP C57BL/6 mouse model of neurotoxicity. *Neurobiology (Bp)* **1999**, *7*, 135–149.

173. Dolezal, P.; Likic, V.; Tachezy, J.; Lithgow, T. Evolution of the molecular machines for protein import into mitochondria. *Science* **2006**, *313*, 314–318.

174. De Marcos-Lousa, C.; Sideris, D.P.; Tokatlidis, K. Translocation of mitochondrial inner-membrane proteins: Conformation matters. *Trends Biochem. Sci.* **2006**, *31*, 259–267.

175. Baker, M.J.; Frazier, A.E.; Gulbis, J.M.; Ryan, M.T. Mitochondrial protein-import machinery: Correlating structure with function. *Trends Cell. Biol.* **2007**, *17*, 456–464.

176. Gottschalk, W.K.; Lutz, M.W.; He, Y.T.; Saunders, A.M.; Burns, D.K.; Roses, A.D.; Chiba-Falek, O. The Broad Impact of TOM40 on Neurodegenerative Diseases in Aging. *J. Parkinsons Dis. Alzheimers Dis.* **2014**, *1*, 12.

177. Bender, A.; Desplats, P.; Spencer, B.; Rockenstein, E.; Adame, A.; Elstner, M.; Laub, C.; Mueller, S.; Koob, A.O.; Mante, M.; *et al.* TOM40 mediates mitochondrial dysfunction induced by alpha-synuclein accumulation in Parkinson's disease. *PLoS ONE* **2013**, *8*, e62277.

178. Okatsu, K.; Kimura, M.; Oka, T.; Tanaka, K.; Matsuda, N. Unconventional PINK1 localization to the outer membrane of depolarized mitochondria drives Parkin recruitment. *J. Cell Sci.* **2015**, *128*, 964–978.

179. Kato, H.; Lu, Q.; Rapaport, D.; Kozjak-Pavlovic, V. Tom70 is essential for PINK1 import into mitochondria. *PLoS ONE* **2013**, *8*, e58435.

180. Cooper, O.; Seo, H.; Andrabi, S.; Guardia-Laguarta, C.; Graziotto, J.; Sundberg, M.; McLean, J.R.; Carrillo-Reid, L.; Xie, Z.; Osborn, T.; *et al.* Pharmacological rescue of mitochondrial deficits in iPSC-derived neural cells from patients with familial Parkinson's disease. *Sci. Transl. Med.* **2012**, *4*, 141ra90.

181. Reinhardt, P.; Schmid, B.; Burbulla, L.F.; Schondorf, D.C.; Wagner, L.; Glatza, M.; Hoing, S.; Hargus, G.; Heck, S.A.; Dhingra, A.; *et al.* Genetic correction of a LRRK2 mutation in human iPSCs links parkinsonian neurodegeneration to ERK-dependent changes in gene expression. *Cell. Stem Cell.* **2013**, *12*, 354–367.

182. Toyoda, Y.; Erkut, C.; Pan-Montojo, F.; Boland, S.; Stewart, M.P.; Muller, D.J.; Wurst, W.; Hyman, A.A.; Kurzchalia, T.V. Products of the Parkinson's disease-related glyoxalase DJ-1, D-lactate and glycolate, support mitochondrial membrane potential and neuronal survival. *Biol. Open* **2014**, *3*, 777–784.

183. Daher, J.P.; Abdelmotilib, H.A.; Hu, X.; Volpicelli-Daley, L.A.; Moehle, M.S.; Fraser, K.B.; Needle, E.; Chen, Y.; Steyn, S.J.; Galatsis, P.; *et al.* LRRK2 Pharmacological Inhibition Abates alpha-Synuclein Induced Neurodegeneration. *J. Biol. Chem.* **2015**, *290*, 19433–19444.

184. Li, T.; He, X.; Thomas, J.M.; Yang, D.; Zhong, S.; Xue, F.; Smith, W.W. A novel GTP-binding inhibitor, FX2149, attenuates LRRK2 toxicity in Parkinson's disease models. *PLoS ONE* **2015**, *10*, e0122461.

185. Chiu, C.C.; Yeh, T.H.; Lai, S.C.; Wu-Chou, Y.H.; Chen, C.H.; Mochly-Rosen, D.; Huang, Y.C.; Chen, Y.J.; Chen, C.L.; Chang, Y.M.; *et al.* Neuroprotective effects of aldehyde dehydrogenase 2 activation in rotenone-induced cellular and animal models of parkinsonism. *Exp. Neurol.* **2015**, *263*, 244–253.

186. Richter, F.; Gao, F.; Medvedeva, V.; Lee, P.; Bove, N.; Fleming, S.M.; Michaud, M.; Lemesre, V.; Patassini, S.; de La Rosa, K.; *et al.* Chronic administration of cholesterol oximes in mice increases transcription of cytoprotective genes and improves transcriptome alterations induced by alpha-synuclein overexpression in nigrostriatal dopaminergic neurons. *Neurobiol. Dis.* **2014**, *69*, 263–275.

187. Rappold, P.M.; Cui, M.; Grima, J.C.; Fan, R.Z.; de Mesy-Bentley, K.L.; Chen, L.; Zhuang, X.; Bowers, W.J.; Tieu, K. Drp1 inhibition attenuates neurotoxicity and dopamine release deficits *in vivo*. *Nat. Commun.* **2014**, *5*, 5244.

188. Mena, N.P.; Garcia-Beltran, O.; Lourido, F.; Urrutia, P.J.; Mena, R.; Castro-Castillo, V.; Cassels, B.K.; Nunez, M.T. The novel mitochondrial iron chelator 5-((methylamino)methyl)-8-hydroxyquinoline protects against mitochondrial-induced oxidative damage and neuronal death. *Biochem. Biophys. Res. Commun.* **2015**, *463*, 787–792.

189. Jiang, J.; Zuo, Y.; Gu, Z. Rapamycin protects the mitochondria against oxidative stress and apoptosis in a rat model of Parkinson's disease. *Int. J. Mol. Med.* **2013**, *31*, 825–832.

190. Xiong, N.; Xiong, J.; Khare, G.; Chen, C.; Huang, J.; Zhao, Y.; Zhang, Z.; Qiao, X.; Feng, Y.; Reesaul, H.; *et al.* Edaravone guards dopamine neurons in a rotenone model for Parkinson's disease. *PLoS ONE* **2011**, *6*, e20677.

191. Dabbeni-Sala, F.; di Santo, S.; Franceschini, D.; Skaper, S.D.; Giusti, P. Melatonin protects against 6-OHDA-induced neurotoxicity in rats: A role for mitochondrial complex I activity. *FASEB J.* **2001**, *15*, 164–170

192. Zaitone, S.A.; Hammad, L.N.; Farag, N.E. *A*ntioxidant potential of melatonin enhances the response to L-dopa in 1-methyl 4-phenyl 1,2,3,6-tetrahydropyridine-parkinsonian mice. *Pharmacol. Rep.* **2013**, *65*, 1213–1226.

193. Karuppagounder, S.S.; Madathil, S.K.; Pandey, M.; Haobam, R.; Rajamma, U.; Mohanakumar, K.P. Quercetin up-regulates mitochondrial complex-I activity to protect against programmed cell death in rotenone model of Parkinson's disease in rats. *Neuroscience* **2013**, *236*, 136–148.

194. Jayaraj, R.L.; Elangovan, N.; Dhanalakshmi, C.; Manivasagam, T.; Essa, M.M. CNB-001, a novel pyrazole derivative mitigates motor impairments associated with neurodegeneration via suppression of neuroinflammatory and apoptotic response in experimental Parkinson's disease mice. *Chem. Biol. Interact.* **2014**, *220*, 149–157.

195. Abdin, A.A.; Sarhan, N.I. Intervention of mitochondrial dysfunction-oxidative stress-dependent apoptosis as a possible neuroprotective mechanism of alpha-lipoic acid against rotenone-induced parkinsonism and L-dopa toxicity. *Neurosci. Res.* **2011**, *71*, 387–395.

196. Kaur, H.; Chauhan, S.; Sandhir, R. Protective effect of lycopene on oxidative stress and cognitive decline in rotenone induced model of Parkinson's disease. *Neurochem. Res.* **2011**, *36*, 1435–1443.

197. Chen, C.H.; Budas, G.R.; Churchill, E.N.; Disatnik, M.H.; Hurley, T.D.; Mochly-Rosen, D. Activation of aldehyde dehydrogenase-2 reduces ischemic damage to the heart. *Science* **2008**, *321*, 1493–1145.

198. Bordet, T.; Buisson, B.; Michaud, M.; Drouot, C.; Galea, P.; Delaage, P.; Akentieva, N.P.; Evers, A.S.; Covey, D.F.; Ostuni, M.A.; *et al.* Identification and characterization of cholest-4-en-3-one, oxime (TRO19622), a novel drug candidate for amyotrophic lateral sclerosis. *J. Pharmacol. Exp. Ther.* **2007**, *322*, 709–720.

199. Gouarne, C.; Tracz, J.; Paoli, M.G.; Deluca, V.; Seimandi, M.; Tardif, G.; Xilouri, M.; Stefanis, L.; Bordet, T.; Pruss, R.M. Protective role of olesoxime against wild-type alpha-synuclein-induced toxicity in human neuronally differentiated SHSY-5Y cells. *Br. J. Pharmacol.* **2015**, *172*, 235–245.

200. Fernandez-Moriano, C.; Gonzalez-Burgos, E.; Gomez-Serranillos, M.P. Mitochondria-Targeted Protective Compounds in Parkinson's and Alzheimer's Diseases. *Oxid. Med. Cell. Longev.* **2015**, *2015*, 408927.

201. Valadas, J.S.; Vos, M.; Verstreken, P. Therapeutic strategies in Parkinson's disease: What we have learned from animal models. *Ann. N. Y. Acad. Sci.* **2015**, *1338*, 16–37.

202. Procaccio, V.; Bris, C.; de la Barca, J.M.C.; Oca, F.; Chevrollier, A.; Amati-Bonneau, P.; Bonneau, D.; Reynier, P. Perspectives of drug-based neuroprotection targeting mitochondria. *Rev. Neurol. (Paris)* **2014**, *170*, 390–400.

203. Yadav, A.; Agarwal, S.; Tiwari, S.K.; Chaturvedi, R.K. Mitochondria: Prospective targets for neuroprotection in Parkinson's disease. *Curr. Pharm. Des.* **2014**, *20*, 5558–5573.

8

Mitochondria Retrograde Signaling and the UPR^{mt}: Where are we in Mammals?

Thierry Arnould [1,*], Sébastien Michel [1,2] and Patricia Renard [1]

1 Laboratory of Biochemistry and Cell Biology (URBC), Namur Research Institute for Life Sciences (NARILIS), University of Namur, Rue de Bruxelles 61, 5000 Namur, Belgium; E-Mails: sebastien.michel@unamur.be (S.M.); patsy.renard@unamur.be (P.R.)

2 Department of Physiology, University of Lausanne, Rue du Bugnon 7, CH-1005 Lausanne, Switzerland

* Author to whom correspondence should be addressed; E-Mail: thierry.arnould@unamur.be

Academic Editors: Lars Olson, Jaime M. Ross and Giuseppe Coppotelli

---

Abstract: Mitochondrial unfolded protein response is a form of retrograde signaling that contributes to ensuring the maintenance of quality control of mitochondria, allowing functional integrity of the mitochondrial proteome. When misfolded proteins or unassembled complexes accumulate beyond the folding capacity, it leads to alteration of proteostasis, damages, and organelle/cell dysfunction. Extensively studied for the ER, it was recently reported that this kind of signaling for mitochondrion would also be able to communicate with the nucleus in response to impaired proteostasis. The mitochondrial unfolded protein response (UPR^{mt}) is activated in response to different types and levels of stress, especially in conditions where unfolded or misfolded mitochondrial proteins accumulate and aggregate. A specific UPR^{mt} could thus be initiated to boost folding and degradation capacity in response to unfolded and aggregated protein accumulation. Although first described in mammals, the UPR^{mt} was mainly studied in *Caenorhabditis elegans*, and accumulating evidence suggests that mechanisms triggered in response to a UPR^{mt} might be different in *C. elegans* and mammals. In this review, we discuss and integrate recent data from the literature to address whether the UPR^{mt} is relevant to mitochondrial homeostasis in mammals and to analyze the putative role of integrated stress response (ISR) activation in response to the inhibition of mtDNA expression and/or accumulation of mitochondrial mis/unfolded proteins.

Keywords: mitochondria; unfolded protein response; cell signaling; gene expression

---

1. Introduction

Mitochondrial, multifunctional, and dynamic organelles (resulting from fusion and fission events of mitochondria fragments, according to cell type and conditions) are linked to pathologies far beyond the *sensu stricto* "mitochondrial diseases" because they regulate metabolism via the Krebs cycle and oxidative phosphorylation (OXPHOS). Linked to their bioenergetics, mitochondria that contain more than 1000 different proteins/peptides, synthesized from both mitochondrial and nuclear genomes in a coordinated manner [1,2] also participate in synthesis (steroids, amino acids, nucleotides), the biogenesis of iron-sulfur centers, calcium homeostasis and redox status [3], control epigenetics marks in the nuclear genome, and constitute an integration platform for signaling and cell death/survival signals, connecting the organelle to apoptosis and autophagy [4–7]. The continued organelle maintenance can be seen as a balance between the biogenesis of the organelle and the mechanisms that provide quality control (involved in the remodeling and mitophagy) that guarantees cell homeostasis and function [8,9]. This function is maintained by the participation of several chaperones, antioxidant enzymes (SOD2, PRDX3, PRDX5, GPX1: human nomenclature according to Uniprot), and quality control proteases that promote protein folding and stability on the mitochondria while performing the degradation of un- or mis-folded proteins that accumulate [10]. Essential for this control is the molecular communication between the mitochondria and nucleus, in which ATP, calcium, and reactive oxygen species have been described to play major roles [1,11–14]. Indeed, upon organelle dysfunction, which can be caused by many events, such as mtDNA depletion, mutations, deletions, oxidative stress, aggregation of misfolded proteins, or dramatic changes in morphology and dynamics, a mitochondrial-to-nucleus communication, known as "retrograde signaling," triggers an orchestrated expression of nuclear genes in an attempt to relieve/resolve the stress and/or to compensate the defect [15]. The importance of these stress response pathways can be highlighted by the various pathophysiological developments associated with their impairment [15]. The aim of the current review is to highlight recent developments in the role of mitochondrial unfolded protein response (UPR^{mt}) in the field of mitochondrial dysfunction and putative connection with other retrograde signaling pathways, with a particular focus on differences between mechanisms retrieved in mammals and other organisms.

1.1. Mitochondrial Retrograde Responses

Several signaling pathways have been described in the mitochondria-to-nucleus communication observed in response to organelle stress and dysfunction. This mitochondrial stress response can be perceived as an attempt to compensate for the metabolic defect by stimulating several biological processes including mitochondria biogenesis. The transcriptional regulation of mitochondrial biogenesis is mediated by a set of transcription factors such as NRF1 and NRF2/GA-binding protein subunit β-1 (nuclear respiratory factors 1 and 2), cyclicAMP-responsive element binding protein1 (CREB1), steroid hormone receptor ERR1 (ERRα), and PPARG (PPARγ). The activity of these transcription factors is coordinated by members of the PGC-1 co-activator family [16]. The retrograde communication often relies on some of these transcriptional regulators to modulate gene expression to adapt mitochondria function in response to the original cue and promote recovery and function to resolve the stress. We now discuss the signaling pathways that regulate the expression, localization, and activity of these transcription factors. Notably, if the following pathways are usually associated with the regulation of mitochondrial protein abundance, lipid content might also be regulated and controlled by retrograde signaling. As with the absence of mtDNA in HeLa cells, mitochondrial DNA absence sensitive (MIDAS) factor increases mitochondrial mass by regulating the expression of enzymes involved in cardiolipin synthesis [17].

Retrograde communication in response to mitochondrial stress comes from seminal studies from Butow's group in the budding yeast *Saccharomyces cerevisiae* depleted of mtDNA [18,19], which showed that electron transport chain (ETC) deficiency leads to the transcription of genes associated with glutamate metabolism (reviewed in [20]). In yeast, there are three retrograde response genes (RTG): Rtg1p and Rtg3p are transcription factors, forming a dimer that translocates from the cytosol to the nucleus to regulate gene expression, while Rtg2p would act/behave as a sensor of mitochondrial stress [21]. Although mammalian orthologs of these proteins have not been found, similar signaling pathways do exist in mammals, as detailed in the next section.

1.1.1. Retrograde Signaling Involving Increased Cytosolic Calcium Concentration

Pioneering works on retrograde signaling in mammalian cells have been described in mtDNA-depleted cells through the pharmacological inhibition of ETC complexes or disruption of the mitochondrial membrane potential and focused on calcium in mouse C2C12 myocytes and human pulmonary carcinoma A549 cells. Avadhani's group showed that mitochondrial dysfunction associated with membrane depolarization leads to increased cytosolic calcium concentration and triggers nuclear factor of activated T-cells (NF-AT) and activating transcription factor 2 (ATF2) translocation to the nucleus in a calcineurin (Cn)- and Mitogen-activated protein kinase 8 (MAPK8)/JNK1-dependent manner, respectively [11,22]. The list of transcriptional regulators activated by calcium, either by a phosphorylation in a JNK1- or calcium/calmodulin-dependent protein kinase type IV (CaMKIV)-dependent manner, or by a dephosphorylation mediated by calcineurin in response to mitochondrial dysfunction has now expanded and includes the transcription factors CREB1, nuclear factor of κ light polypeptide gene enhancer in B-cells 1 (NF-κB), cellular tumor antigen p53 (p53), myocyte-specific enhancer factor 2A (MEF-2), and co-activators peroxisome proliferator-activated receptor γ coactivator 1-α (PGC-1α) and heterogenous nuclear ribonucleoprotein A2 (hnRNPA2) [13,22–25]. Altogether, data about calcium-dependent mitochondrial retrograde response represent a good example of functional compensation after organelle stress. Overexpressed nuclear genes include proteins associated with calcium homeostasis, such as calsequestrin and calreticulin, OXPHOS subunits for stabilizing energy production, and chloride intracellular channel protein 4 (Clic 4), which contribute to mitochondrial membrane potential [26–28].

1.1.2. Retrograde Response Associated with ROS Production and Signaling

Reactive oxygen species (ROS) are by-products of ETC complexes, and while they represent important physiological second messengers, their excessive production can lead to protein, lipid, and DNA damages and, ultimately, cell death. When antioxidant defenses are overwhelmed, a transcriptional program regulated by the NRF2/GA-binding protein subunit β-1 is initiated [29]. Indeed, excessive production of ROS triggers the stabilization and translocation of the factor to the nucleus, where the transcription of target genes containing an antioxidant-response element (ARE) in their promoter is increased, including NRF1, which regulates mitochondrial gene expression [29,30]. It has also been suggested that NRF1 might be directly regulated by redox signaling in phosphatidylinositol 3-kinase (PI3K) and RAC-α serine/threonine-protein kinase/Akt-dependent manners to regulate mtDNA transcription and replication by increasing mitochondrial transcription factor A (TFAM) protein expression [31]. PGC-1α is also activated by AMP-activated protein kinase (AMPK) in response to mitochondrial ROS to promote organelle biogenesis [32]. As for calcium, ROS have been the focus of numerous studies on retrograde communication, and additional transcription factors, such as NF-κB, p53, and AP-1, are activated by oxidants [33–35]. Surprisingly, while several studies have demonstrated enhanced biogenesis in response to mitochondrial stress [33,36–38], recent biological analysis of retrograde signaling by Chae and collaborators demonstrated the role of ROS in the down-regulation of OXPHOS enzymes [39]. These authors reported that, among the 72 transcription factors, for which activity was modified in response to mtDNA mutation, the abundance of RXRα Retinoic acid receptor RXR α) decreased after ROS-dependent JNK1 activation. As a direct consequence, the interaction between RXRα and PGC-1α was reduced, which attenuated the expression of gene encoding of OXPHOS enzymes as well as mitochondrial ribosomal proteins—two phenomena that further exacerbate mitochondrial dysfunction [39].

1.1.3. Energy Deprivation and Retrograde Response

The most obvious retrograde signaling might be the stress response associated with energy deprivation. AMPK is a sentinel activated in response to an increase in the AMP/ATP ratio, which triggers allosteric activation and the serine/threonine-protein kinase STK11 (LKB1)-dependent Thr172 phosphorylation of AMPK. Once activated, AMPK increases ATP production by inducing mitochondrial biogenesis, thus stimulating fatty acid β-oxidation and glycolysis, and shuts down energy consuming processes such as cell growth and lipid and protein synthesis [40–43]. AMPK regulates PGC-1α activity and abundance as well as its binding activity to transcription factors, such as NRF1, at the promoter of nuclear gene encoding mitochondrial proteins [36,42]. Moreover, AMPK is also well-known for regulating the serine/threonine-protein kinase mammalian target of rapamycin (mTOR) through control of tuberous sclerosis 2 (Tuberin) activity [44]. In mammals, two complexes have been described: mTORC1 and mTORC2. While mTORC1 regulates processes, such as transcription, translation, autophagy, and metabolism, mTORC2 is mainly associated with cell survival, proliferation, and metabolism. The activities of mTOR complexes are negatively regulated by Hamartin/Tuberin (tuberous sclerosis 1/2) via phosphorylation. In response to mitochondrial stress and energy deprivation, AMPK decreases mTOR activity and downstream processes, such as protein synthesis, by a mechanism involving either direct phosphorylation of the raptor or the activating phosphorylation of Tuberin [41,43,44].

Finally, the NAD^+/NADH ratio and acetyl-coenzyme A (acetyl-CoA) are important mitochondrial coenzymes and metabolites, respectively, which are associated with retrograde response [45]. However, the reduction of NAD^+ into NADH during the Krebs cycle and its oxidation by the NADH dehydrogenase complex represent the basis of mitochondrial respiration. Because NAD^+ is also a substrate used by enzymes, such as poly[ADP-ribose] polymerase-1 (PARP-1) and Sirtuins, the NAD^+/NADH ratio, which not only results from the redox status of the molecule, but also a balance between its synthesis and its degradation, might modulate gene expression [46]. Conversely, acetyl-CoA, resulting from pyruvate metabolism and fatty acid β-oxidation, is a substrate for lysine acetylation by acetyltransferases that controls the acetylation status of the transcription factors of the forkhead box protein O family members [47], histone acetylation, chromatin remodeling, and thus gene expression [48,49]. Yet, even if both NAD^+/NADH and acetyl-CoA are associated with the activity of post-translational modifying enzymes, and thus regulation of signaling pathways or gene expression, their direct involvement in retrograde signaling has not been extensively characterized to date.

1.1.4. Beyond Regulation by Transcription Factors

Beyond this level of retrograde communication, which regulates nuclear gene expression by the activation of transcription factors, it has been shown that the mtDNA copy number can also lead to epigenetic modifications. It is known that ATP, acetyl-CoA concentration, NAD^+/NADH ratio, and SAM (s-adenosylmethionine) are cofactors and metabolites whose concentrations can be modulated by mitochondria. Several of them, such as acetyl-CoA and SAM, are important acetyl- or methyl-group donors for DNA, RNA, and protein modifications that could also control gene expression and contribute to adaptation to mitochondria dysfunction [50]. These molecules can thus be seen as messengers for retrograde signaling responses [50].

Indeed, it has been demonstrated that mitochondrial depletion leads to increased promoter methylation of nuclear genes, which can be reverted, at least partly, by mtDNA repletion [7]. In addition, in conditions of mitochondrial stress, mitochondria-to-nucleus signaling is also know to confer a long-term adaptive response to the cell, which can protect it from future insults related or not to the initial stress. This concept, called mitochondrial hormesis or mitohormesis, provides a long-term increase in cytoprotective function after mild and transient mitochondrial stress such as ROS production. Mitohormesis has been particularly highlighted in the context of increased lifespan in various organisms such as yeast, worms, flies, and mice [51]. Moreover, it has been suggested that mitochondrial stress can generate a non-cell-autonomous response, meaning the activation of a stress response in a cell/tissue that is not affected by the initial triggered stress event. For instance, in response to OXPHOS deficiency, muscle fibers secrete the cytokine fibroblast growth factor 21 (FGF21) in the bloodstream, which results in chronic lipid recruitment and mobilization from adipose tissue and triggers ketogenesis in the liver [52,53]. It seems that skeletal muscle increases FGF21 expression in mitochondrial disorders to compensate for metabolic insufficiency by activating the mTOR Transcriptional repressor protein YY1-PGC-1α pathway [54]. Durieux and co-workers have described another example in *C. elegans*, in which stressed mitochondria, via electron chain manipulations in key tissues, release soluble signals (the so-called mitokines), which trigger expression of cytoprotective genes in distal tissues and have a beneficial effect on longevity of the worm [55]. Mitokines have also been unveiled in the context of a recent retrograde response, which we discuss in the next sections of this review: the mitochondrial unfolded protein response (UPR^{mt}). Following the example of the endoplasmic reticulum unfolded protein response (UPR^{er}), when damaged and/or unfolded proteins accumulate within mitochondria, this stress response coordinates expression of chaperones and proteases in a positive feedback loop and results in a first attempt to resolve the stress and bring cytoprotection [56].

1.2. The Mitochondrial Unfolded Protein Response (UPR^{mt})

Protein homeostasis relies on the equilibrium between the number of unfolded proteins and the folding capacity of a compartment. The mitochondria possess their own arsenal of chaperones and proteases. When unfolded, misfolded, or unassembled proteins accumulate beyond folding capacity, this leads to damage and organelle/cell dysfunction. As seen previously, mitochondrion is able to communicate with the nucleus in response to organelle dysfunction and bioenergetic impairment. As with the endoplasmic reticulum, for which the different signaling pathways and branches of the unfolded protein response are well described [57], and affecting the regulation of autophagy and the biology of mitochondria [58,59], a specific UPR^{mt} could also be initiated to boost folding and degradation capacity in response to unfolded and aggregated protein accumulations in mitochondria. Thus, mitochondrial protein homeostasis—or proteostasis—is preserved by retrograde communication to coordinate transcriptional activation of nuclear-encoded mitochondrial chaperones and proteases. Although its existence was initially reported in mammals almost 20 years ago [60], later studies related to the UPR^{mt} mainly focused in *Caenorhabditis elegans*, with mechanisms mainly revealed in the model organism and linked to aging, healthy lifespan, and longevity [61–63]. The UPR^{mt} is now considered as a mechanism that allows synchronization of nuclear and mitochondrial genomes [64].

1.2.1. The UPR^{mt} in *C. elegans*

Stress Models for the Study of UPR^{mt}

The worm *C. elegans* has been a useful model to identify stresses that activate a UPR^{mt} and to decipher signaling from proteotoxic stress to expression of stress-responsive genes. Most of the time, the induction of the UPR^{mt} can be easily assayed in worms that express green fluorescent protein (GFP) under the control of the promoter that drives the expression of gene encoding chaperones and orthologues of the mammalian mtHSP70 and heat shock 60 kDa protein 1 (HSPD1), respectively [61–63]. Ethidium bromide, an inhibitor of the mitochondrial DNA polymerase γ (POLG) was the first UPR^{mt} stressor described in *C. elegans* [65]. Because mitochondrial ETC complexes rely on a precise stoichiometry between nuclear- and mtDNA-encoded proteins, it is reasonable to think that the inhibition of mitochondrial transcription and replication by ethidium bromide, as transcriptions in mitochondria, clearly depend on mtDNA replication [66], which will lead to a protein imbalance and an increase in unassembled components. Similar mechanisms were later used to induce a UPR^{mt} with other molecules. On the one hand, decreasing mtDNA-encoded proteins through interference with mitochondrial translation, either with inhibitors, such as doxycycline, or with siRNA, allowing the silencing of mitoribosomal proteins, triggers a UPR^{mt} [46]. On the other hand, the stimulation of organelle biogenesis, by NAD$^+$ supplementation or mTOR inhibition by rapamycin, has similar effects [67]. Additional stresses, which induce a UPR^{mt}, following impaired assembly of mitochondrial complexes, defective protein folding, or processing, have also been described in a genome-wide RNAi screen [65].

Signaling Pathway of the UPR^{mt}

Ten years of research using the UPR^{mt} reporter in worms and large-scale experiments with siRNA were needed to shed light on a currently well-accepted model of retrograde signaling that followed mitochondrial proteotoxic stress in *C. elegans*. Briefly, accumulating unfolded proteins within mitochondria are degraded by the ATP-dependent Clp protease proteolytic subunit (ClpP) into small peptides, which are actively exported across the inner mitochondrial membrane (IMM) by HAF-1, a matrix peptide exporter belonging to the ATP-binding cassette (ABC) transporters [68,69]. Passive diffusion then brings peptides through the outer mitochondrial membrane (OMM) to the cytosol that triggers nuclear translocation of activating transcription factor associated with stress-1 (ATFS-1), a transcription factor that orchestrates expression of mitochondrial chaperones and proteases as well as other genes associated with ROS detoxification, mitochondrial protein import, and glycolysis [70]. Experiments in which the expression of ClpP and HAF-1 was silenced confirmed that both proteins were essential for the nuclear translocation of ATFS-1 and the activation of a UPR^{mt} [68,69]. The mechanism that controls the cellular localization of ATFS-1 was recently discovered [70]. In addition to its nuclear localization sequence (NLS), the transcription factor also contains a mitochondrial targeting sequence (MTS). Under basal conditions, ATFS-1 is constitutively imported in mitochondria and degraded by the Lon protease [71]. However, upon mitochondrial proteotoxic stress conditions, mitochondrial import efficiency decreases, and a fraction of ATFS-1 proteins accumulates in the cytosol and translocates to the nucleus. Very recently, Haynes and colleagues showed that the ATFS-1 transcription factor not only induces chaperones, OXPHOS assembly factor, and glycolysis genes, but also directly regulates OXPHOS gene promoters by limiting the accumulation of OXPHOS transcripts. Altogether, ATFS-1 coordinates the abundance of transcripts involved in OXPHOS expression and assembly factors to the protein-folding capacity of mitochondria [72]. The importance of the efficiency of the import machinery of mitochondrial proteins in the re-localization of ATFS-1 is supported by the fact that the disruption of mitochondrial import in response to the silencing of the mitochondrial import inner membrane translocase subunit Tim23 is sufficient to trigger ATFS-1 accumulation in the nucleus. It can also activate the GFP reporter construct driven by the promoter of HSPD1 encoding HSP60 [70], a mitochondrial matrix chaperonin crucial for the folding and assembly of newly imported mitochondrial proteins [73].

Interestingly, Rainbolt and colleagues obtained similar results and showed an increase in the expression of HSP60-GFP in response to the silencing of Tim17, a response that can be prevented in ATFS-1 but not HAF-1 mutant worms [74]. In addition, these authors noted that silencing of either Tim23 or Tim17 conferred increased resistance to paraquat-induced oxidative stress, although the protection was rather modest in mammalian cells. However, increased resistance to oxidative stress was not disrupted in the ATFS-1 mutant worm, thereby questioning the active participation of the UPR^{mt} in this protection. Finally, while the reduced protein import was able to induce stress-responsive genes independently of mitochondrial matrix stress (HAF-1 mutant), in the context of a UPR^{mt}, both the reduced mitochondrial protein import and the ATFS-1 nuclear translocation relied on HAF-1 and peptides efflux; however, the precise mechanism and tight regulation are still unclear [70].

In addition to ATFS-1, in *C. elegans*, the transcription factors DVE-1 and ubiquitin-like protein 5 (UBL-5) appear to be involved in the transcriptional regulation of the UPR^{mt}. During the onset of mitochondrial stress, both proteins redistribute into the nucleus and form a complex that binds the promoter of gene encoding chaperones such as mtHSP70 and HSPD1. As for ATFS-1, ClpP is also required for DVE-1 nuclear translocation, but HAF-1 would be dispensable [68]. While these proteins might help in chromatin remodeling to facilitate ATFS-1 access to the promoters of the specific target genes of the UPR^{mt}, future research on the importance of the contribution of DVE-1/UBL-5 complex to the UPR^{mt} is still needed. In addition, it is also very likely that the activation of a UPR^{mt} could be accompanied by signals that trigger and connect the activation of other cell signaling pathways.

Crosstalk between the UPR^{mt} and Other Stress Pathways in *C. elegans*

The UPR^{mt} relies on more than the over-expression of chaperones and proteases to maintain mitochondria proteostasis; several genes that are directly or indirectly controlled by ATFS-1 have been identified by comparing the relative abundance of transcripts between wild type and ATFS-1 mutant worms raised either under basal or UPR^{mt}-inducing conditions. In these conditions, among 685 genes differentially expressed in response to the UPR^{mt}, only 391 genes (encoding important components of ROS detoxification and antioxidant enzymes, glycolytic enzymes, and proteins of the mitochondrial protein import machinery) seemed to be dependent on ATFS-1, suggesting the activation of other pathways [70].

Other studies described the UPR^{mt} as interconnected cell signaling with other stress-activated pathways such as the integrative stress response (ISR) [75] and the antioxidant response [67]. Baker and co-workers demonstrated that, during mitochondrial stress, eIF2α is phosphorylated by GCN2 in a ROS-dependent manner, attenuating the protein synthesis within the cytosol. This is similar to what happens in the well-characterized UPR^{er} [75]. These two pathways are complementary and maintain the activity of mitochondria. These authors show that, in a worm genetic model of UPR^{mt}, RNAi-mediated silencing of eukaryotic translation initiation factor 2-α kinase 4/GCN2-like protein (GCN2) is associated with the accumulation of carbonylated proteins and decreased oxygen consumption, despite enhanced activity of the mtHSP70-GFP reporter plasmid. Conversely, upon mitochondrial stress, the inhibition of the UPR^{mt} increases the load on the ISR, as demonstrated by an increased eIF2α phosphorylation. In addition, concomitant alterations of both pathways result in growth defects, even in the absence of any stress [75].

In addition, boosting the NAD$^+$ level not only activates Sirtuins and, especially, mitochondrial Sirtuin 3 [76], but also induces mitonuclear protein imbalances and activates UPR^{mt} [67]. Importantly, these authors revealed that, in mice and *C. elegans*, in time-course experiments, treatments with nicotinamide riboside (NR), a NAD$^+$ precursor led to a protective response in a two-step process. First, during the early phase response (one day of treatment), the UPR^{mt} was activated, with the associated increase in ClpP and HSP6 expression. Then, after three days of treatment with NAD$^+$ booster/NR, with the UPR^{mt} markers still overexpressed, an antioxidant response was also triggered, as demonstrated by enhanced activation of the promoter of the gene encoding SOD3 and nuclear localization of the transcription factor daf-16 (FOXO3A orthologue in mammals). Interestingly, it appears that both pathways might be interconnected; the induction of Extracellular superoxide dismutase [Cu/Zn] (SOD3) was dependent on UBL-5, a transcriptional regulator activated during the UPR^{mt} [67].

1.2.2. The UPR^{mt} Models in Mammals

In mammals, two different models of UPR^{mt} have been described (Figure 1). While Hoogenraad's group described a DDIT3 (DNA damage-inducible transcript 3 protein)/CHOP-10-dependent UPR^{mt}

induced by the overexpression of a truncated form of a mitochondrial matrix enzyme [77], Germain and colleagues reported a CHOP-10-independent UPR^{mt} model in which protein aggregates accumulated within the internal membrane space (IMS) when a mutant (catalytically inactive enzyme) form of endonuclease G (N174A) was overexpressed in breast adenocarcinoma MCF-7 cells [78].

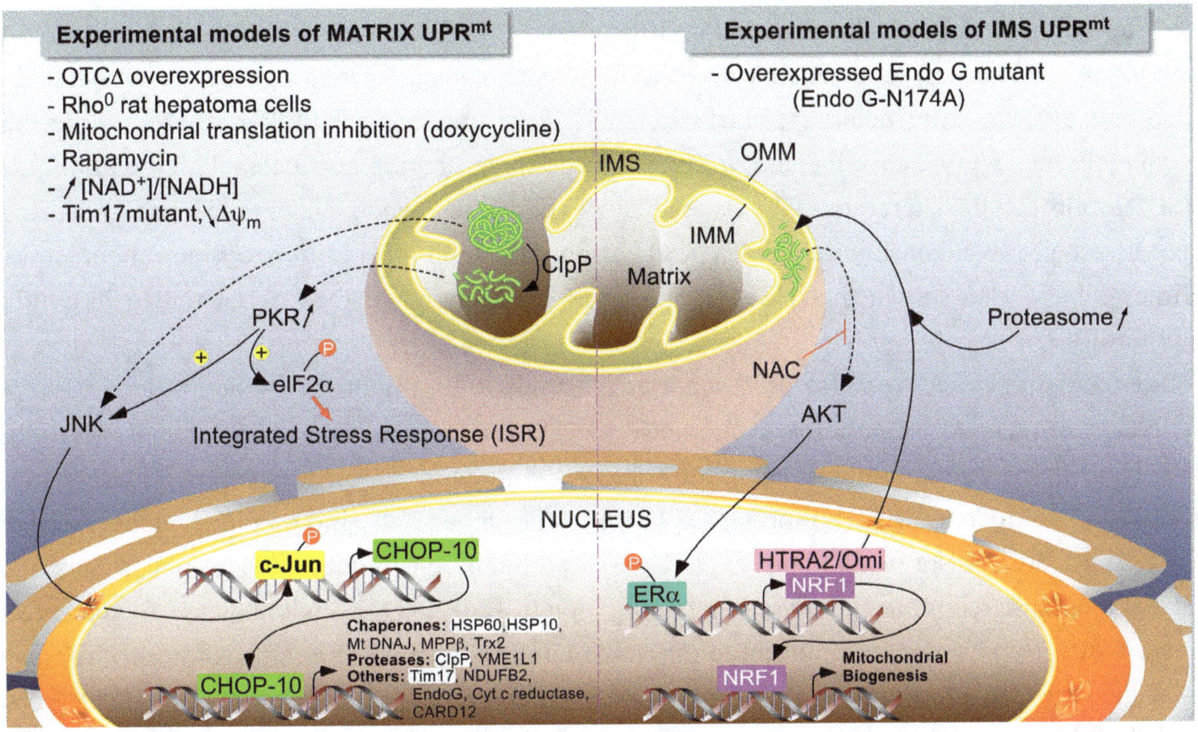

Figure 1. The UPR^{mt} models in mammalians. Two major independent models of UPR^{mt} have been described. When unfolded proteins accumulate in the mitochondrial matrix, they are first cleaved by ClpP proteases. Peptides exit mitochondria by unknown mechanisms and trigger a signaling pathway, leading to the activation of c-Jun N-terminal kinase (JNK) and PKR, which phosphorylate as c-Jun (part of AP-1 transcription factor) and eIF2α, respectively. The phosphorylation of eIF2α can also be mediated by GCN2 in response to mitochondrial translation inhibition (not illustrated), and it turns on the integrated stress response (see Figure 2). PKR could also activate JNK. The activation of c-Jun triggers CHOP-10 expression. In turn, the transcription factor regulates the expression of stress-resolving genes such as mitochondrial proteases and chaperones (left side of the chart). When unfolded proteins accumulate in the inter-membrane space, the UPR^{mt} is mediated by the activation of ERα, triggered by an AKT-dependent phosphorylation. The activation of AKT would then be mediated by oxidative stress, as inhibited by *N*-acetylcysteine (NAC). Mitochondrial stress resolves by an increase in the expression of HTRA2/Omi protease and enhanced proteasome activity, while the biogenesis of mitochondria is enhanced and under the control of NRF1, a target gene of ERα. The highlighted acronyms represent genes for which an endogenous differential expression (RT-qPCR or Western blot analysis) in response to UPR^{mt} has been experimentally demonstrated to be dependent on CHOP-10; this is in addition to analysis using reporter constructs. The non-highlighted genes refer to genes controlled by CHOP-10 in response to a UPR^{mt}, but only demonstrated using reporter constructs.

Accumulation of Unfolded Proteins within the Mitochondrial Matrix

Pioneering work by Hoogenraad and collaborators first showed the selective induction of mitochondrial chaperones HSPD1 and HSPE1 in response to complete mitochondrial genome depletion (ρ°) in rat hepatoma H4 cells [60]. Later on, they set up a UPR^{mt} model based on the overexpression of a truncated form of the mitochondrial ornithine carbamoyltransferase (OTC) enzyme (OTCΔ). The truncated form of the enzyme, while correctly localized in the organelle matrix, formed aggregates associated with the insoluble fraction after detergent extraction [79]. First clues of a UPR^{mt} have been described in the monkey kidney COS-7 cell line, as aggregated protein elimination correlates with the up-regulation of HSPD1 and ClpP. Therefore, OTCΔ abundance decreases when chaperones and proteases are overexpressed, yet abundance of the wild type OTC remains stable. In addition, co-immunoprecipitation experiments have also shown that HSPD1 and ClpP were bound to OTCΔ but not to its wild type counterpart [79].

Hoogenraad and colleagues also used reporter plasmids to highlight the induction of several genes that encoded mitochondrial proteins in response to OTCΔ aggregation such as chaperones (HSP60, HSP10, DnaJ homolog subfamily B member 1/HSP40, mitochondrial-processing peptidase subunit β, mitochondrial thioredoxin/MTRX), proteases (ClpP, ATP-dependent zinc metalloprotease YME1L1), and other mitochondrial proteins (Tim17, NADH dehydrogenase (ubiquinone) 1β subcomplex subunit 2/NDUFB2, caspase recruitment domain-containing protein 12/CARD12, endonuclease G, cytochrome C reductase) [80]. However, it is important to emphasize that increased expression of most of these target genes using artificial reporters has not been confirmed for endogenous-related genes [81] (personal unpublished data)—an observation that could be explained by differential epigenetic regulation and chromatin remodeling between a naked DNA promoter in the reporter construct and the DNA of authentic endogenous promoters [81]. In addition, the response seems to be specific; no change has been detected in the abundance for the UPR^{er} gene markers (BiP/GRP-78/78 kDa glucose-regulated protein, endoplasmin/GRP-94/94 kDa glucose-regulated protein), for cytosolic chaperones (Heat shock cognate 71 kDa protein, heat shock-related 70 kDa protein 2), and the requirement for mitochondrial localization. Demonstrated as the induction of nuclear gene encoding, mitochondrial proteins were not observed when cells were transfected with OTCΔ lacking the mitochondrial signal peptide [77,79].

Since these works, few advances have been made in the comprehension of mechanisms by which a UPR^{mt} is set up in mammals. However, the pathological relevance of the protein kinase RNA-activated (PKR) and the UPR^{mt} emerged in two murine models of colitis as well as in patients with inflammatory bowel diseases (IBDs) [82]. The expression of PKR increases in Mode K intestinal epithelial cells that overexpress the truncated form of OTC (OTCΔ). Silencing experiments pointed out that PKR might play a role in the early phase of the UPR^{mt}; the silencing of the kinase expression using siRNA prevents the phosphorylation of c-Jun and the induction of CHOP-10 [82]. In addition, these authors showed a PKR-dependent phosphorylation of eIF2α, suggesting the concomitant induction of ISR and UPR^{mt}, as previously described in *C. elegans* [82]. Various strategies were used to disrupt the stoichiometric equilibrium between nuclear- and mitochondrial-encoded ETC (Electron Transport Chain) subunits to trigger a UPR^{mt}. Among them, doxycycline, an inhibitor of mitochondrial translation, triggered the activation of a UPR^{mt} activation in AML12 cells (a mouse hepatic cell line), as demonstrated by an increased expression of HSPD1 protein and the activation of a reporter gene driven by the ClpP promoter [45]. Similar results were obtained when mitochondrial protein abundance was increased by either increasing the NAD$^+$ concentration with two precursors of its synthesis (nicotinamide mononucleotide (NAM) and nicotinamide riboside (NR)) known to trigger the expression of several nuclear gene encoding mitochondrial proteins in an NAD-dependent protein deacetylase Sirtuin-1 (SIRT1)-dependent manner [83] or by boosting mitochondrial biogenesis by mTOR inhibition with rapamycin [67]. In primary murine hepatocytes, it was also confirmed that an antioxidant response mediated by superoxide dismutase 2 (SOD2) was associated with the UPR^{mt}, and that both responses were dependent on the Sirtuin-1 deacetylase [67]. Interestingly, at least one other Sirtuin is involved in the resolution of mitochondrial proteotoxic stress. Indeed, NAD-dependent protein deacetylase Sirtuin-3 (SIRT3) has been shown to be activated in response to a large panel of mitochondrial stresses known to induce the accumulation of protein aggregates and/or ROS, including antimycin A, rotenone, an inhibitor of heat shock protein HSP90 and the overexpression of mutated EndoG (ΔMLS-Endo G-His), mutated SOD1, and OTCΔ. These mitochondrial stresses activated an antioxidant response and mitophagy in a SIRT3-dependent manner, but the possible SIRT3-dependency of the UPR^{mt} (evaluated by CHOP-10 and HSP60 mRNA levels) was not reported. Interestingly, the proteotoxic stress resulted in a heterogeneous mitochondrial phenotype, a fraction of the mitochondria under severe stress. The cell viability under proteotoxic stress was compromised in the absence of SIRT3. Altogether, this set of data suggests that cells undergoing proteotoxic stress present a heterogeneous adaptive response with the induction of a UPR^{mt} to resolve the stress in mildly affected mitochondria and a SIRT3-orchestrated mitophagy to remove irreversibly damaged organelles [84].

Despite the requirement for HAF-1 to observe the activation of a UPR^{mt} in *C. elegans*, no clear mammalian orthologue of this protein has been identified thus far. However, Rainbolt and colleagues confirmed that, in mammalian cells, the expression of the worm Tim17 mutant and the reduction of mitochondrial imported induce UPR^{mt}. In addition, the knock down of Tim17 in HEK293 cells resulted in a slight increase in HSPD1 and YME1L1 mRNA abundance [74]. Interestingly, these authors proposed a stress pathway based on the phosphorylation of eIF2α and the attenuation of translation that leads to a down-regulation of Tim17 abundance, which is further enhanced by an increased YME1L1-dependent degradation of the protein. Ultimately, a lower abundance of Tim17 and, as a consequence, the reduced mitochondrial protein import might thus facilitate the activation of the UPR^{mt} [74].

As for HAF-1, no mammalian orthologue of ATFS-1 has been identified to date. Detailed mechanisms as well as several actors of the UPR^{mt} are thus still likely missing in mammals, especially regarding the effectors involved in the transcriptional regulation of target genes. However, both DVE-1 and UBL5 have mammalian orthologues: SATB2 and UBL5, respectively [68]. Interestingly, SATB2 is a global chromatin organizer, and this protein is also able to form a complex with UBL5; however, no evidence of involvement of SATB2/UBL5 in the mammalian UPR^{mt} has been experimentally demonstrated thus far [68].

The analysis of the HSPD1/HSPE1 bidirectional promoter revealed the existence of a CHOP-10-C/EBPβ binding element crucial for the activation of the UPR^{mt}-responsive genes. In addition, the abundance of these two transcription factors increased upon the overexpression of OTCΔ [79,80]. As in *C. elegans*, the current mammalian model of the UPR^{mt}-induced gene expression would be a two-step process. It would firstly require the expression of transcription factors, such as CHOP-10, which would in turn activate the expression of gene encoding of mitochondrial chaperones/chaperonins and proteases.

In terms of upstream signaling pathways, it seems that the phosphorylation of c-Jun by JNK2 and/or PKR is necessary to activate the expression of these transcription factors [77,82]. Phosphorylated c-Jun would bind to an AP-1 binding site within the CHOP-10 promoter, allowing the regulation of a UPR^{mt}-specific gene expression. These genes do not contain the well-described endoplasmic reticulum stress response element (ERSE) required for the classic genes regulated during the UPR^{er} [80]. However, considering the multiple cell responses involving CHOP-10 and the putative 3522 promoters of nuclear genes predicted to contain the binding consensus element for this transcription factor, it is likely that additional factors are required to specifically regulate gene expression during the UPR^{mt} [80]. The sequence alignment of the 1000 bp promoter sequences of the UPR^{mt}-induced genes revealed two well-conserved elements (except for the HSPD1/HSPE1 bidirectional promoter) surrounding the CHOP-10 binding site at a constant distance, named MURE1/2 (Mitochondrial unfolded protein response element 1/2) [80]. The requirement of these regulatory elements was demonstrated by reporter plasmid experiments as mutations within MURE1 or MURE2 to reduce the transcriptional activity of the UPR^{mt}-target genes triggered by the overexpression of OTCΔ [80]. However, the relevance of the MURE1-CHOP-10-MURE2 element might be questionable because (i) it has not been confirmed since then; (ii) this motif was widespread in the genome and was not restricted to the UPR^{mt} responsive genes (personal unpublished data); and (iii) the ligands putatively associated with these two regulatory elements have not yet been identified.

IMS-Associated mtUPR

In addition to the UPR^{mt} initiated by events occurring in the mitochondrial matrix, Germain and Papa described another UPR^{mt} model triggered by mitochondrial IMS protein aggregate accumulations, which caused the overexpression of a mutant form of endonuclease G (EndoG-N174A) in breast cancer MCF-7 cells, a condition that increases HTRA2/OMI protein abundance (a mitochondria-located serine protease that can be released by mitochondria during apoptosis) [85], NRF1 expression, and enhanced proteasome activity [78]. To contrast with OTCΔ-induced UPR^{mt}, the IMS stress response is not dependent on CHOP-10 expression, but rather relies on the ligand-independent activation of estrogen receptor α (ERα). It has been clearly shown that the overexpression of EndoG-N174A induces the RAC-α serine/threonine-protein kinase/AKT/PKB-dependent phosphorylation of ERα phosphorylation (Ser167). Additionally, AKT activation is dependent on ROS production triggered by the IMS stress as a treatment, with the antioxidant *N*-acetyl cysteine (NAC), completely abolishing the activation of ERα [78]. More recently, this research group deciphered events associated with IMS stress in triple negative cancer cells (lacking ERα) and highlighted the involvement of the mitochondrial SIRT3 in the regulation of the antioxidant response and mitophagy [84]. Strikingly, while these authors had previously shown that EndoG-N174A overexpression in MCF-7 cells did not induce a cell signaling response comparable to the one observed in response to the accumulation of truncated OTC in the mitochondrial matrix, in ERα-deficient breast cancer cells, they recently revealed that IMS stress also increases the expression of CHOP-10 and HSPD1 [84].

In conclusion, it seems that different retrograde responses that are dependent on the mitochondrial compartment and location at the origin of the stress could be initiated. Indeed, as we have seen, different quality control factors have been described, depending on the organelle compartment, such as ClpP and

HSPD1, for matrix proteins or serine protease HTRA2/Omi and the proteasome for IMS proteins. Thus, because initiating stress and responding effectors to recover protein homeostasis are different between the matrix and IMS stress, this may explain the existence of independent signaling pathways such as those described with CHOP-10 and OTCΔ, or ERα and EndoG-N174A. In a recent study, we showed that, the transcription factor CHOP-10 was systematically overexpressed when the expression of mtDNA was inhibited by different means and in different cell types, while the UPR^{mt}-related gene markers HSPD1 and ClpP were not induced [86]. Furthermore, increased expression of this transcription factor correlated with the activation of another stress-responsive pathway called the integrated stress response (ISR) [86].

1.3. The Integrated Stress Response (ISR)

Although the integrated stress response (ISR) is not usually described as a typical retrograde response, several studies reported its activation in response to mitochondrial dysfunction [86–90]. The next paragraph presents the main features of ISR as well as its crosstalk with mitochondrial stress because it is a complementary pathway to the UPR^{mt}.

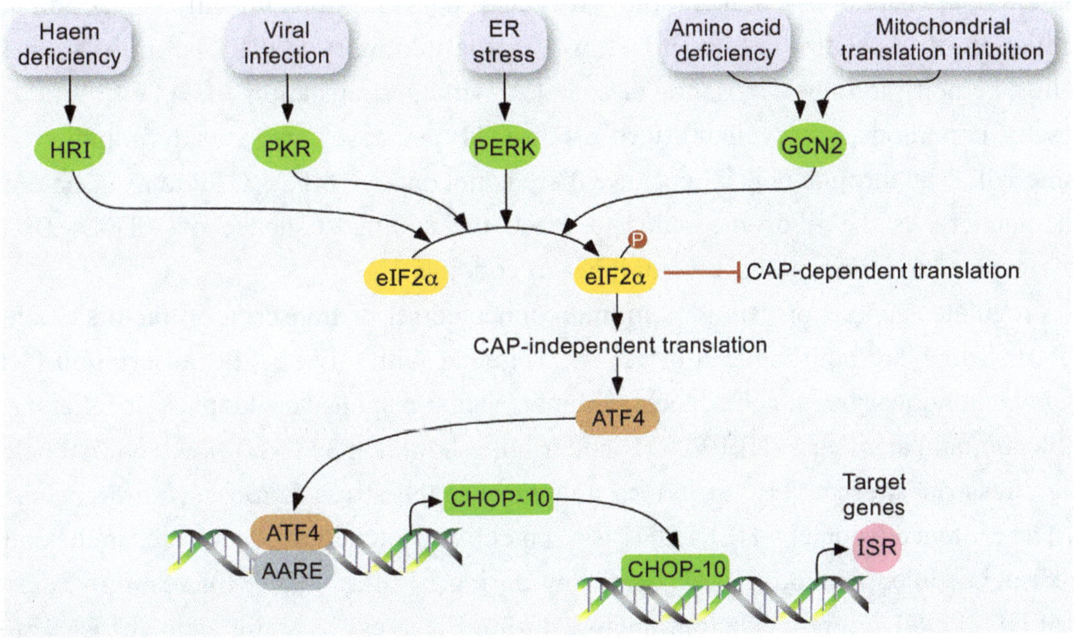

Figure 2. Hypothetical model for stress response to mitochondrial dysfunction. In response to various stresses, several kinases, such as HRI, PKR, PERK, and GCN2, are activated and converge to phosphorylate the translation initiation factor eIF2α. In turn, the phosphorylation of eIF2α has at least two consequences. While the cytosolic translation is globally attenuated as it inhibits CAP-dependent translation, mRNAs containing a uORF are preferentially translated, such as the transcription factor ATF4, which controls the expression of CHOP-10. The profit of this GCN2-eIF2α-ATF4 pathway might be stress attenuation gained by relieving the load of proteins imported into the mitochondria and increasing the expression of integrated stress responsive genes.

The ISR is a stress adaptive pathway conserved by evolution as recovered from yeast to human. In mammals, molecular effectors of this signaling pathway (Figure 2) are activated during ER stress, amino acid depletion [91], virus infection, oxidative stress, heme deprivation, or UV irradiation [92].

These stressors are sensed by four different kinases: Eukaryotic translation initiation factor 2-α kinase 1/heme-regulated inhibitor (HRI), PKR, RNA-dependent protein kinase-like endoplasmic reticulum kinase (PERK), and general control non-derepressible 2 (GCN2) that converge to the phosphorylation of eIF2α eukaryotic translation Initiation Factor 2α at serine 51 [93]. Phosphorylation of eIF2α is a central event of the ISR; it integrates upstream signals to reduce the ATP-consuming events, such as cytosolic translation, and to regulate the specific expression of a set of nuclear genes. Phosphorylated eIF2α (P-eIF2α) inhibits recycling of its GTP-bound form (active) by eIF2B, which is required for delivery of the methionyl-tRNA to the ribosome during translation initiation [94]. In addition, P-eIF2α also elicits preferential translation of mRNAs containing an upstream open reading frame (uORF) in their 5′-leader sequence such as the bZIP transcription factor activating transcription factor (4ATF4). Under unstressed conditions, translation initiation occurs at the start codon of the first uORF (uORF1) within the 5′-leader of ATF4 mRNA. While the ribosome scans the sequence, after translation termination at the uORF1, the ribosome does not dissociate from the mRNA and resumes scanning to reach uORF2. When eIF2α is not phosphorylated, it is rapidly recycled in its eIF2α-GTP form—a reaction that allows rapid re-initiation of translation at the site. However, the inhibitory uORF2 is out of frame from the ATF4 coding region, and the ribosome dissociates without translation of ATF4. When cells are under stressful conditions, the availability of eIF2α-GTP decreases and after termination at uORF1, the ribosome will scan through uORF2 because there is not enough eIF2α-GTP to re-initiate translation in a rapid manner. As the ribosome scans to reach the coding sequence of ATF4, eIF2α-GTP is sufficiently recycled to initiate translation at the start codon [92].

ISR also regulates gene expression by an interconnected set of transcription factors, such as ATF3 and CHOP-10, which are both induced by ATF4. Together with other bZIP transcription factors, they regulate many other downstream genes, such as homocysteine-responsive endoplasmic reticulum-resident ubiquitin-like domain member 1 (HERPUD1) and tribbles homolog 3 (TRIB3), which first help to either resolve the stress or, alternatively, to induce apoptosis if the stress is too severe or persists for too long [91]. The protein encoded by HERPUD1 is a protective protein that participates in the endoplasmic reticulum-associated degradation (ERAD) pathway during ER stress [95]. Interestingly, these authors showed that HERPUD1 might be neuroprotective during ER stress by stabilization of Ca^{2+} storage and maintenance of mitochondria function [95]. The tribbles homolog 3 protein is a pseudo-kinase because, despite having a region of high homology with kinase domains, it lacks a consensus sequence for protein phosphorylation. So far, TRIB3 is known to directly interact with different proteins, such as AKT, C/EBPs (CCAAT/enhancer binding proteins), and ATF4/CHOP-10, thereby regulating the biological process and responses such as insulin signaling, autophagy, adipogenesis, and induction of apoptosis, respectively [96].

The ISR is interconnected with several pathways, and some effectors are shared by other signaling such as the UPR^{er} or the UPR^{mt} [82]. In addition, the GCN2 kinase can be activated by amino acid depletion, glucose depletion, UV irradiation, proteasome inhibition, and oxidative stress [93]. The mechanism of activation upon amino acid starvation is well described. Under normal conditions, the kinase is kept inactive by auto-inhibitory molecular interactions. Upon binding of unloaded tRNA to the histidyl-tRNA synthetase-like domain, an allosteric rearrangement creates an open conformation of the protein, and GCN2 is activated by auto-phosphorylation (Thr898 and Thr903) within the activation loop [97].

This kinase is a good example to illustrate divergent signaling pathways of the cell responses following eIF2α phosphorylation. Indeed, while the signaling pathway downstream the P-eIF2α triggered by the UV-irradiation enhances the activation of NF-κB and cell survival, the phosphorylation of eIF2α in response to the inhibition of proteasome leads to the expression of CHOP-10 [98,99]. As mentioned, CHOP-10 is a pleiotropic transcription factor that can form heterodimers with other members of the bZIP family transcription factors, such as C/EBPs and ATF4, to function either as a trans-activator or a trans-repressor of target genes [100–102]. Moreover, while CHOP-10 has been associated with apoptosis in many different conditions, paradoxically, recent studies have revealed that it can also promote cell survival during ER stress and amino acid starvation by regulating the expression of several autophagy-related genes such as p62, ATG5, ATG7, and ATG10 [103,104]. The resulting pro- or anti-apoptotic response is most likely dependent on the integration and balance of multiple signals that are activated in parallel with the cell nuclear background, together with the type/origin, nature, severity, and duration of the stress.

In addition to this coordinated expression of chaperones and proteases, when the defect is too severe, the next level of quality control is the degradation of the entire organelle by specific autophagy called mitophagy [105]. We have seen clearly that mitochondria display several lines of quality control mechanisms: mitochondria-specific chaperones and proteases protect against misfolded proteins at the molecular level, and fission/fusion and mitophagy segregate and eliminate damage at the organelle level [9]. Therefore, if an increase in unfolded proteins in mitochondria activates a UPR^{mt} to increase chaperone production, the mitochondrial serine/threonine-protein kinase PINK1 and the E3 ubiquitin-protein ligase parkin, whose mutations cause familial Parkinson's disease, remove depolarized mitochondria through mitophagy [106]. This research group also showed that both phenomenon were linked; the expression of unfolded proteins in the matrix causes the accumulation of PINK1 on energetically healthy mitochondria, resulting in mitochondrial translocation of Parkin and mitophagy and subsequent reduction of unfolded protein load, a response that is strongly enhanced by the knockdown of the mitochondrial Lon protease homolog [105].

1.4. What Is the (Patho)Physiological Relevance of a UPR^{mt} in Mammals?

1.4.1. The UPR^{mt} and Neurodegenerative Diseases

The accumulation of unfolded or aggregated proteins is a hallmark of a number of neurodegenerative diseases, including Alzheimer's disease (AD) and Parkinson's disease (PD), although it is currently unclear if these protein aggregates participate or are the consequence of such pathologies. For instance, extracellular amyloid β deposits are characteristic of AD, but they are also present in mitochondria [107]. Amyloid β precursor protein is cleaved by the mitochondrial HTRA2/Omi serine protease [108] and the latter has been shown to delay the aggregation of the amyloid β 1–42 peptide [109]. Although HTRA2 is a target gene of IMS-induced UPR^{mt} (Figure 1), a putative role of the UPR^{mt} in this pathology has not been demonstrated clearly. Regarding PD, accumulation of α-synuclein has been demonstrated in the mitochondria of post-mortem brains of PD patients as well as in cell models of PD [110,111]. Interestingly, a higher level of misfolded respiratory proteins was detected in post-mortem brains of PD patients, which correlates with an elevated abundance of HSP60 [112], suggesting the occurrence of a UPR^{mt}. However, the UPR^{mt} is still poorly characterized in PD models, contrary to mitophagy, because PINK and/or Parkin deficiencies are strongly associated with PD [113]. Clearly, investigating the relationship between the UPR^{mt} and neurodegenerative diseases would be of great interest. In addition, the development of new tools to better monitor the UPR^{mt} *in vivo* in individual cells would be welcome, as the mitochondrial proteotoxic stress generates a heterogeneous phenotype associated with a UPR^{mt} and/or mitophagy [84].

1.4.2. The UPR^{mt} in Aging

One of the pioneering studies reporting a link between aging and the UPR^{mt} was the silencing of cco-1, a subunit of Cytochrome c oxidase, in neuronal tissue of *C. elegans*, which increases lifespan and induces a UPR^{mt} not only in neurons, but also in the intestine [55], suggesting an inter-tissue signaling that results in a UPR^{mt} activation. Since then, the UPR^{mt} has been implicated in increased lifespan in worms, flies, and mice, highlighting the importance of maintaining mitochondrial proteostasis for longevity. This aspect will not be developed here; several reviews have been recently devoted to UPR^{mt} and aging [56,62,114].

1.4.3. The UPR^{mt} and Stem Cells

Recently, Mohrin and co-workers highlighted a SIRT7-NRF1 regulatory axis in the UPR^{mt} and linked the UPR^{mt} to hematopoietic stem cells' (HSC) quiescence and aging [115]. They first demonstrated that the NAD$^+$-dependent deacetylase SIRT7 is induced in response to a UPR^{mt}. SIRT7 directly binds to the transcription factor NRF1, specifically on the promoter of gene encoding mitochondrial ribosomal proteins and mitochondrial translation factors, to reduce, indirectly, their expression. SIRT7 induction thus contributes to attenuating the mitochondrial translation, thereby decreasing the proteotoxic stress imposed on the organelle. In addition, SIRT7 deficiency results in a constitutive UPR^{mt}. Because stem cells undergo an oxidative metabolic shift when entering into differentiation, it is not surprising that SIRT7, by limiting mitochondria biogenesis and the UPR^{mt}, contributes to HSCs' quiescence and prevents differentiation. Furthermore, the activation of a UPR^{mt} observed in aged HSCs has been linked to SIRT7 decreasing in these cells, contributing to their decline. Interestingly, reintroduction of SIRT7 in aged HSCs reduces the UPR^{mt} and improves their regenerative capacities. Altogether, these authors concluded that the UPR^{mt} might be an aged-sensitive metabolic checkpoint to regulate HSCs' renewal properties.

1.4.4. The UPR^{mt} in Innate Immunity

Several bacterial pathogens target the mitochondria, disrupting essential functions, such as calcium and redox homeostasis, or mitochondrial morphology [116]. However, the impact of pathogens on the UPR^{mt} is still poorly described, especially in mammals. In *C. elegans*, several bacteria strains, and particularly *P. aeruginosa*, activate a UPR^{mt}, as shown by a specific chaperone induction in the intestine [117]. Because the worm is more susceptible to *P. aeruginosa* exposure when ATFS-1 expression is down-regulated, the UPR^{mt} might be protective against bacterial infection. In addition, the knockdown of an mtDNA-encoded ATP synthase subunit not only activates the UPR^{mt}, but also the expression of innate immune genes and makes the worm more resistant to *P. aeruginosa* exposure. Finally, these authors showed that human cells undergoing a UPR^{mt} also induce the transcription of antimicrobial peptides (reviewed in [113]). In line with these observations, PKR-mediated UPR^{mt} has been highlighted in two murine models of colitis, as well as in patients with inflammatory bowel diseases under inflammatory conditions [82]. The UPR^{mt} might thus represent one of the mechanisms that senses bacterial pathogens in the intestinal mucosae and adapts the cell functions and immune response (reviewed in [118]). However, it is still currently unclear whether PKR-dependent induction of a UPR^{mt} in colitis models is detrimental or beneficial due to controversial data found in the literature [82,119].

Importantly, the sensitivity to the UPR^{mt} may vary according to different parameters, including tissue-specificity and stress intensity. This has been demonstrated by Dogan and co-workers, who studied a mouse model that lacked a protein directly involved in mitochondrial translation, such as the gene DARS2 encoding the mitochondrial aspartate-tRNA ligase, specifically in heart and skeletal muscle tissues [120]. While both tissues presented a comparable mitochondrial respiratory deficiency, the strategies developed to cope with this proteotoxic stress were different, depending on the tissue. DARS2-deficient hearts from 6-week-old mice presented a mitochondrial cardiopathy, with an increase in mitochondria biogenesis and activation of a UPR^{mt}, accompanied by reduced autophagy markers. However, 3-week-old mice showed no sign of cardiopathy, suggesting that the mitochondrial stress imposed by the conditional loss of DARS2 had not reached a sufficient level to trigger these stress responses. Conversely, these stress responses were not observed in the skeletal muscles of 6-week-old mice despite a 60% to 80% decrease in the activity of respiratory chain complexes, presumably because a low turnover of mitochondrial transcripts was the preferred strategy to cope with mitochondrial stress in this tissue [120].

2. Conclusions

2.1. The UPR^{mt} versus ISR or Both in Response to a Mitochondrial Stress?

While the number of papers reporting on the UPR^{mt} is ever growing, it is interesting to note that this stress response is still poorly defined in terms of target genes and signaling pathways, especially in mammalians. This is probably reinforced by the fact that some researchers tend to designate UPR^{mt} as any kind of retrograde response triggered by a mitochondrial stress, even if the stress has not been associated with the accumulation of unfolded protein aggregates. Although the pioneering work performed by the group of Hoogenraad [79] established a large list of target genes specifically induced by the UPR^{mt} (and not by the UPR^{er}), these interesting findings were mainly obtained with reporter assays. Only a few of these target genes were shown later to be induced at the endogenous level by one or several of the mitochondria stressing conditions listed in Figure 1. These endogenous UPR^{mt} targets were CHOP-10 [79], HSP60 [78,79,82,115], ClpP [115], and YME1L1 [74], although the last three UPR^{mt} markers were generally induced in a modest manner. However, it is interesting to point out that other types of mitochondrial stresses, such as the inhibition of mitochondrial translation or the depletion of mtDNA [86], MELAS and NARP mtDNA point mutations [88], inhibition of the respiratory chain by rotenone and antimycin A [89], or loss of HTRA2 expression [90] do not induce HSP60 and ClpP expression, but rather activate the integrated stress response (ISR) characterized by an attenuation of the cytosolic translation achieved through the phosphorylation of eIF2α and by the induction of another set of target genes including CHOP-10, TRIB3, and HERPUD1 [86]. Importantly, Rainbolt and co-workers revealed that arsenite, an ETC inhibitor, triggered the ISR, with a transient phosphorylation of eIF2α, accompanied by the induction of HSP60 and Tim17 at the transcript level, but not at the protein level [74]. Tim17 protein abundance strongly decreased in response to arsenite due to both a decreased biogenesis and an increased YME1L1-dependent degradation, thereby reducing the mitochondrial protein import. Interestingly, depletion of Tim17 by RNA interference not only reduced mitochondrial import, but also modestly increased (1.5-fold) the transcript level of HSP60 and YME1L1 [74]. Altogether, this suggests a two-wave response to the arsenite-induced mitochondrial stress: the first wave activated the ISR, with a transient phosphorylation of eIF2α that reduces Tim17 synthesis. The lower mitochondrial import capacity associated with reduced Tim17 abundance in turn induced the transcription of some UPR^{mt} target genes. It is important to know whether cooperation between the ISR and the UPR^{mt} activation could be generalized to other types of mitochondrial stresses, which still needs to be addressed, although Haller's group has already shown that the UPR^{mt} observed in human intestinal epithelial cells overexpressing OTCΔ is dependent on PKR-mediated ISR [82]. Researchers interested in mitochondria-induced stress responses should therefore be encouraged to monitor the kinetics—systematically—of both the UPR^{mt} and the ISR markers, including the efficiency of the mitochondrial import capacity.

It is important to emphasize the fact that, while the ISR and the UPR^{mt} were first discriminated based on the source of mitochondrial stress generated (depletion of mtDNA *versus* accumulation of misfolded proteins) in experimental models, it is most likely that a certain level of overlap exists between these two signaling pathways, especially as several effectors are common to both cell responses. One can easily imagine that mtDNA depletion and mutations affecting ETC that are known to trigger ISR could also eventually lead to ROS-dependent oxidation of mitochondrial proteins and lipids, causing defects in folding or misassembling of mitochondrial protein complexes, initiating a UPR^{mt}.

2.2. Future Directions

Important questions remain unanswered regarding the UPR^{mt}, as well as mitochondria-induced ISR. First, while the link between a mitochondria proteotoxic stress and the cytosolic UPR^{mt} cascade is relatively clear in *C. elegans*, it remains poorly understood in mammals. Does the mammalian UPR^{mt} signaling depend on peptide efflux? If so, what are the peptide transporters and the primary cytosolic targets?

Second, the attenuation of cap-dependent translation, achieved through phosphorylation of eIF2α, is a shared and common feature between the UPR^{er}, the UPR^{mt}, and the ISR (Figures 1 and 2). These three pathways are powerful mechanisms for alleviating the proteotoxic stress imposed on organelles or cytosol. However, an intriguing feature is the kinetics of eIF2α phosphorylation: while transient in the case of a UPR^{er}, it is sustained (up to 30–48 h) in the case of a UPR^{mt} triggered by ΔOTC overexpression [82], and in the case of ISR, it is induced by the inhibition of mitochondrial translation [86].

In response to a UPR^{er}, it is known that the induction of encoding the protein phosphatase 1 regulatory subunit 15A/Growth arrest and DNA damage-inducible protein (PPP1R15A/GADD34) contributes to dephosphorylation of the phosphorylated form of eIF2α and thus resumes cap-dependent translation. One might wonder how cells deal with a reduced cap-dependent translation for as long as 2–3 days, a phenomenon accompanied by cell growth arrest in some cell types, but not all of them (Sébastien Michel, University of Namur, Belgium. Personal communication, 2014). During this period, several ISR-target genes are induced. Even if the presence of internal ribosome entry sites (IRES) has been described for at least some of them, one cannot exclude a specific regulation at the translation level on these ISR-responsive transcripts, following the mechanism of RNA regulon control [121]. In line with this hypothesis, a recent multi-layered "omics" dissection of mitochondrial activity in the liver proteome from 40 strains of the BXD mouse genetic reference population revealed that six members of the UPR^{mt} pathway are coordinately regulated; this occurs more strongly at the protein level than at the transcript level [122].

Third, the specificity of CHOP-10-dependent cell responses needs to be investigated. Indeed, CHOP-10 is induced in response to number of cell stress conditions, such as nutrient deprivation, viral infection, hypoxia, or accumulation of protein aggregates in organelles such as the endoplasmic reticulum and mitochondria. This transcription factor is undoubtedly a key actor for coordinating the adequate cell response that ranges from apoptosis, the UPR^{er}, and the UPR^{mt} to the ISR. In addition, CHOP-10 was also shown to reduce the proliferation of intestinal cells in models of acute and chronic colitis [123]. However, the molecular determinants that confer the specificity of CHOP-10-dependent stress-response are still largely unclear and might involve specific post-transcriptional modifications of CHOP-10 as well as the activation of additional transcriptional regulators.

Fourth, while inducing chaperones and proteases to restore proteostasis makes sense, the biological significance of some ISR target genes in the context of a mitochondrial proteotoxic stress remains undetermined. For instance, it has been shown that the induction of TRIB3 participates in a negative feedback loop to repress most of the ATF4-dependent ISR target genes [124]. However, the biological role of other ISR-induced genes, and particularly the induction of several aminoacyl-tRNA synthetases in response to mitochondrial stresses [88,124], still requires analysis. Briefly, in all eukaryotes, aminoacyl-tRNA synthetases are sequestrated in large multimeric complexes, where they ensure their primary role of tRNA aminoacylation. It has been shown that upon release from these complexes, they can carry out alternative functions. When yeasts are submitted to the diauxic transition from anaerobic glycolysis to oxidative respiration, two aminoacyl-tRNA synthetases, cERS and cMRS, dissociate from their cytosolic anchor complex, relocating to the mitochondria and the nucleus, respectively. While cERS stimulates mitochondrial translation, cMRS triggers the transcription of nuclear gene encoding of ATP synthase subunits, enabling synchronization of the expression of mitochondria-encoded and nuclear-encoded subunits of the complex V [125]. The fact that non-canonical functions have been recently highlighted for two tRNA synthetases in the yeast could pave the way for new researches needed in the future to fully understand signaling and effectors of the UPR^{mt} in mammalians.

Acknowledgments

Sebastien Michel was a recipient of a doctoral fellowship from the Fonds pour la Recherche dans l'Industrie et l'Agriculture (FRIA, Brussels, Belgium). The authors also thank the Belgian Association for Muscular Diseases for their support (ABMM, La Louvière, Belgium, grant 2013-08) and Michel Savels for the Figure layout.

Author Contributions

Thierry Arnould initiated the writing and mainly contributed to the writing of the mitochondrial retrograde signaling and UPR^{mt}. Sébastien Michel was involved in the different aspects of UPR^{mt} signaling in various species and Patricia Renard contributed to the physiological relevance of UPR^{mt} and drafted the figures. All authors contributed actively to the writing and the revision of the review.

References

1. Goffart, S.; Wiesner, R.J. Regulation and co-ordination of nuclear gene expression during mitochondrial biogenesis. *Exp. Physiol.* **2003**, *88*, 33–40.
2. Ryan, M.T.; Hoogenraad, N.J. Mitochondrial-nuclear communications. *Annu. Rev. Biochem.* **2007**, *76*, 701–722.
3. Collins, Y.; Chouchani, E.T.; James, A.M.; Menger, K.E.; Cocheme, H.M.; Murphy, M.P. Mitochondrial redox signalling at a glance. *J. Cell Sci.* **2012**, *125*, 801–806.
4. Lackner, L.L. Shaping the dynamic mitochondrial network. *BMC Biol.* **2014**, *12*, 35.
5. Van Vliet, A.R.; Verfaillie, T.; Agostinis, P. New functions of mitochondria associated membranes in cellular signaling. *Biochim. Biophys. Acta* **2014**, *1843*, 2253–2262.
6. Milane, L.; Trivedi, M.; Singh, A.; Talekar, M.; Amiji, M. Mitochondrial biology, targets, and drug delivery. *J. Control. Release* **2015**, *207*, 40–58.
7. Smiraglia, D.J.; Kulawiec, M.; Bistulfi, G.L.; Gupta, S.G.; Singh, K.K. A novel role for mitochondria in regulating epigenetic modification in the nucleus. *Cancer Biol. Ther.* **2008**, *7*, 1182–1190.
8. Kotiadis, V.N.; Duchen, M.R.; Osellame, L.D. Mitochondrial quality control and communications with the nucleus are important in maintaining mitochondrial function and cell health. *Biochim. Biophys. Acta* **2014**, *1840*, 1254–1265.
9. Michel, S.; Wanet, A.; de Pauw, A.; Rommelaere, G.; Arnould, T.; Renard, P. Crosstalk between mitochondrial (dys)function and mitochondrial abundance. *J. Cell. Physiol.* **2012**, *227*, 2297–2310.
10. Rugarli, E.I.; Langer, T. Mitochondrial quality control: A matter of life and death for neurons. *EMBO J.* **2012**, *31*, 1336–1349.

11. Biswas, G.; Adebanjo, O.A.; Freedman, B.D.; Anandatheerthavarada, H.K.; Vijayasarathy, C.; Zaidi, M.; Kotlikoff, M.; Avadhani, N.G. Retrograde Ca^{2+} signaling in C2C12 skeletal myocytes in response to mitochondrial genetic and metabolic stress: A novel mode of inter-organelle crosstalk. *EMBO J.* **1999**, *18*, 522–533.

12. Amuthan, G.; Biswas, G.; Ananadatheerthavarada, H.K.; Vijayasarathy, C.; Shephard, H.M.; Avadhani, N.G. Mitochondrial stress-induced calcium signaling, phenotypic changes and invasive behavior in human lung carcinoma A549 cells. *Oncogene* **2002**, *21*, 7839–7849.

13. Arnould, T.; Vankoningsloo, S.; Renard, P.; Houbion, A.; Ninane, N.; Demazy, C.; Remacle, J.; Raes, M. CREB activation induced by mitochondrial dysfunction is a new signaling pathway that impairs cell proliferation. *EMBO J.* **2002**, *21*, 53–63.

14. Whelan, S.P.; Zuckerbraun, B.S. Mitochondrial signaling: Forwards, backwards, and in between. *Oxid. Med. Cell. Longev.* **2013**, *2013*, 351613.

15. Cagin, U.; Enriquez, J.A. The complex crosstalk between mitochondria and the nucleus: What goes in between? *Int. J. Biochem. Cell Biol.* **2015**, *63*, 10–15.

16. Hock, M.B.; Kralli, A. Transcriptional control of mitochondrial biogenesis and function. *Annu. Rev. Physiol.* **2009**, *71*, 177–203.

17. Nakashima-Kamimura, N.; Asoh, S.; Ishibashi, Y.; Mukai, Y.; Shidara, Y.; Oda, H.; Munakata, K.; Goto, Y.; Ohta, S. MIDAS/GPP34, a nuclear gene product, regulates total mitochondrial mass in response to mitochondrial dysfunction. *J. Cell Sci.* **2005**, *118*, 5357–5367.

18. Parikh, V.S.; Morgan, M.M.; Scott, R.; Clements, L.S.; Butow, R.A. The mitochondrial genotype can influence nuclear gene expression in yeast. *Science* **1987**, *235*, 576–580.

19. Parikh, V.S.; Conrad-Webb, H.; Docherty, R.; Butow, R.A. Interaction between the yeast mitochondrial and nuclear genomes influences the abundance of novel transcripts derived from the spacer region of the nuclear ribosomal DNA repeat. *Mol. Cell. Biol.* **1989**, *9*, 1897–1907.

20. Jazwinski, S.M. The retrograde response: When mitochondrial quality control is not enough. *Biochim. Biophys. Acta* **2013**, *1833*, 400–409.

21. Liao, X.; Butow, R.A. *RTG1* and *RTG2*: Two yeast genes required for a novel path of communication from mitochondria to the nucleus. *Cell* **1993**, *72*, 61–71.

22. Biswas, G.; Anandatheerthavarada, H.K.; Zaidi, M.; Avadhani, N.G. Mitochondria to nucleus stress signaling: A distinctive mechanism of NF-κB/Rel activation through calcineurin-mediated inactivation of IκBβ. *J. Cell Biol.* **2003**, *161*, 507–519.

23. Vankoningsloo, S.; de Pauw, A.; Houbion, A.; Tejerina, S.; Demazy, C.; de Longueville, F.; Bertholet, V.; Renard, P.; Remacle, J.; Holvoet, P.; *et al.* CREB activation induced by mitochondrial dysfunction triggers triglyceride accumulation in 3T3-L1 preadipocytes. *J. Cell Sci.* **2006**, *119*, 1266–1282.

24. Guha, M.; Pan, H.; Fang, J.K.; Avadhani, N.G. Heterogeneous nuclear ribonucleoprotein A2 is a common transcriptional coactivator in the nuclear transcription response to mitochondrial respiratory stress. *Mol. Biol. Cell* **2009**, *20*, 4107–4119.

25. Acharya, P.; Engel, J.C.; Correia, M.A. Hepatic CYP3A suppression by high concentrations of proteasomal inhibitors: A consequence of endoplasmic reticulum (ER) stress induction, activation of RNA-dependent protein kinase-like ER-bound eukaryotic initiation factor 2α (eIF2α)-kinase (PERK) and general control nonderepressible-2 eIF2α kinase (GCN2), and global translational shutoff. *Mol. Pharmacol.* **2009**, *76*, 503–515.
26. Arnould, T.; Mercy, L.; Houbion, A.; Vankoningsloo, S.; Renard, P.; Pascal, T.; Ninane, N.; Demazy, C.; Raes, M. mtCLIC is up-regulated and maintains a mitochondrial membrane potential in mtDNA-depleted L929 cells. *FASEB J.* **2003**, *17*, 2145–2147.
27. Rohas, L.M.; St-Pierre, J.; Uldry, M.; Jager, S.; Handschin, C.; Spiegelman, B.M. A fundamental system of cellular energy homeostasis regulated by PGC-1α. *Proc. Natl. Acad. Sci. USA* **2007**, *104*, 7933–7938.
28. Guha, M.; Tang, W.; Sondheimer, N.; Avadhani, N.G. Role of calcineurin, hnRNPA2 and Akt in mitochondrial respiratory stress-mediated transcription activation of nuclear gene targets. *Biochim. Biophys. Acta* **2010**, *1797*, 1055–1065.
29. Itoh, K.; Wakabayashi, N.; Katoh, Y.; Ishii, T.; Igarashi, K.; Engel, J.D.; Yamamoto, M. Keap1 represses nuclear activation of antioxidant responsive elements by Nrf2 through binding to the amino-terminal Neh2 domain. *Genes Dev.* **1999**, *13*, 76–86.
30. Lerner, C.; Bitto, A.; Pulliam, D.; Nacarelli, T.; Konigsberg, M.; van Remmen, H.; Torres, C.; Sell, C. Reduced mammalian target of rapamycin activity facilitates mitochondrial retrograde signaling and increases life span in normal human fibroblasts. *Aging Cell* **2013**, *12*, 966–977.
31. Piantadosi, C.A.; Suliman, H.B. Mitochondrial transcription factor a induction by redox activation of nuclear respiratory factor 1. *J. Biol. Chem.* **2006**, *281*, 324–333.
32. Favre, C.; Zhdanov, A.; Leahy, M.; Papkovsky, D.; O'Connor, R. Mitochondrial pyrimidine nucleotide carrier (PNC1) regulates mitochondrial biogenesis and the invasive phenotype of cancer cells. *Oncogene* **2010**, *29*, 3964–3976.
33. Perez-de-Arce, K.; Foncea, R.; Leighton, F. Reactive oxygen species mediates homocysteine-induced mitochondrial biogenesis in human endothelial cells: Modulation by antioxidants. *Biochem. Biophys. Res. Commun.* **2005**, *338*, 1103–1109.
34. Storz, P.; Doppler, H.; Toker, A. Protein kinase D mediates mitochondrion-to-nucleus signaling and detoxification from mitochondrial reactive oxygen species. *Mol. Cell. Biol.* **2005**, *25*, 8520–8530.
35. Liu, B.; Chen, Y.; St Clair, D.K. ROS and p53: A versatile partnership. *Free Radic. Biol. Med.* **2008**, *44*, 1529–1535.
36. Bergeron, R.; Ren, J.M.; Cadman, K.S.; Moore, I.K.; Perret, P.; Pypaert, M.; Young, L.H.; Semenkovich, C.F.; Shulman, G.I. Chronic activation of AMP kinase results in NRF-1 activation and mitochondrial biogenesis. *Am. J. Physiol. Endocrinol. Metab.* **2001**, *281*, E1340–E1346.
37. Holmuhamedov, E.; Jahangir, A.; Bienengraeber, M.; Lewis, L.D.; Terzic, A. Deletion of mtDNA disrupts mitochondrial function and structure, but not biogenesis. *Mitochondrion* **2003**, *3*, 13–19.
38. Mercy, L.; Pauw, A.; Payen, L.; Tejerina, S.; Houbion, A.; Demazy, C.; Raes, M.; Renard, P.; Arnould, T. Mitochondrial biogenesis in mtDNA-depleted cells involves a Ca^{2+}-dependent pathway and a reduced mitochondrial protein import. *FEBS J.* **2005**, *272*, 5031–5055.

39. Chae, S.; Ahn, B.Y.; Byun, K.; Cho, Y.M.; Yu, M.H.; Lee, B.; Hwang, D.; Park, K.S. A systems approach for decoding mitochondrial retrograde signaling pathways. *Sci. Signal.* **2013**, *6*, rs4.
40. Hardie, D.G. The AMP-activated protein kinase pathway—New players upstream and downstream. *J. Cell Sci.* **2004**, *117*, 5479–5487.
41. Bolster, D.R.; Crozier, S.J.; Kimball, S.R.; Jefferson, L.S. AMP-activated protein kinase suppresses protein synthesis in rat skeletal muscle through down-regulated mammalian target of rapamycin (mTOR) signaling. *J. Biol. Chem.* **2002**, *277*, 23977–23980.
42. Jager, S.; Handschin, C.; St-Pierre, J.; Spiegelman, B.M. AMP-activated protein kinase (AMPK) action in skeletal muscle via direct phosphorylation of PGC-1α. *Proc. Natl. Acad. Sci. USA* **2007**, *104*, 12017–12022.
43. Krause, U.; Bertrand, L.; Hue, L. Control of p70 ribosomal protein S6 kinase and acetyl-CoA carboxylase by AMP-activated protein kinase and protein phosphatases in isolated hepatocytes. *Eur. J. Biochem.* **2002**, *269*, 3751–3759.
44. Inoki, K.; Zhu, T.; Guan, K.L. TSC2 mediates cellular energy response to control cell growth and survival. *Cell* **2003**, *115*, 577–590.
45. Houtkooper, R.H.; Mouchiroud, L.; Ryu, D.; Moullan, N.; Katsyuba, E.; Knott, G.; Williams, R.W.; Auwerx, J. Mitonuclear protein imbalance as a conserved longevity mechanism. *Nature* **2013**, *497*, 451–457.
46. Houtkooper, R.H.; Canto, C.; Wanders, R.J.; Auwerx, J. The secret life of NAD^+: An old metabolite controlling new metabolic signaling pathways. *Endocr. Rev.* **2010**, *31*, 194–223.
47. Daitoku, H.; Sakamaki, J.; Fukamizu, A. Regulation of FoxO transcription factors by acetylation and protein–protein interactions. *Biochim. Biophys. Acta* **2011**, *1813*, 1954–1960.
48. Madiraju, P.; Pande, S.V.; Prentki, M.; Madiraju, S.R. Mitochondrial acetylcarnitine provides acetyl groups for nuclear histone acetylation. *Epigenetics* **2009**, *4*, 399–403.
49. Spange, S.; Wagner, T.; Heinzel, T.; Kramer, O.H. Acetylation of non-histone proteins modulates cellular signalling at multiple levels. *Int. J. Biochem. Cell Biol.* **2009**, *41*, 185–198.
50. Wallace, D.C.; Fan, W. Energetics, epigenetics, mitochondrial genetics. *Mitochondrion* **2010**, *10*, 12–31.
51. Yun, J.; Finkel, T. Mitohormesis. *Cell Metab.* **2014**, *19*, 757–766.
52. Nunnari, J.; Suomalainen, A. Mitochondria: In sickness and in health. *Cell* **2012**, *148*, 1145–1159.
53. Ribas, F.; Villarroya, J.; Hondares, E.; Giralt, M.; Villarroya, F. FGF21 expression and release in muscle cells: Involvement of MyoD and regulation by mitochondria-driven signalling. *Biochem. J.* **2014**, *463*, 191–199.
54. Ji, K.; Zheng, J.; Lv, J.; Xu, J.; Ji, X.; Luo, Y.B.; Li, W.; Zhao, Y.; Yan, C. Skeletal muscle increase FGF21 expression in mitochondrial disorder to compensate for the energy metabolic insufficiency by activating mTOR-YY1-PGC1α pathway. *Free Radic. Biol. Med.* **2015**, *84*, 161–170.
55. Durieux, J.; Wolff, S.; Dillin, A. The cell-non-autonomous nature of electron transport chain-mediated longevity. *Cell* **2011**, *144*, 79–91.
56. Schulz, A.M.; Haynes, C.M. UPR^{mt}-mediated cytoprotection and organismal aging. *Biochim. Biophys. Acta* **2015**, doi:10.1016/j.bbabio.2015.03.008.
57. Chakrabarti, A.; Chen, A.W.; Varner, J.D. A review of the mammalian unfolded protein response. *Biotechnol. Bioeng.* **2011**, *108*, 2777–2793.

58. Vannuvel, K.; Renard, P.; Raes, M.; Arnould, T. Functional and morphological impact of ER stress on mitochondria. *J. Cell. Physiol.* **2013**, *228*, 1802–1818.
59. Senft, D.; Ronai, Z.A. UPR, autophagy, and mitochondria crosstalk underlies the ER stress response. *Trends Biochem. Sci.* **2015**, *40*, 141–148.
60. Martinus, R.D.; Garth, G.P.; Webster, T.L.; Cartwright, P.; Naylor, D.J.; Hoj, P.B.; Hoogenraad, N.J. Selective induction of mitochondrial chaperones in response to loss of the mitochondrial genome. *Eur. J. Biochem.* **1996**, *240*, 98–103.
61. Bennett, C.F.; Kaeberlein, M. The mitochondrial unfolded protein response and increased longevity: Cause, consequence, or correlation? *Exp. Gerontol.* **2014**, *56*, 142–146.
62. Jensen, M.B.; Jasper, H. Mitochondrial proteostasis in the control of aging and longevity. *Cell Metab.* **2014**, *20*, 214–225.
63. Jovaisaite, V.; Mouchiroud, L.; Auwerx, J. The mitochondrial unfolded protein response, a conserved stress response pathway with implications in health and disease. *J. Exp. Biol.* **2014**, *217*, 137–143.
64. Jovaisaite, V.; Auwerx, J. The mitochondrial unfolded protein response-synchronizing genomes. *Curr. Opin. Cell Biol.* **2015**, *33*, 74–81.
65. Yoneda, T.; Benedetti, C.; Urano, F.; Clark, S.G.; Harding, H.P.; Ron, D. Compartment-specific perturbation of protein handling activates genes encoding mitochondrial chaperones. *J. Cell Sci.* **2004**, *117*, 4055–4066.
66. Taanman, J.W. The mitochondrial genome: Structure, transcription, translation and replication. *Biochim. Biophys. Acta* **1999**, *1410*, 103–123.
67. Mouchiroud, L.; Houtkooper, R.H.; Moullan, N.; Katsyuba, E.; Ryu, D.; Canto, C.; Mottis, A.; Jo, Y.S.; Viswanathan, M.; Schoonjans, K.; *et al.* The NAD^+/sirtuin pathway modulates longevity through activation of mitochondrial UPR and FoxO signaling. *Cell* **2013**, *154*, 430–441.
68. Haynes, C.M.; Petrova, K.; Benedetti, C.; Yang, Y.; Ron, D. ClpP mediates activation of a mitochondrial unfolded protein response in *C. elegans*. *Dev. Cell* **2007**, *13*, 467–480.
69. Haynes, C.M.; Yang, Y.; Blais, S.P.; Neubert, T.A.; Ron, D. The matrix peptide exporter HAF-1 signals a mitochondrial UPR by activating the transcription factor ZC376.7 in *C. Elegans*. *Mol. Cell* **2010**, *37*, 529–540.
70. Nargund, A.M.; Pellegrino, M.W.; Fiorese, C.J.; Baker, B.M.; Haynes, C.M. Mitochondrial import efficiency of ATFS-1 regulates mitochondrial UPR activation. *Science* **2012**, *337*, 587–590.
71. Harbauer, A.B.; Zahedi, R.P.; Sickmann, A.; Pfanner, N.; Meisinger, C. The protein import machinery of mitochondria—A regulatory hub in metabolism, stress, and disease. *Cell Metab.* **2014**, *19*, 357–372.
72. Nargund, A.M.; Fiorese, C.J.; Pellegrino, M.W.; Deng, P.; Haynes, C.M. Mitochondrial and nuclear accumulation of the transcription factor ATFS-1 promotes OXPHOS recovery during the UPR^{mt}. *Mol. Cell* **2015**, *58*, 123–133.
73. Cheng, M.Y.; Hartl, F.U.; Horwich, A.L. The mitochondrial chaperonin HSP60 is required for its own assembly. *Nature* **1990**, *348*, 455–458.
74. Rainbolt, T.K.; Atanassova, N.; Genereux, J.C.; Wiseman, R.L. Stress-regulated translational attenuation adapts mitochondrial protein import through Tim17A degradation. *Cell Metab.* **2013**, *18*, 908–919.

75. Baker, B.M.; Nargund, A.M.; Sun, T.; Haynes, C.M. Protective coupling of mitochondrial function and protein synthesis via the eIF2α kinase GCN-2. *PLoS Genet.* **2012**, *8*, e1002760.
76. Kincaid, B.; Bossy-Wetzel, E. Forever young: SIRT3 a shield against mitochondrial meltdown, aging, and neurodegeneration. *Front. Aging Neurosci.* **2013**, *5*, 48.
77. Horibe, T.; Hoogenraad, N.J. The *chop* gene contains an element for the positive regulation of the mitochondrial unfolded protein response. *PLoS ONE* **2007**, *2*, e835.
78. Papa, L.; Germain, D. Estrogen receptor mediates a distinct mitochondrial unfolded protein response. *J. Cell Sci.* **2011**, *124*, 1396–1402.
79. Zhao, Q.; Wang, J.; Levichkin, I.V.; Stasinopoulos, S.; Ryan, M.T.; Hoogenraad, N.J. A mitochondrial specific stress response in mammalian cells. *EMBO J.* **2002**, *21*, 4411–4419.
80. Aldridge, J.E.; Horibe, T.; Hoogenraad, N.J. Discovery of genes activated by the mitochondrial unfolded protein response (mtUPR) and cognate promoter elements. *PLoS ONE* **2007**, *2*, e874.
81. Cairns, B.R. The logic of chromatin architecture and remodelling at promoters. *Nature* **2009**, *461*, 193–198.
82. Rath, E.; Berger, E.; Messlik, A.; Nunes, T.; Liu, B.; Kim, S.C.; Hoogenraad, N.; Sans, M.; Sartor, R.B.; Haller, D. Induction of dsRNA-activated protein kinase links mitochondrial unfolded protein response to the pathogenesis of intestinal inflammation. *Gut* **2012**, *61*, 1269–1278.
83. Brenmoehl, J.; Hoeflich, A. Dual control of mitochondrial biogenesis by sirtuin 1 and sirtuin 3. *Mitochondrion* **2013**, *13*, 755–761.
84. Papa, L.; Germain, D. SirT3 regulates the mitochondrial unfolded protein response. *Mol. Cell. Biol.* **2014**, *34*, 699–710.
85. Hegde, R.; Srinivasula, S.M.; Zhang, Z.; Wassell, R.; Mukattash, R.; Cilenti, L.; DuBois, G.; Lazebnik, Y.; Zervos, A.S.; Fernandes-Alnemri, T.; *et al.* Identification of Omi/HtrA2 as a mitochondrial apoptotic serine protease that disrupts inhibitor of apoptosis protein-caspase interaction. *J. Biol. Chem.* **2002**, *277*, 432–438.
86. Michel, S.; Canonne, M.; Arnould, T.; Renard, P. Inhibition of mitochondrial genome expression triggers the activation of chop-10 by a cell signaling dependent on the integrated stress response but not the mitochondrial unfolded protein response. *Mitochondrion* **2015**, *21*, 58–68.
87. Cortopassi, G.; Danielson, S.; Alemi, M.; Zhan, S.S.; Tong, W.; Carelli, V.; Martinuzzi, A.; Marzuki, S.; Majamaa, K.; Wong, A. Mitochondrial disease activates transcripts of the unfolded protein response and cell cycle and inhibits vesicular secretion and oligodendrocyte-specific transcripts. *Mitochondrion* **2006**, *6*, 161–175.
88. Fujita, Y.; Ito, M.; Nozawa, Y.; Yoneda, M.; Oshida, Y.; Tanaka, M. CHOP (C/EBP homologous protein) and ASNS (asparagine synthetase) induction in cybrid cells harboring MELAS and NARP mitochondrial DNA mutations. *Mitochondrion* **2007**, *7*, 80–88.
89. Ishikawa, F.; Akimoto, T.; Yamamoto, H.; Araki, Y.; Yoshie, T.; Mori, K.; Hayashi, H.; Nose, K.; Shibanuma, M. Gene expression profiling identifies a role for CHOP during inhibition of the mitochondrial respiratory chain. *J. Biochem.* **2009**, *146*, 123–132.
90. Moisoi, N.; Klupsch, K.; Fedele, V.; East, P.; Sharma, S.; Renton, A.; Plun-Favreau, H.; Edwards, R.E.; Teismann, P.; Esposti, M.D.; *et al.* Mitochondrial dysfunction triggered by loss of HtrA2 results in the activation of a brain-specific transcriptional stress response. *Cell Death Differ.* **2009**, *16*, 449–464.

91. Harding, H.P.; Zhang, Y.; Zeng, H.; Novoa, I.; Lu, P.D.; Calfon, M.; Sadri, N.; Yun, C.; Popko, B.; Paules, R.; *et al.* An integrated stress response regulates amino acid metabolism and resistance to oxidative stress. *Mol. Cell* **2003**, *11*, 619–633.
92. Wek, R.C.; Jiang, H.Y.; Anthony, T.G. Coping with stress: eIF2 kinases and translational control. *Biochem. Soc. Trans.* **2006**, *34*, 7–11.
93. Donnelly, N.; Gorman, A.M.; Gupta, S.; Samali, A. The eIF2α kinases: Their structures and functions. *Cell Mol. Life Sci.* **2013**, *70*, 3493–3511.
94. Baird, T.D.; Wek, R.C. Eukaryotic initiation factor 2 phosphorylation and translational control in metabolism. *Adv. Nutr.* **2012**, *3*, 307–321.
95. Chan, S.L.; Fu, W.; Zhang, P.; Cheng, A.; Lee, J.; Kokame, K.; Mattson, M.P. Herp stabilizes neuronal Ca^{2+} homeostasis and mitochondrial function during endoplasmic reticulum stress. *J. Biol. Chem.* **2004**, *279*, 28733–28743.
96. Prudente, S.; Sesti, G.; Pandolfi, A.; Andreozzi, F.; Consoli, A.; Trischitta, V. The mammalian tribbles homolog TRIB3, glucose homeostasis, and cardiovascular diseases. *Endocr. Rev.* **2012**, *33*, 526–546.
97. Padyana, A.K.; Qiu, H.; Roll-Mecak, A.; Hinnebusch, A.G.; Burley, S.K. Structural basis for autoinhibition and mutational activation of eukaryotic initiation factor 2α protein kinase GCN2. *J. Biol. Chem.* **2005**, *280*, 29289–29299.
98. Jiang, H.Y.; Wek, R.C. GCN2 phosphorylation of eIF2α activates NF-κB in response to UV irradiation. *Biochem. J.* **2005**, *385*, 371–380.
99. Jiang, H.Y.; Wek, R.C. Phosphorylation of the α-subunit of the eukaryotic initiation factor-2 (eIF2α) reduces protein synthesis and enhances apoptosis in response to proteasome inhibition. *J. Biol. Chem.* **2005**, *280*, 14189–14202.
100. Ron, D.; Habener, J.F. CHOP, a novel developmentally regulated nuclear protein that dimerizes with transcription factors C/EBP and LAP and functions as a dominant-negative inhibitor of gene transcription. *Genes Dev.* **1992**, *6*, 439–453.
101. Ohoka, N.; Yoshii, S.; Hattori, T.; Onozaki, K.; Hayashi, H. TRB3, a novel ER stress-inducible gene, is induced via ATF4-CHOP pathway and is involved in cell death. *EMBO J.* **2005**, *24*, 1243–1255.
102. Su, N.; Kilberg, M.S. C/EBP homology protein (CHOP) interacts with activating transcription factor 4 (ATF4) and negatively regulates the stress-dependent induction of the asparagine synthetase gene. *J. Biol. Chem.* **2008**, *283*, 35106–35117.
103. B'Chir, W.; Maurin, A.C.; Carraro, V.; Averous, J.; Jousse, C.; Muranishi, Y.; Parry, L.; Stepien, G.; Fafournoux, P.; Bruhat, A. The eIF2α/ATF4 pathway is essential for stress-induced autophagy gene expression. *Nucleic Acids Res.* **2013**, *41*, 7683–7699.
104. B'Chir, W.; Chaveroux, C.; Carraro, V.; Averous, J.; Maurin, A.C.; Jousse, C.; Muranishi, Y.; Parry, L.; Fafournoux, P.; Bruhat, A. Dual role for CHOP in the crosstalk between autophagy and apoptosis to determine cell fate in response to amino acid deprivation. *Cell Signal.* **2014**, *26*, 1385–1391.
105. Jin, S.M.; Youle, R.J. The accumulation of misfolded proteins in the mitochondrial matrix is sensed by PINK1 to induce PARK2/Parkin-mediated mitophagy of polarized mitochondria. *Autophagy* **2013**, *9*, 1750–1757.

106. Jin, S.M.; Lazarou, M.; Wang, C.; Kane, L.A.; Narendra, D.P.; Youle, R.J. Mitochondrial membrane potential regulates PINK1 import and proteolytic destabilization by PARL. *J. Cell Biol.* **2010**, *191*, 933–942.

107. Caspersen, C.; Wang, N.; Yao, J.; Sosunov, A.; Chen, X.; Lustbader, J.W.; Xu, H.W.; Stern, D.; McKhann, G.; Yan, S.D. Mitochondrial Aβ: A potential focal point for neuronal metabolic dysfunction in Alzheimer's disease. *FASEB J.* **2005**, *19*, 2040–2041.

108. Park, H.J.; Kim, S.S.; Seong, Y.M.; Kim, K.H.; Goo, H.G.; Yoon, E.J.; Min, D.S.; Kang, S.; Rhim, H. β-Amyloid precursor protein is a direct cleavage target of HtrA2 serine protease. Implications for the physiological function of HtrA2 in the mitochondria. *J. Biol. Chem.* **2006**, *281*, 34277–34287.

109. Kooistra, J.; Milojevic, J.; Melacini, G.; Ortega, J. A new function of human HtrA2 as an amyloid-β oligomerization inhibitor. *J. Alzheimer's Dis.* **2009**, *17*, 281–294.

110. Devi, L.; Raghavendran, V.; Prabhu, B.M.; Avadhani, N.G.; Anandatheerthavarada, H.K. Mitochondrial import and accumulation of α-synuclein impair complex I in human dopaminergic neuronal cultures and Parkinson disease brain. *J. Biol. Chem.* **2008**, *283*, 9089–9100.

111. Shavali, S.; Brown-Borg, H.M.; Ebadi, M.; Porter, J. Mitochondrial localization of α-synuclein protein in α-synuclein overexpressing cells. *Neurosci. Lett.* **2008**, *439*, 125–128.

112. De Castro, I.P.; Costa, A.C.; Lam, D.; Tufi, R.; Fedele, V.; Moisoi, N.; Dinsdale, D.; Deas, E.; Loh, S.H.; Martins, L.M. Genetic analysis of mitochondrial protein misfolding in *Drosophila melanogaster*. *Cell Death Differ.* **2012**, *19*, 1308–1316.

113. Pellegrino, M.W.; Haynes, C.M. Mitophagy and the mitochondrial unfolded protein response in neurodegeneration and bacterial infection. *BMC Biol.* **2015**, *13*, 22.

114. Hill, S.; van Remmen, H. Mitochondrial stress signaling in longevity: A new role for mitochondrial function in aging. *Redox Biol.* **2014**, *2*, 936–944.

115. Mohrin, M.; Shin, J.; Liu, Y.; Brown, K.; Luo, H.; Xi, Y.; Haynes, C.M.; Chen, D. Stem cell aging. A mitochondrial UPR-mediated metabolic checkpoint regulates hematopoietic stem cell aging. *Science* **2015**, *347*, 1374–1377.

116. Lobet, E.; Letesson, J.J.; Arnould, T. Mitochondria: A target for bacteria. *Biochem. Pharmacol.* **2015**, *94*, 173–185.

117. Pellegrino, M.W.; Nargund, A.M.; Kirienko, N.V.; Gillis, R.; Fiorese, C.J.; Haynes, C.M. Mitochondrial UPR-regulated innate immunity provides resistance to pathogen infection. *Nature* **2014**, *516*, 414–417.

118. Rath, E.; Haller, D. Unfolded protein responses in the intestinal epithelium: Sensors for the microbial and metabolic environment. *J. Clin. Gastroenterol.* **2012**, *46*, S3–S5.

119. Cao, S.S.; Song, B.; Kaufman, R.J. PKR protects colonic epithelium against colitis through the unfolded protein response and prosurvival signaling. *Inflamm. Bowel Dis.* **2012**, *18*, 1735–1742.

120. Dogan, S.A.; Pujol, C.; Maiti, P.; Kukat, A.; Wang, S.; Hermans, S.; Senft, K.; Wibom, R.; Rugarli, E.I.; Trifunovic, A. Tissue-specific loss of DARS2 activates stress responses independently of respiratory chain deficiency in the heart. *Cell Metab.* **2014**, *19*, 458–469.

121. Blackinton, J.G.; Keene, J.D. Post-transcriptional RNA regulons affecting cell cycle and proliferation. *Semin. Cell Dev. Biol.* **2014**, *34*, 44–54.

122. Wu, Y.; Williams, E.G.; Dubuis, S.; Mottis, A.; Jovaisaite, V.; Houten, S.M.; Argmann, C.A.; Faridi, P.; Wolski, W.; Kutalik, Z.; *et al.* Multilayered genetic and omics dissection of mitochondrial activity in a mouse reference population. *Cell* **2014**, *158*, 1415–1430.
123. Waldschmitt, N.; Berger, E.; Rath, E.; Sartor, R.B.; Weigmann, B.; Heikenwalder, M.; Gerhard, M.; Janssen, K.P.; Haller, D. C/EBP homologous protein inhibits tissue repair in response to gut injury and is inversely regulated with chronic inflammation. *Mucosal Immunol.* **2014**, *7*, 1452–1466.
124. Jousse, C.; Deval, C.; Maurin, A.C.; Parry, L.; Cherasse, Y.; Chaveroux, C.; Lefloch, R.; Lenormand, P.; Bruhat, A.; Fafournoux, P. TRB3 inhibits the transcriptional activation of stress-regulated genes by a negative feedback on the ATF4 pathway. *J. Biol. Chem.* **2007**, *282*, 15851–15861.
125. Frechin, M.; Enkler, L.; Tetaud, E.; Laporte, D.; Senger, B.; Blancard, C.; Hammann, P.; Bader, G.; Clauder-Munster, S.; Steinmetz, L.M.; *et al.* Expression of nuclear and mitochondrial genes encoding ATP synthase is synchronized by disassembly of a multisynthetase complex. *Mol. Cell* **2014**, *56*, 763–776.

9

Mitochondrial Mechanisms in Septic Cardiomyopathy

María Cecilia Cimolai [1,2], Silvia Alvarez [2], Christoph Bode [1] and Heiko Bugger [1,*]

[1] Department of Cardiology and Angiology, Heart Center Freiburg University, Hugstetter Str. 55, 79106 Freiburg, Germany; E-Mails: cecilia.cimolai@universitaets-herzzentrum.de (M.C.C.); christoph.bode@universitaets-herzzentrum.de (C.B.)

[2] Institute of Biochemistry and Molecular Medicine, School of Pharmacy and Biochemistry, University of Buenos Aires-National Scientific and Technical Research Council (UBA-CONICET), Junín 956, C1113AAD Buenos Aires, Argentina; E-Mail: salvarez@ffyb.uba.ar

* Author to whom correspondence should be addressed;
E-Mail: heiko.bugger@universitaets-herzzentrum.de

Academic Editors: Giuseppe Coppotelli and Jaime M. Ross

Abstract: Sepsis is the manifestation of the immune and inflammatory response to infection that may ultimately result in multi organ failure. Despite the therapeutic strategies that have been used up to now, sepsis and septic shock remain a leading cause of death in critically ill patients. Myocardial dysfunction is a well-described complication of severe sepsis, also referred to as septic cardiomyopathy, which may progress to right and left ventricular pump failure. Many substances and mechanisms seem to be involved in myocardial dysfunction in sepsis, including toxins, cytokines, nitric oxide, complement activation, apoptosis and energy metabolic derangements. Nevertheless, the precise underlying molecular mechanisms as well as their significance in the pathogenesis of septic cardiomyopathy remain incompletely understood. A well-investigated abnormality in septic cardiomyopathy is mitochondrial dysfunction, which likely contributes to cardiac dysfunction by causing myocardial energy depletion. A number of mechanisms have been proposed to cause mitochondrial dysfunction in septic cardiomyopathy, although it remains controversially discussed whether some mechanisms impair mitochondrial function or serve to restore mitochondrial function. The purpose of this review is to discuss mitochondrial mechanisms that may causally contribute to mitochondrial dysfunction and/or may represent adaptive responses to mitochondrial dysfunction in septic cardiomyopathy.

Keywords: septic cardiomyopathy; mitochondrial dysfunction; heart; bioenergetics

1. Introduction

Sepsis is responsible for millions of deaths worldwide each year and is a frequent cause of death in people who have been hospitalized [1]. In the United States, 3 in 1000 people suffer from sepsis, and severe sepsis contributes to more than 200,000 deaths per year [2]. Sepsis is the manifestation of the immune and inflammatory response to infection that may ultimately result in multi organ failure. It is believed that the release of pro-inflammatory mediators is not answered by an appropriate anti-inflammatory response and overwhelms the immune system, resulting in an uncontrolled excessive inflammatory state and an inability to neutralize pathogens [3]. Both effects of bacterial endotoxins and exotoxins, as well as an excessive release of a large number of cytokines contribute to organ dysfunction and failure during sepsis and septic shock. Guidelines recommend early treatment with antibiotics, to restore fluid deficits, and vasopressor treatment if necessary, actions that are critical for patient outcomes and overall survival. Besides specifically neutralizing the pathogen with antibiotics, the currently recommended therapeutic strategies are symptomatic, and no specific treatment is available to treat the consequences of specific organ damage in sepsis.

End-organ damage and organ failure in sepsis affects the most significant organs of the body, including the heart. Myocardial dysfunction is a well-described complication of severe sepsis, also referred to as septic cardiomyopathy, which includes both systolic and diastolic dysfunction [4]. Adequate O_2 supply in sepsis suggests that myocardial depression is not related to tissue hypoperfusion but to the presence of circulating depressant factors, or other mechanisms [5]. A number of mechanisms have been proposed to be involved in myocardial dysfunction in sepsis, including toxins, cytokines, nitric oxide, complement activation, apoptosis and energy metabolic derangements [6,7]. The latter two mechanisms imply that mitochondrial function may be compromised in septic cardiomyopathy. Already a century ago, it had been recognized that low arterial oxygen tension, low circulating hemoglobin levels and/or tissue hypoperfusion may lead to tissue hypoxia during sepsis, thereby contributing to impaired ATP generation [8–10]. Others demonstrated that these parameters were not impaired or even improved during sepsis [11,12]. Instead, more recent studies strongly suggest that organ dysfunction in sepsis, including myocardial depression, is rather related to the presence of circulating depressant factors or other mechanisms, leading to cellular energy depletion [5].

Indeed, a number of studies identified myocardial mitochondrial dysfunction in septic conditions both in animal models and humans [13–18]. Since the heart is highly dependent on continuous delivery of ATP to maintain contractile function, impairment in mitochondrial function is energetically detrimental for the heart. For a number of cardiac pathologies, mitochondrial dysfunction and energy depletion have been proposed to significantly contribute to cardiac dysfunction, including prevalent cardiac diseases such as ischemia reperfusion injury, diabetic cardiomyopathy and heart failure [19–21]. While a number of mechanisms have been proposed to causally contribute to myocardial mitochondrial dysfunction in sepsis, a number of other mitochondria-related alterations occur in septic cardiomyopathy the purpose of which may be to restore mitochondrial function. In the following sections, we will review the evidence of mitochondrial dysfunction in septic cardiomyopathy and discuss mitochondrial mechanisms that may causally contribute to mitochondrial dysfunction and/or may represent adaptive responses to mitochondrial dysfunction in septic cardiomyopathy.

2. Myocardial Mitochondrial Dysfunction in Sepsis

In the heart, high-energy phosphates are mainly generated from the oxidation of fatty acids, while glucose and other substrates contribute to myocardial ATP regeneration to a lesser extent. Substrate oxidation yields mainly NADH (reduced nicotinamide adenine dinucleotide) and $FADH_2$ (reduced flavin adenine dinucleotide), which feed electrons into the electron transport chain of oxidative phosphorylation (OXPHOS) via complex I and II. Electron transport results in the reduction of O_2 to H_2O, which is coupled to ATP regeneration by the F_0F_1-ATPase (ATP synthase). ATP exits the mitochondrion and is mainly utilized by myosin ATPases to maintain contractile function, whereas a minor part of ATP is used to maintain cellular ion homeostasis. Of note, the entire pool of myocardial ATP is turned over completely every few seconds, emphasizing that mitochondrial energy substrate metabolism is critically important to maintain ATP availability and thus cardiac function.

Mitochondrial dysfunction is widely discussed as a crucial mechanism of organ dysfunction in sepsis, including the heart, although only few data on mitochondrial function are available from human septic hearts. Takasu and colleagues reported edema of the mitochondrial matrix, associated with cystic alterations of the cristae and collapse into small myelin-like clusters in hearts of septic patients [22]. Soriano *et al.* observed that patients that did not survive sepsis presented a more severe degree of cardiac dysfunction compared to survivors, and non-survivors also showed derangements in mitochondrial cristae [13]. In contrast, numerous animal studies reported myocardial mitochondrial dysfunction in sepsis. The most commonly used animal models of sepsis are the induction of sepsis by injection of an exogenous bacterial toxin, mostly lipopolysaccharide (LPS), or the induction of intestinal leakage by cecal ligation and puncture (CLP) or colon ascendant stent peritonitis (CASP) [23]. Using these models, it was well documented by Smeding and colleagues [24] that, in the majority of the cases, the impairment in cardiac mitochondrial function correlated with decreased cardiac contractility, both measured *in vivo* by echochardiography or *ex vivo* in the isolated heart perfusion. Mitochondrial dysfunction was characterized by decreased rates of State 3 respiration and ATP synthesis, decreased respiratory control ratios and membrane potential, decreased activities of mitochondrial OXPHOS Complexes, increased rates of State 4 respiration, and increased mitochondrial size and fragility [14,25–29]. Several authors reported a decrease in State 3 respiration in the heart using glutamate as a Complex I substrate, whereas an impairment in Complex II-mediated respiration was less consistent [29–31]. Decreased activity or expression of individual OXPHOS Complexes was also widely reported, in particular Complex I. Decreased State 3 respiration was frequently associated with a decreased respiratory control ratio (RCR), although the lower RCR was sometimes related to increased State 4 respiration [26], which could be indicative of mitochondrial uncoupling. In addition, a decrease of the mitochondrial transmembrane potential ($\Delta\psi$) or increased mitochondrial permeability transition was often observed in animal models of sepsis. Functional impairment of cardiac mitochondria was often associated with structural defects of mitochondria, such as cristae derangement, cleared matrix and mitochondrial swelling in animal models, thereby confirming the findings in human septic hearts as described above. Finally, increased mitochondrial reactive oxygen species (ROS) generation also confirms dysfunction of mitochondria in septic cardiomyopathy [32].

Although mitochondrial dysfunction in septic cardiomyopathy is associated with cardiac dysfunction, the underlying mechanisms of mitochondrial dysfunction are still incompletely elucidated. In addition, a number of mitochondrial mechanisms are triggered in septic cardiomyopathy, which may rather serve to restore than impair mitochondrial function, although controversially discussed. In the following sections, we will discuss mitochondrial mechanisms that may causally contribute to mitochondrial dysfunction and/or may represent adaptive responses to mitochondrial dysfunction in septic cardiomyopathy (illustrated in Figure 1).

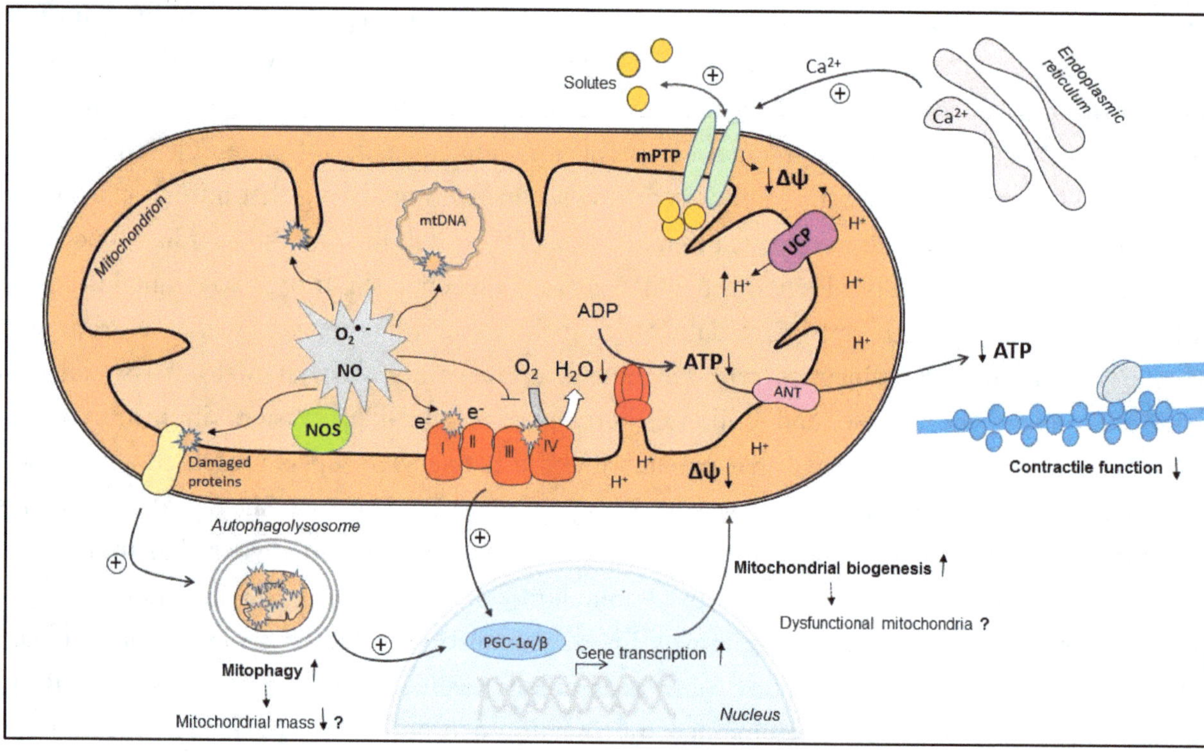

Figure 1. Mitochondrial mechanisms in septic cardiomyopathy: Increased superoxide ($O_2\cdot^-$) and nitric oxide (NO) production can cause direct oxidative or nitrosative damage and inhibition of oxidative phosphorylation (OXPHOS) complexes, leading to decreased O_2 consumption and decreased mitochondrial membrane potential ($\Delta\psi$). In addition, $\Delta\psi$ may drop due to increased uncoupling protein (UCP)-mediated proton leak, increased Ca^{2+}-induced mitochondrial permeability transition pore (mPTP) opening and direct oxidative damage of the inner mitochondrial membrane. As a consequence, mitochondrial ATP regeneration is compromised, and energy depletion may contribute to cardiac contractile dysfunction. Increased mitophagy may eliminate dysfunctional mitochondria, which may be replaced by increased mitochondrial biogenesis, mediated by activation of peroxisome proliferator-activated receptor γ coactivator 1α/β (PGC-1α/β). However, if uncoordinatedly activated, mitophagy and mitochondrial biogenesis may lead to decreased mitochondrial mass and dysfunctional mitochondria (upward pointing arrow: increase; downward pointing arrow: decrease; dashed arrow: possible consequence).

3. Mechanisms of Myocardial Mitochondrial Dysfunction in Sepsis

3.1. Mitochondrial NO Production and Oxidative Stress

During respiration, mitochondria may reduce O_2 univalently thereby producing $O_2^{\cdot-}$ (superoxide anion) as a normal metabolite. Under physiological conditions, only a small percentage (1%–2%) of the O_2 utilized is driven to $O_2^{\cdot-}$ formation, which is dismutated to H_2O_2 by Mn-superoxide dismutase. However, situations in which steady-state concentrations of these reactive oxygen species (ROS) are increased can lead to a diverse array of reversible and irreversible toxic modifications on biomolecules, such as protein carbonylation or lipid peroxidation. The continuous mitochondrial production of free radicals and the significant amount of mitochondria within cells makes mitochondria the most important source of these harmful species within the cell. At the same time, mitochondrial enzymes, function and mtDNA are particularly sensitive to ROS-induced damage [33,34]. Mitochondria also produce nitric oxide (NO) through the activity of mitochondrial NO synthase (mtNOS), which physiologically regulates mitochondrial respiration by inhibition of cytochrome c oxidase [35].

Sepsis is an excessive, systemic inflammatory response characterized by massive increases in ROS, NO and inflammatory cytokines [36]. A large body of evidence strongly suggests that excessive mitochondrial production of NO and ROS contribute to mitochondrial dysfunction in sepsis in various tissues, including the heart. Studies demonstrated that NO production, production of $O_2^{\cdot-}$ and H_2O_2, global protein nitration, nitrotyrosine content, protein carbonylation and lipid peroxidation are increased in cardiac mitochondria of animals subjected to endotoxemia [14,26,37–39]. Moreover, the antioxidant systems, aimed to scavenge these reactive species, seem to be inhibited as shown by decreased activity of Mn-superoxide dismutase and glutathione peroxidase, and depletion of gutathione [40,41]. Peroxynitrite ($ONOO^-$) is a particularly powerful oxidant, which results from a diffusion-controlled reaction of NO with $O_2^{\cdot-}$. In a model of CLP, Escames *et al.* showed increased activity of the inducible mitochondrial nitric oxide synthase (i-mtNOS) in septic mice, which leads to increased mitochondrial $ONOO^-$ levels [42,43]. Increased oxidative stress, impairment in OXPHOS function and a decrease in ATP production were restored by genetic deletion of iNOS (iNOS$^{-/-}$ mice), which also includes the mitochondrial isoform, suggesting a significant role of $ONOO^-$ in myocardial mitochondrial dysfunction in sepsis [43]. This argument is further supported by the fact that treatment with melatonin, an inhibitor of iNOS, prevented the impairment of mitochondrial homeostasis after sepsis, restored ATP production and improved survival. Some studies also showed an improvement of cardiac function by pharmacological inhibition or genetic deletion of NOS in septic animal models [37,44]. One of these studies showed that recovery of NO production and content of NO metabolites in mitochondria was associated with partial improvement of left ventricular (LV) developed pressure. A causal relationship between mitochondrial NO production and cardiac dysfunction needs to be considered carefully though, since NO is produced by several NOS isoforms (not only mitochondrial) at different intracellular locations and in different cell types, and since the role of NO in cardiovascular function and regulation of vascular tone is indisputably contributing to cardiac dysfunction in sepsis.

While increased ROS levels in sepsis may origin from ROS production at different cellular sites, the importance of mitochondrial oxidative stress in animal models of sepsis has been underlined by studies demonstrating beneficial effects by treatment with mitochondria-targeted antioxidants. Treatment with α-lipoic acid, an antioxidant that is reduced in the mitochondria to a more active form (dihyrolipoic acid), led to full recovery of cardiac mitochondrial function during endotoxemia [45]. Furthermore, administration of Coenzyme Q10 that was conjugated to triphenylphosphonium (MitoQ), a powerful mitochondria-targeted antioxidant, not only restored mitochondrial function but also ameliorated myocardial dysfunction in mice and rats treated with LPS [46]. Finally, treatment with vitamin E conjugated to triphenylphosphonium (Mito-VitE) in a rat pneumonia-related sepsis model prevented mitochondrial ROS production and ROS-related damage in the heart, with recovery of mitochondrial structure and function [32].

3.2. Ca^{2+} Handling and Mitochondrial Permeability Transition

On the basis of the chemiosmotic model of energy transduction, the impermeability of the inner mitochondrial membrane is essential to allow the electron transport chain to pump protons across the inner mitochondrial membrane (IMM) and to build up a proton gradient that can be used by the F_0F_1-ATPase to regenerate ATP. Mitochondrial permeability transition is a sudden increase of IMM permeability for solutes with a molecular mass up to 1500 Da, and it is believed that this permeability transition is mediated by a voltage- and Ca^{2+}-dependent, cyclosporin A (CsA)-sensitive, high-conductance channel located in the inner mitochondrial membrane termed mitochondrial permeability transition pore (mPTP). The consequence of mPTP opening is loss of $\Delta\psi$, a decrease in State 3 respiration, mitochondrial swelling and rupture of the outer mitochondrial membrane, ultimately leading to activation of pro-apoptotic pathways and necrotic cell death. Ca^{2+} overload is the primary trigger of mPTP opening, but the sensitivity of the pore is dependent on prevailing conditions, such as oxidative stress, adenine nucleotide depletion, increased inorganic phosphate concentrations and mitochondrial depolarization [47]. Using the LPS-induced endotoxemia model in rats, Hassoun *et al.* observed decreased Ca^{2+} uptake into the sarcoplasmic reticulum and increased Ca^{2+} leak from this organelle, which was associated with increased mitochondrial Ca^{2+} content. Administration of the sarcoplasmic reticulum Ca^{2+} leak inhibitor dantrolene prevented mitochondrial Ca^{2+} overload, and improved mitochondrial dysfunction and cardiac contractility defects, suggesting that mitochondrial Ca^{2+} overload could be an important mechanism contributing to myocardial mitochondrial dysfunction in sepsis [30]. Larche and colleagues reported decreased survival, myocardial dysfunction, decreased State 3 and increased State 4 respiration, and increased opening of the mPTP in CLP-induced sepsis. Inhibition of the mPTP by CsA treatment improved cardiac function, survival, and also attenuated mitochondrial dysfunction in this study. In addition, the amount of Ca^{2+} required to increase mPTP opening in CLP-treated mice was less than 25% of the Ca^{2+} needed in control animals, indicative of increased sensitivity for Ca^{2+}-induced mPTP opening of the mitochondria in CLP-treated mice [48]. In a different study by Fauvel *et al.*, endotoxemic mice showed decreased LV developed pressure in isolated perfused hearts, and this contractile defect was rescued by *in vivo* treatment with CsA. Treatment with CsA also decreased caspase activation and cytochrome c release, suggestive of prevention of mPTP opening [49]. Chopra *et al.* observed severe morphological deformations that could be indicative of pore formation in rats treated with cecal inoculum, accompanied by a mitochondrial $\Delta\psi$ collapse. Restorative effects were observed in both parameters by treatment with

5-hydroxydecanoate (5-HD), a blocker of mitoK$_{ATP}$ (mitochondrial ATP-sensitive K^+ channel). Moreover, sepsis-related decreases in cardiac output and ejection fraction were recovered after inhibition of the mitoK$_{ATP}$ [28]. The primary function of mitoK$_{ATP}$ channels is thought to be the regulation of mitochondrial volume; mitoK$_{ATP}$ channel opening results in mitochondrial uptake of potassium and associated inorganic phosphates, anions, and water [50], and is believed to sensitize outer membrane permeabilization and rupture in this sepsis model. Thus, it is tempting to speculate that in some models of sepsis, an impairment in cytosolic calcium handling may lead to mitochondrial Ca^{2+} overload, which triggers opening of a hypersensitive mPTP, thereby contributing to mitochondrial and contractile dysfunction. Further evidence is required to test whether activation of mitoK$_{ATP}$ contributes to increased sensitivity for mPTP opening. It remains to be mentioned though that some studies reported no change of mPTP opening using the mitochondrial swelling assay, although Ca^{2+} handling was found to be impaired in isolated cardiomyocytes [51].

3.3. Mitochondrial Uncoupling

Mitchell's chemiosmotic theory of energy transduction states that the free energy generated by electron transport through the OXPHOS chain and ultimately transfer on O_2 is used to set up an electrochemical H^+ gradient across the inner mitochondrial membrane. Protons return into the matrix via the F_0 subunit of the F_0F_1-ATPase, thereby regenerating ATP from ADP. Under physiological conditions, some H^+ return into the matrix but bypass the F_0F_1-ATPase (e.g., via uncoupling proteins (UCPs)), resulting in "mitochondrial uncoupling" of ATP synthesis from O_2 consumption. Physiologic uncoupling is used for heat generation in brown adipose tissue or to decrease mitochondrial ROS production [52]. Pathologic conditions with increased uncoupling include diabetic cardiomyopathy, where increased utilization of long-chain fatty acids results in ROS-induced activation of UCPs, resulting in decreased amounts of ATP per oxygen consumed and therefore impaired cardiac efficiency (cardiac work/myocardial O_2 consumption) [53–55]. Similarly, an increase in UCP3 expression may increase mitochondrial uncoupling and decrease cardiac efficiency in chronically infarcted failing rat hearts [56].

The role of UCPs in sepsis remains controversial. Twelve hours following CLP, Roshon and colleagues found a 35% reduction in cardiac efficiency in isolated perfused hearts, probably due to a decrease in cardiac work and associated with increased UCP2 mRNA levels [57]. Similarly, myocardial mRNA content of both UCP2 and UCP3 were increased in LPS-induced endotoxemia in rats, and increased myocardial mRNA and protein expression of UCP2 was observed in a canine model of endotoxin-induced shock associated with decreased phosphocreatine/ATP ratios [41,58]. Thus, UCP-mediated uncoupling may impair cardiac efficiency under septic conditions. In contrast, a very recent study reported that treatment of rat embryonic cardiomyoblasts (H9C2) with LPS plus peptidoglycan G led to increased mRNA expression of UCP2, associated with decreased membrane potential, decreased ATP content, increased ROS and depletion of mtDNA. These effects were aggravated by additional silencing of UCP2, suggesting that increased UCP2 expression may actually exert protective effects [59]. Less and also controversial evidence is available on UCP3. Aguirre *et al.* showed increased myocardial UCP3 levels in LPS-induced endotoxemia [60]. These authors report decreased State 3 respiration and Complex IV activity in wildtype and UCP3$^{-/-}$ mice treated with LPS.

Proton conductance was found unchanged after 24 h of LPS treatment, regardless of the higher levels of UCP3. In contrast, UCP3 levels were increased with a concomitant loss of membrane potential and ATP content in neonatal cardiomyocytes treated with LPS [61].

3.4. Mitochondrial Biogenesis

Mitochondrial biogenesis can be described as growth and division of mitochondria that requires the proper functioning of a number of cellular processes including the coordinated synthesis and import of proteins encoded by the nuclear genome, mtDNA replication, synthesis of proteins encoded by the mitochondrial genome and fusion–fission processes [62]. Mitochondrial biogenesis is triggered by different cellular stressors, such as increased levels of NO, oxidative stress, and an elevated AMP/ATP ratio, among others, resulting in the activation of the peroxisome proliferator-activated receptor y (PPARγ) coactivator (PGC) family of transcriptional coactivators, most importantly PGC-1α and PGC-1β [63,64]. PGC-1α coactivates and/or increases the expression of transcription factors, including ERRα/γ, NRF1/2 and Tfam, that mediate the transcription of nuclear and mitochondrially encoded OXPHOS subunits, nuclear proteins necessary for mtDNA transcription and replication, OXPHOS assembly factors, and components of mitochondrial protein import [65]. By regulating ERRα and PPARα, PGC-1α/β also stimulate the expression of genes encoding for proteins and enzymes of fatty acid oxidation, the main metabolic pathway of fuel oxidation in cardiac tissue.

Several studies have illustrated that mitochondrial biogenesis or the process of generating new mitochondria improves survival in sepsis, and inhibition of mitochondrial biogenesis worsens outcome [25,66]. It is generally thought that loss and/or removal of damaged mitochondria in septic organs is compensated by generation of new mitochondria by increased mitochondrial biogenesis, which should be responsible for increasing survival and reversing organ damage. In the heart, Vanasco *et al.* recently reported increased myocardial levels of PGC1-α and Tfam, accompanied by increased mitochondrial mass and recovery of initially decreased amounts of mtDNA 24 h following LPS treatment [14]. Electron microscopy revealed an increased amount of structurally damaged mitochondria, including swelling, loss and/or disruption of cristae, cleared matrix, and internal vesicles. Despite evidence of mitochondrial biogenesis, mitochondrial State 3 respiration related to Complex I substrates and activity of Complexes I and II were impaired. Others also reported increased expression of mediators of mitochondrial biogenesis, including PGC-1α, NRF-1, and NRF-2, as well as Tfam and the mitochondrial DNA polymerase, associated with mitochondrial structural changes such as swelling and loss of cristae [41]. In neonatal cardiomyocytes, LPS treatment stimulated an increase in Tfam, nuclear accumulation of NRF-1, and expression of PGC-1α, and the changes correlated with signs of mitochondrial dysfunction (decreased mitochondrial membrane potential and ATP content) and markers of autophagy [61]. Taken together, it appears that mitochondrial biogenesis signaling is activated in endotoxemia, which however may not necessarily result in improvement or recovery of mitochondrial function. Russell and colleagues demonstrated that cardiomyocyte-specific overexpression of PGC-1α resulted in a markedly increased mitochondrial biogenesis but also heart failure [67]. Possibly, intensive stimulation of mitochondrial biogenesis might have perturbed the complex process of gene transcription and mitochondrial dynamics in this study, resulting in mitochondria with altered biochemical composition and stoichiometry and thus functional deficits. Thus, while being a compensatory mechanism to overcome loss of dysfunctional mitochondria, uncontrolled triggering of mitochondrial biogenesis in endotoxemia may also result in dysfunctional mitochondria *per se*. Alternatively, mitochondrial biogenesis may simply not be sufficient to compensate for mitochondrial defects in septic conditions.

In contrast to increased mitochondrial biogenesis signaling, Schilling *et al.* showed that LPS injection led to a rapid downregulation of PGC-1α and PGC-1β mRNA levels, accompanied by decreased levels of PPARα, ERRα, and NRF1. Cardiac function was depressed 6 h after the onset of endotoxemia, associated with a decrease in palmitate oxidation in the isolated working heart and increased myocyte lipid accumulation [68]. The total mtDNA copy number and mRNA levels of several enzymes were decreased, and the content of respiratory Complexes I and IV, assessed by 2D-gel electrophoresis and MALDI-TOF mass spectrometry, was reduced in treated rats [41]. Thus, time points and models investigated may affect initiation of mitochondrial biogenesis, and it remains controversial whether activation of mitochondrial biogenesis is actually adaptive or maladaptive under septic and endotoxemic conditions. Of interest, cells overexpressing PGC-1β and mice overexpressing PGC-1β selectively in cardiomyocytes were resistant to the LPS-induced decrease of the expression of PPARα, ERRα, and enzymes of fatty acid oxidation. PGC-1β transgenic mice showed improvement in palmitate oxidation rates in isolated working hearts after treatment with LPS, and echocardiography revealed only a modest decrease in cardiac function [68]. These data suggests that impairment in mitochondrial energetics may importantly contribute to cardiac dysfunction, and that activation of mitochondrial biogenesis may have positive effects in endotoxemia.

When assessing the role of PPARα in endotoxemia, Drosatos *et al.* found that LPS administration reduced the mRNA expression of PGC-1α/β, PPARα, and metabolic PPARα downstream targets. Treatment with a PPARα agonist, however, did not improve cardiac function or alter the mRNA levels of these enzymes after endotoxemia. Interestingly, treatment with rosiglitazone, a PPARγ agonist, led to an improvement of cardiac dysfunction together with improved expression of PGC-1 molecules and Tfam. Moreover, αMHC-PPARγ mice that express PPARγ constitutively in cardiomyocytes showed no disturbances in cardiac function and lipid accumulation after LPS treatment [16]. The exact mechanism by which PPARγ agonism improves or maintains cardiac function in endotoxemia, and in how far these mechanisms are related to mitochondrial biogenesis, remains to be determined.

3.5. Mitophagy

Recovery of organ function in sepsis is dependent on removal of dysfunctional mitochondria and generation of new, functionally competent mitochondria, as may be achieved by stimulation of mitochondrial biogenesis. Removal of dysfunctional mitochondria occurs via autophagy of mitochondria, or simply "mitophagy", which is a quality control mechanism by which cells eliminate dysfunctional mitochondria [69]. Damaged mitochondria are selectively sequestered in autophagosomes and ultimately degraded following fusion with lysosomes. Different pathways have been described to recognize and eliminate the dysfunctional mitochondria. Activation of the PINK1/Parkin (PARK2) pathway leads to ubiquitinylation of special substrates of mitochondria that can be recognized by the adaptor protein p62, which in turn binds to microtubule-associated protein 1A/1B-light chain 3 (LC3). An ubiquitin-independent pathway involves direct binding of autophagy-related protein 8 (ATG8) family proteins to autophagy receptors, such as NIX (Nip3-like protein X) [63,69]. Mitophagy may serve to eliminate dysfunctional mitochondria before they induce cellular damage or cell death, but may also serve as a cell survival pathway by suppressing apoptosis or as a back-up mechanism when the process of apoptosis is defective [70].

Several studies have shown that autophagy and mitophagy are increased in various organs in sepsis [71,72]. In the heart, several studies observed increased myocardial mRNA or protein levels of ATG3, ATG5, ATG7, LC3 II and p62 in LPS-induced endotoxemia, suggestive of increased rates of autophagy [14,16,73]. In some cases, a reduced number of cardiomyocyte mitochondria was reported in LPS-treated animals [16,25,41] potentially suggesting increased mitochondrial degradation by mitophagy, although unchanged mitochondrial mass was observed in other studies [41,74]. An interesting study was published by Piquereau *et al.* who addressed the role of Parkin/PARK2 in mitochondrial dysfunction in endotoxemia. LPS treatment led to a 50% decrease in cardiac output and stroke volume in wildtype and PARK2$^{-/-}$ mice. While wildtype mice completely recovered cardiac function after 48 h, cardiac function remained impaired in PARK2$^{-/-}$ mice. Similarly, a decrease in OXPHOS complex activities, mitochondrial O_2 consumption rates, and increased susceptibility to mPTP opening were normalized 48 h after endotoxemia induction in wildtype but not PARK2$^{-/-}$ mice. Thus, PARK2-dependent autophagy/mitophagy is required to recover mitochondrial and cardiac function following LPS treatment. In mitochondrial fractions of LPS-treated wildtype mice, levels of polyubiquitinated proteins and PARK2 were increased, and proteins involved in mitochondrial quality control and autophagy (SQTSTM1, LC3 II, BNIP3L/NIX, and BNIP3) were recruited to mitochondria both in wildtype and PARK2$^{-/-}$ mice, showing that septic mitochondria can be targeted for degradation by mitophagy, even in the absence of PARK2. Using electron microscopy, the presence of autophagosome-like structures was confirmed as well [31]. Thus, mitophagy may be increased in endotoxemia and may serve a compensatory role in the recovery of mitochondrial function by eliminating defective mitochondria. The occurrence of mitochondrial biogenesis after the early activation of mitophagy during the course of endotoxemia suggests that an increase in mitochondrial biogenesis may serve to recover an appropriate volume density of mitochondria to restore the full mitochondrial energy-producing capacity of the heart [14,75,76]. Recent studies suggest that the degradation of mitochondria via mitophagy promotes mitochondrial biogenesis directly via a Toll-like receptor 9 (TLR9)-dependent signaling, although the mechanistic links between autophagy and mitochondrial biogenesis are not well characterized yet [72]. In contrast to a beneficial role of autophagy, the controlled elimination of organelles may also be part of apoptosis. Interactions among signaling components of the two pathways (autophagy and apoptosis) have been reported and indicate a complex cross-talk [77]. Thus, it remains unclear whether autophagy is actually an attempt to remove damaged mitochondria or part of a cellular program leading to apoptosis during sepsis. More studies are needed to further define the role of mitophagy in septic cardiomyopathy and to define how mitophagy may directly regulate mitochondrial biogenesis.

4. Conclusions

Cardiac dysfunction is common in patients suffering from severe sepsis and septic shock. At present, treatment mainly includes management of the infectious focus and hemodynamic support. However, recent progress in sepsis research increased our understanding of underlying molecular alterations contributing to organ dysfunction, including mitochondrial dysfunction. In the heart, a number of mechanisms have been proposed to underlie myocardial mitochondrial dysfunction, including excessive production of mitochondrial NO and ROS, increased mPTP opening, and increased mitochondrial uncoupling. Several reports showed recovery of mitochondrial function which was associated with improvement or normalization of cardiac pump function following treatment with drugs that target different mechanisms within mitochondria but that had no influence on the inflammatory response, supporting the idea that impaired myocardial energetics rather than cardiac inflammation *per se* are critical for cardiac dysfunction in sepsis [11,23,63]. We now need more studies using pharmacological modulators to prevent or reverse specific mitochondrial mechanisms to further evaluate the significance of mitochondrial mechanisms in septic cardiomyopathy, and to test whether such treatments may be a promising and feasible therapeutic approach that could potentially be transferred from bench to bedside. Such approaches may, for example, include inhibition of mitochondrial ROS and NO production and inhibition of mPTP opening. In addition, it would be very informative to perform such studies in the different models of sepsis and endotoxemia, which may not only reveal differences in the underlying disease mechanisms but may also allow evaluating whether the treatment may be broadly applicable and effective irrespective of the sepsis model. The increasing interest in sepsis research and the availability of novel research tools promises that effective specific treatments may be available in the future to complement our current treatment options of severe sepsis and septic shock to improve outcome for this highly jeopardized patient cohort.

Acknowledgments

This work was supported by research grants from the German Research Foundation (DFG) and the argentinean National Scientific and Technical Research Council (CONICET).

Author Contributions

María Cecilia Cimolai, Silvia Alvarez, Christoph Bode and Heiko Bugger contributed to the writing of this review.

References

1. Deutschman, C.S.; Tracey, K.J. Sepsis: Current dogma and new perspectives. *Immunity* **2014**, *40*, 463–475.
2. Soong, J.; Soni, N. Sepsis: Recognition and treatment. *Clin. Med.* **2012**, *12*, 276–280.
3. Sagy, M.; Al-Qaqaa, Y.; Kim, P. Definitions and pathophysiology of sepsis. *Curr. Prob. Pediatr. Adolesc. Health Care* **2013**, *43*, 260–263.
4. Court, O.; Kumar, A.; Parrillo, J.E.; Kumar, A. Clinical review: Myocardial depression in sepsis and septic shock. *Crit. Care* **2002**, *6*, 500–508.

5. Cunnion, R.E.; Schaer, G.L.; Parker, M.M.; Natanson, C.; Parrillo, J.E. The coronary circulation in human septic shock. *Circulation* **1986**, *73*, 637–644.
6. Romero-Bermejo, F.J.; Ruiz-Bailen, M.; Gil-Cebrian, J.; Huertos-Ranchal, M.J. Sepsis-induced cardiomyopathy. *Curr. Cardiol. Rev.* **2011**, *7*, 163–183.
7. Rudiger, A.; Singer, M. Mechanisms of sepsis-induced cardiac dysfunction. *Crit. Care Med.* **2007**, *35*, 1599–1608.
8. Vallet, B.; Lund, N.; Curtis, S.E.; Kelly, D.; Cain, S.M. Gut and muscle tissue PO2 in endotoxemic dogs during shock and resuscitation. *J. Appl. Physiol.* **1994**, *76*, 793–800.
9. Joseph, B. On anoxaemia. *Lancet* **1920**, 485–489.
10. Sair, M.; Etherington, P.J.; Curzen, N.P.; Winlove, C.P.; Evans, T.W. Tissue oxygenation and perfusion in endotoxemia. *Am. J. Physiol.* **1996**, *271*, H1620–H1625.
11. Hotchkiss, R.S.; Rust, R.S.; Dence, C.S.; Wasserman, T.H.; Song, S.K.; Hwang, D.R.; Karl, I.E.; Welch, M.J. Evaluation of the role of cellular hypoxia in sepsis by the hypoxic marker [18F] fluoromisonidazole. *Am. J. Physiol.* **1991**, *261*, R965–R972.
12. Fink, M.P. Cytopathic hypoxia. Mitochondrial dysfunction as mechanism contributing to organ dysfunction in sepsis. *Crit. Care Clin.* **2001**, *17*, 219–237.
13. Soriano, F.G.; Nogueira, A.C.; Caldini, E.G.; Lins, M.H.; Teixeira, A.C.; Cappi, S.B.; Lotufo, P.A.; Bernik, M.M.; Zsengeller, Z.; Chen, M.; *et al.* Potential role of poly(adenosine 5′-diphosphate-ribose) polymerase activation in the pathogenesis of myocardial contractile dysfunction associated with human septic shock. *Crit. Care Med.* **2006**, *34*, 1073–1079.
14. Vanasco, V.; Saez, T.; Magnani, N.D.; Pereyra, L.; Marchini, T.; Corach, A.; Vaccaro, M.I.; Corach, D.; Evelson, P.; Alvarez, S. Cardiac mitochondrial biogenesis in endotoxemia is not accompanied by mitochondrial function recovery. *Free Radic. Biol. Med.* **2014**, *77*, 1–9.
15. Tavener, S.A.; Long, E.M.; Robbins, S.M.; McRae, K.M.; van Remmen, H.; Kubes, P. Immune cell toll-like receptor 4 is required for cardiac myocyte impairment during endotoxemia. *Circ. Res.* **2004**, *95*, 700–707.
16. Drosatos, K.; Khan, R.S.; Trent, C.M.; Jiang, H.; Son, N.H.; Blaner, W.S.; Homma, S.; Schulze, P.C.; Goldberg, I.J. Peroxisome proliferator-activated receptor-γ activation prevents sepsis-related cardiac dysfunction and mortality in mice. *Circ. Heart Fail.* **2013**, *6*, 550–562.
17. Escames, G.; Leon, J.; Macias, M.; Khaldy, H.; Acuna-Castroviejo, D. Melatonin counteracts lipopolysaccharide-induced expression and activity of mitochondrial nitric oxide synthase in rats. *FASEB J.* **2003**, *17*, 932–934.
18. Alvarez, S.; Boveris, A. Mitochondrial nitric oxide metabolism in rat muscle during endotoxemia. *Free Radic. Biol. Med.* **2004**, *37*, 1472–1478.
19. Bugger, H.; Abel, E.D. Mitochondria in the diabetic heart. *Cardiovasc. Res.* **2010**, *88*, 229–240.
20. Bugger, H.; Boudina, S.; Hu, X.X.; Tuinei, J.; Zaha, V.G.; Theobald, H.A.; Yun, U.J.; McQueen, A.P.; Wayment, B.; Litwin, S.E.; *et al.* Type 1 diabetic akita mouse hearts are insulin sensitive but manifest structurally abnormal mitochondria that remain coupled despite increased uncoupling protein 3. *Diabetes* **2008**' *57*, 2924–2932.

21. Lesnefsky, E.J.; Chen, Q.; Slabe, T.J.; Stoll, M.S.; Minkler, P.E.; Hassan, M.O.; Tandler, B.; Hoppel, C.L. Ischemia, rather than reperfusion, inhibits respiration through cytochrome oxidase in the isolated, perfused rabbit heart: Role of cardiolipin. *Am. J. Physiol. Heart Circ. Physiol.* **2004**, *287*, H258–H267.
22. Takasu, O.; Gaut, J.P.; Watanabe, E.; To, K.; Fagley, R.E.; Sato, B.; Jarman, S.; Efimov, I.R.; Janks, D.L.; Srivastava, A.; *et al.* Mechanisms of cardiac and renal dysfunction in patients dying of sepsis. *Am. J. Respir. Crit. Care Med.* **2013**, *187*, 509–517.
23. Doi, K.; Leelahavanichkul, A.; Yuen, P.S.; Star, R.A. Animal models of sepsis and sepsis-induced kidney injury. *J. Clin. Investig.* **2009**, *119*, 2868–2878.
24. Smeding, L.; Plotz, F.B.; Groeneveld, A.B.; Kneyber, M.C. Structural changes of the heart during severe sepsis or septic shock. *Shock* **2012**, *37*, 449–456.
25. Reynolds, C.M.; Suliman, H.B.; Hollingsworth, J.W.; Welty-Wolf, K.E.; Carraway, M.S.; Piantadosi, C.A. Nitric oxide synthase-2 induction optimizes cardiac mitochondrial biogenesis after endotoxemia. *Free Radic. Biol. Med.* **2009**, *46*, 564–572.
26. Joshi, M.S.; Julian, M.W.; Huff, J.E.; Bauer, J.A.; Xia, Y.; Crouser, E.D. Calcineurin regulates myocardial function during acute endotoxemia. *Am. J. Respir. Crit. Care Med.* **2006**, *173*, 999–1007.
27. An, J.; Du, J.; Wei, N.; Guan, T.; Camara, A.K.; Shi, Y. Differential sensitivity to LPS-induced myocardial dysfunction in the isolated brown norway and Dahl S rat hearts: Roles of mitochondrial function, NF-κB activation, and TNF-α production. *Shock* **2012**, *37*, 325–332.
28. Chopra, M.; Golden, H.B.; Mullapudi, S.; Dowhan, W.; Dostal, D.E.; Sharma, A.C. Modulation of myocardial mitochondrial mechanisms during severe polymicrobial sepsis in the rat. *PLoS ONE* **2011**, *6*, e21285.
29. Kozlov, A.V.; Staniek, K.; Haindl, S.; Piskernik, C.; Ohlinger, W.; Gille, L.; Nohl, H.; Bahrami, S.; Redl, H. Different effects of endotoxic shock on the respiratory function of liver and heart mitochondria in rats. *Am. J. Physiol. Gastrointest. Liver Physiol.* **2006**, *290*, G543–G549.
30. Hassoun, S.M.; Marechal, X.; Montaigne, D.; Bouazza, Y.; Decoster, B.; Lancel, S.; Neviere, R. Prevention of endotoxin-induced sarcoplasmic reticulum calcium leak improves mitochondrial and myocardial dysfunction. *Crit. Care Med.* **2008**, *36*, 2590–2596.
31. Piquereau, J.; Godin, R.; Deschenes, S.; Bessi, V.L.; Mofarrahi, M.; Hussain, S.N.; Burelle, Y. Protective role of PARK2/Parkin in sepsis-induced cardiac contractile and mitochondrial dysfunction. *Autophagy* **2013**, *9*, 1837–1851.
32. Zang, Q.S.; Sadek, H.; Maass, D.L.; Martinez, B.; Ma, L.; Kilgore, J.A.; Williams, N.S.; Frantz, D.E.; Wigginton, J.G.; Nwariaku, F.E.; *et al.* Specific inhibition of mitochondrial oxidative stress suppresses inflammation and improves cardiac function in a rat pneumonia-related sepsis model. *Am. J. Physiol. Heart Circ. Physiol.* **2012**, *302*, H1847–H1859.
33. Alvarez, S.; Evelson, P.; Cimolai, M.C. Oxygen and nitric oxide metabolism in sepsi. In *Free Radical Pathophysiology*; Alvarez, S., Ed.; Transworld Research Network: Kerala, India, 2008; pp. 223–236.
34. Wallace, D.C.; Shoffner, J.M.; Watts, R.L.; Juncos, J.L.; Torroni, A. Mitochondrial oxidative phosphorylation defects in Parkinson's disease. *Ann. Neurol.* **1992**, *32*, 113–114.

35. Kanai, A.J.; Pearce, L.L.; Clemens, P.R.; Birder, L.A.; VanBibber, M.M.; Choi, S.Y.; de Groat, W.C.; Peterson, J. Identification of a neuronal nitric oxide synthase in isolated cardiac mitochondria using electrochemical detection. *Proc. Natl. Acad. Sci. USA* **2001**, *98*, 14126–14131.
36. Galley, H.F. Oxidative stress and mitochondrial dysfunction in sepsis. *Br. J. Anaesth.* **2011**, *107*, 57–64.
37. Van de Sandt, A.M.; Windler, R.; Godecke, A.; Ohlig, J.; Zander, S.; Reinartz, M.; Graf, J.; van Faassen, E.E.; Rassaf, T.; Schrader, J.; *et al.* Endothelial nos (NOS3) impairs myocardial function in developing sepsis. *Basic Res. Cardiol.* **2013**, *108*, 330.
38. Vanasco, V.; Magnani, N.D.; Cimolai, M.C.; Valdez, L.B.; Evelson, P.; Boveris, A.; Alvarez, S. Endotoxemia impairs heart mitochondrial function by decreasing electron transfer, ATP synthesis and atp content without affecting membrane potential. *J. Bioenerg. Biomembr.* **2012**, *44*, 243–252.
39. Burgoyne, J.R.; Rudyk, O.; Mayr, M.; Eaton, P. Nitrosative protein oxidation is modulated during early endotoxemia. *Nitric Oxide* **2011**, *25*, 118–124.
40. Zang, Q.; Maass, D.L.; Tsai, S.J.; Horton, J.W. Cardiac mitochondrial damage and inflammation responses in sepsis. *Surg. Infect.* **2007**, *8*, 41–54.
41. Suliman, H.B.; Welty-Wolf, K.E.; Carraway, M.; Tatro, L.; Piantadosi, C.A. Lipopolysaccharide induces oxidative cardiac mitochondrial damage and biogenesis. *Cardiovasc. Res.* **2004**, *64*, 279–288.
42. Boveris, A.; Alvarez, S.; Navarro, A. The role of mitochondrial nitric oxide synthase in inflammation and septic shock. *Free Radic. Biol. Med.* **2002**, *33*, 1186–1193.
43. Escames, G.; Lopez, L.C.; Ortiz, F.; Lopez, A.; Garcia, J.A.; Ros, E.; Acuna-Castroviejo, D. Attenuation of cardiac mitochondrial dysfunction by melatonin in septic mice. *FEBS J.* **2007**, *274*, 2135–2147.
44. Xu, C.; Yi, C.; Wang, H.; Bruce, I.C.; Xia, Q. Mitochondrial nitric oxide synthase participates in septic shock myocardial depression by nitric oxide overproduction and mitochondrial permeability transition pore opening. *Shock* **2012**, *37*, 110–115.
45. Vanasco, V.; Cimolai, M.C.; Evelson, P.; Alvarez, S. The oxidative stress and the mitochondrial dysfunction caused by endotoxemia are prevented by α-lipoic acid. *Free Radic. Res.* **2008**, *42*, 815–823.
46. Supinski, G.S.; Murphy, M.P.; Callahan, L.A. MitoQ administration prevents endotoxin-induced cardiac dysfunction. *Am. J. Physiol. Regul. Integr. Comp. Physiol.* **2009**, *297*, R1095–R1102.
47. Halestrap, A.P.; McStay, G.P.; Clarke, S.J. The permeability transition pore complex: Another view. *Biochimie* **2002**, *84*, 153–166.
48. Larche, J.; Lancel, S.; Hassoun, S.M.; Favory, R.; Decoster, B.; Marchetti, P.; Chopin, C.; Neviere, R. Inhibition of mitochondrial permeability transition prevents sepsis-induced myocardial dysfunction and mortality. *J. Am. Coll. Cardiol.* **2006**, *48*, 377–385.
49. Fauvel, H.; Marchetti, P.; Obert, G.; Joulain, O.; Chopin, C.; Formstecher, P.; Neviere, R. Protective effects of cyclosporin a from endotoxin-induced myocardial dysfunction and apoptosis in rats. *Am. J. Respir. Crit. Rare Red.* **2002**, *165*, 449–455.
50. Costa, A.D.; Quinlan, C.L.; Andrukhiv, A.; West, I.C.; Jaburek, M.; Garlid, K.D. The direct physiological effects of mitoKATP opening on heart mitochondria. *Am. J. Physiol. Heart Circ. Physiol.* **2006**, *290*, H406–H415.

51. Bougaki, M.; Searles, R.J.; Kida, K.; Yu, J.; Buys, E.S.; Ichinose, F. Nos3 protects against systemic inflammation and myocardial dysfunction in murine polymicrobial sepsis. *Shock* **2010**, *34*, 281–290.
52. Ricquier, D.; Bouillaud, F. Mitochondrial uncoupling proteins: From mitochondria to the regulation of energy balance. *J. Physiol.* **2000**, *1*, 3–10.
53. Boudina, S.; Sena, S.; Theobald, H.; Sheng, X.; Wright, J.J.; Hu, X.X.; Aziz, S.; Johnson, J.I.; Bugger, H.; Zaha, V.G.; *et al.* Mitochondrial energetics in the heart in obesity-related diabetes: Direct evidence for increased uncoupled respiration and activation of uncoupling proteins. *Diabetes* **2007**, *56*, 2457–2466.
54. Bugger, H.; Abel, E.D. Molecular mechanisms of diabetic cardiomyopathy. *Diabetologia* **2014**, *57*, 660–671.
55. Konig, A.; Bode, C.; Bugger, H. Diabetes mellitus and myocardial mitochondrial dysfunction: Bench to bedside. *Heart Fail. Clin.* **2012**, *8*, 551–561.
56. Murray, A.J.; Cole, M.A.; Lygate, C.A.; Carr, C.A.; Stuckey, D.J.; Little, S.E.; Neubauer, S.; Clarke, K. Increased mitochondrial uncoupling proteins, respiratory uncoupling and decreased efficiency in the chronically infarcted rat heart. *J. Mol. Cell. Cardiol.* **2008**, *44*, 694–700.
57. Roshon, M.J.; Kline, J.A.; Thornton, L.R.; Watts, J.A. Cardiac UCP2 expression and myocardial oxidative metabolism during acute septic shock in the rat. *Shock* **2003**, *19*, 570–576.
58. Wang, X.; Liu, D.; Chai, W.; Long, Y.; Su, L.; Yang, R. The role of uncoupling protein-2 (UCP2) during myocardial dysfunction in a canine model of endotoxin shock. *Shock* **2014**, *3*, 292–297.
59. Zheng, G.; Lyu, J.; Liu, S.; Huang, J.; Liu, C.; Xiang, D.; Xie, M.; Zeng, Q. Silencing of uncoupling protein 2 by small interfering rna aggravates mitochondrial dysfunction in cardiomyocytes under septic conditions. *Int. J. Mol. Med.* **2015**, *35*, 1525–1536.
60. Aguirre, E.; Cadenas, S. GDP and carboxyatractylate inhibit 4-hydroxynonenal-activated proton conductance to differing degrees in mitochondria from skeletal muscle and heart. *Biochim. Biophys. Acta* **2010**, *1797*, 1716–1726.
61. Hickson-Bick, D.L.; Jones, C.; Buja, L.M. Stimulation of mitochondrial biogenesis and autophagy by lipopolysaccharide in the neonatal rat cardiomyocyte protects against programmed cell death. *J. Mol. Cell. Cardiol.* **2008**, *44*, 411–418.
62. Jornayvaz, F.R.; Shulman, G.I. Regulation of mitochondrial biogenesis. *Essays Biochem.* **2010**, *47*, 69–84.
63. Gottlieb, R.A.; Gustafsson, A.B. Mitochondrial turnover in the heart. *Biochim. Biophys. Acta* **2011**, *1813*, 1295–1301.
64. Wenz, T. Regulation of mitochondrial biogenesis and PGC-1α under cellular stress. *Mitochondrion* **2013**, *13*, 134–142.
65. Kelly, D.P.; Scarpulla, R.C. Transcriptional regulatory circuits controlling mitochondrial biogenesis and function. *Genes Dev.* **2004**, *18*, 357–368.
66. Lancel, S.; Hassoun, S.M.; Favory, R.; Decoster, B.; Motterlini, R.; Neviere, R. Carbon monoxide rescues mice from lethal sepsis by supporting mitochondrial energetic metabolism and activating mitochondrial biogenesis. *J. Pharmacol. Exp. Ther.* **2009**, *329*, 641–648.

67. Russell, L.K.; Mansfield, C.M.; Lehman, J.J.; Kovacs, A.; Courtois, M.; Saffitz, J.E.; Medeiros, D.M.; Valencik, M.L.; McDonald, J.A.; Kelly, D.P. Cardiac-specific induction of the transcriptional coactivator peroxisome proliferator-activated receptor γ coactivator-1α promotes mitochondrial biogenesis and reversible cardiomyopathy in a developmental stage-dependent manner. *Circ. Res.* **2004**, *94*, 525–533.
68. Schilling, J.; Lai, L.; Sambandam, N.; Dey, C.E.; Leone, T.C.; Kelly, D.P. Toll-like receptor-mediated inflammatory signaling reprograms cardiac energy metabolism by repressing peroxisome proliferator-activated receptor gamma coactivator-1 signaling. *Circ. Heart Fail.* **2011**, *4*, 474–482.
69. Kubli, D.A.; Gustafsson, A.B. Mitochondria and mitophagy: The yin and yang of cell death control. *Circ. Res.* **2012**, *111*, 1208–1221.
70. Eisenberg-Lerner, A.; Bialik, S.; Simon, H.U.; Kimchi, A. Life and death partners: Apoptosis, autophagy and the cross-talk between them. *Cell Death Differ.* **2009**, *16*, 966–975.
71. Carchman, E.H.; Rao, J.; Loughran, P.A.; Rosengart, M.R.; Zuckerbraun, B.S. Heme oxygenase-1-mediated autophagy protects against hepatocyte cell death and hepatic injury from infection/sepsis in mice. *Hepatology* **2011**, *53*, 2053–2062.
72. Carchman, E.H.; Whelan, S.; Loughran, P.; Mollen, K.; Stratamirovic, S.; Shiva, S.; Rosengart, M.R.; Zuckerbraun, B.S. Experimental sepsis-induced mitochondrial biogenesis is dependent on autophagy, TLR4, and TLR9 signaling in liver. *FASEB J.* **2013**, *27*, 4703–4711.
73. Ceylan-Isik, A.F.; Zhao, P.; Zhang, B.; Xiao, X.; Su, G.; Ren, J. Cardiac overexpression of metallothionein rescues cardiac contractile dysfunction and endoplasmic reticulum stress but not autophagy in sepsis. *J. Mol. Cell. Cardiol.* **2010**, *48*, 367–378.
74. Suliman, H.B.; Welty-Wolf, K.E.; Carraway, M.S.; Schwartz, D.A.; Hollingsworth, J.W.; Piantadosi, C.A. Toll-like receptor 4 mediates mitochondrial DNA damage and biogenic responses after heat-inactivated *E. coli. FASEB J.* **2005**, *19*, 1531–1533.
75. Crouser, E.D.; Julian, M.W.; Huff, J.E.; Struck, J.; Cook, C.H. Carbamoyl phosphate synthase-1: A marker of mitochondrial damage and depletion in the liver during sepsis. *Crit. Care Med.* **2006**, *34*, 2439–2446.
76. Yuan, H.; Perry, C.N.; Huang, C.; Iwai-Kanai, E.; Carreira, R.S.; Glembotski, C.C.; Gottlieb, R.A. LPS-induced autophagy is mediated by oxidative signaling in cardiomyocytes and is associated with cytoprotection. *Am. J. Physiol. Heart Circ. Physiol.* **2009**, *296*, H470–H479.
77. Nikoletopoulou, V.; Markaki, M.; Palikaras, K.; Tavernarakis, N. Crosstalk between apoptosis, necrosis and autophagy. *Biochim. Biophys. Acta* **2013**, *1833*, 3448–3459.

10

Autophagy as a Regulatory Component of Erythropoiesis

Jieying Zhang [1,2], Kunlu Wu [1], Xiaojuan Xiao [1], Jiling Liao [1], Qikang Hu [1], Huiyong Chen [1], Jing Liu [1,*] and Xiuli An [3,4]

[1] State Key Laboratory of Medical Genetics & School of Life Sciences, Central South University, Changsha 410078, China; E-Mails: jzhang@nybloodcenter.org (J.Z.); wukunlu@163.com (K.W.); xiaojuan_xiao0809@163.com (X.X.); liaojilingcsu@163.com (J.L.); huqikang1001@163.com (Q.H.); chenhy5000@126.com (H.C.)

[2] Red Cell Physiology Laboratory, New York Blood Center, New York, NY 10065, USA

[3] Laboratory of Membrane Biology, New York Blood Center, New York, NY 10065, USA; E-Mail: xan@nybloodcenter.org

[4] College of Life Science, Zhengzhou University, Zhengzhou 450001, China

* Author to whom correspondence should be addressed; E-Mail: liujing2@sklmg.edu.cn

Academic Editors: Jaime M. Ross and Giuseppe Coppotelli

Abstract: Autophagy is a process that leads to the degradation of unnecessary or dysfunctional cellular components and long-lived protein aggregates. Erythropoiesis is a branch of hematopoietic differentiation by which mature red blood cells (RBCs) are generated from multi-potential hematopoietic stem cells (HSCs). Autophagy plays a critical role in the elimination of mitochondria, ribosomes and other organelles during erythroid terminal differentiation. Here, the modulators of autophagy that regulate erythroid differentiation were summarized, including autophagy-related (Atg) genes, the B-cell lymphoma 2 (Bcl-2) family member Bcl-2/adenovirus E1B 19 kDa interacting protein 3-like (Nix/Binp3L), transcription factors globin transcription factor 1 (GATA1) and forkhead box O3 (FoxO3), intermediary factor KRAB-associated protein1 (KAP1), and other modulators, such as focal adhesion kinase family-interacting protein of 200-kDa (FIP200), Ca^{2+} and 15-lipoxygenase. Understanding the modulators of autophagy in erythropoiesis will benefit the autophagy research field and facilitate the prevention and treatment of autophagy-related red blood cell disorders.

Keywords: autophagy; modulators; erythropoiesis

1. Introduction

Erythropoiesis is a continuous and dynamic process by which erythrocytes are generated from multipotent hematopoietic stem cells (HSCs). Erythropoiesis is mainly divided into two stages, early erythroid progenitor proliferation and terminal erythroid differentiation. HSCs proliferate and differentiate into the earliest erythroid progenitors: burst-forming-unit erythroid (BFU-E) cells, and then, colony-forming-unit erythroid (CFU-E) cells. Subsequently, terminal erythroid differentiation starts with proerythroblasts, which undergo three mitoses to produce basophilic, polychromatic, and orthochromatic erythroblasts. Eventually, orthochromatic erythroblasts expel their nucleus and become reticulocytes, which subsequently become mature erythrocytes [1]. Noticeable changes in cellular composition and structure occur during terminal erythroid differentiation, including the filling of the cells with abundant hemoglobin and the clearance of all intracellular organelles from the cells, such as mitochondria and ribosomes [2].

A study performed in 1962 revealed an abundance of membranous structures in murine hepatocytes following treatment with glucagon and found that the mitochondria are degraded by lysosomes [3]. Deter and de Duve first proposed the biological concept of autophagy in an international forum [4]. Autophagy is a key cellular catabolic pathway that can be divided into macroautophagy, microautophagy and chaperone-mediated autophagy, according to the various enveloped substances and transport methods. Macroautophagy is comprised of two types: selective and non-selective. Selective autophagy includes mitophagy and pexophagy, and non-selective autophagy plays an important role in cell starvation [5,6].

Lemasters *et al.* [7] formally proposed the concept of mitophagy in 2005. This group observed that decreasing mitochondrial membrane potentials and the opening of the conductance permeability transition pores of the mitochondrial inner membrane cause mitophagy. The mitochondrion is the powerhouse of the cell, providing almost all energy for cellular activities and generating reactive oxygen species (ROS). ROS may cause damage to mitochondria, releasing apoptosis-inducing factors and leading to cell death. As the major site of biosynthesis (hemoglobin and lipid), the mitochondrion participates in the regulation of intracellular calcium. Therefore, the timely sequestration of damaged mitochondria is important for the normal growth of cells and the maintenance of a stable cellular environment [8,9].

Mitochondrial clearance from reticulocytes occurs through a special process that is regulated by multi-domain autophagy-related protein. The programmed removal of the mitochondria that occurs in reticulocytes represents a physiological model for studying the molecular mechanisms involved in mitophagy [10]. A hemin-induced human myeloid leukemia cell line (K562) has been shown to possess the capacity for erythroid differentiation *in vitro*. Multi-vesicular bodies and autophagy have been observed during K562 cell erythroid maturation [11]. In this review, we summarize the relevant modulators of autophagy involved in the regulation of erythroid differentiation under physiological and pathological conditions.

2. Autophagy Regulators and Erythroid Maturation

The targeted deletion of genes related to autophagy has been shown to cause anemia, indicating the presence of defective erythrocyte maturation and impaired mitophagy during terminal erythroid differentiation. The reported autophagy-related regulators that act during erythropoiesis have been summarized in Table 1.

Table 1. Autophagy-related modulators in erythropoiesis.

Modulators	Interactions with Other Molecules or Targets	Functions	References
Atg1/Ulk1	Atg13, Hsp90-Cdc37	Regulation of mitochondrial and ribosomal clearance	[12–14]
Atg4	-	Fusion of autophagosomes with lysosomes	[15]
Atg7	Atg5	Regulation of mitochondrial removal	[16–21]
Nix/Bnip3L	LC3, Atg8, miRNA	Modulation of mitochondrial clearance and autophagosome formation	[22–27]
GATA1	FoxO3, LC3-I	Direct regulation of autophagy genes	[28–33]
KRAB/KAP1-miRNA	Nix/Bnip3L, Ulk1	Participation in cascade controlling mitophagy	[34]
FIP200	Ulk1, Atg13	Essential autophagy gene in hematopoietic cells	[35,36]
Ca^{2+} and 15-lipoxygenase	-	Ca^{2+} promotes binding of 15-lipoxygenase to modulate the clearance of mitochondria	[37–39]

2.1. Autophagy-Related Gene (Atg) Family

Many autophagy-related genes have been identified that are critical for selective and/or nonselective autophagic regulatory mechanisms [40]. Atg1 (Ulk1), Atg13 and Atg17 are serine-threonine kinase complexes that regulate the cell cycle and cell growth and proliferation; E1-like enzyme Atg7 can activate Atg12 and conjugate Atg5 and E2-like protein Atg10 to form the preautophagosomal structure. Atg7 also mediates the conjugation of Atg12 to Atg5 and of Atg8 to phosphatidylethanolamine (PE), which participates in the extension of autophagy vesicles [41,42]. Among these genes, Ulk1, Atg4 and Atg7 are reported to play important roles during erythropoiesis.

2.2. Uncoordinated 51-Like Autophagy Activating Kinase 1 (Ulk1)

Ulk1, which is a homolog of yeast Atg1, is critical for mitochondrial and ribosomal clearance during erythroid terminal differentiation. The number of reticulocytes, mean cell volume (MCV), mean corpuscular hemoglobin level (MCH), and relative distribution width (RDW) of mature erythroid cells are increased in $Ulk1^{-/-}$ mice. $Ulk1^{-/-}$ reticulocytes exhibit the delayed removal of mitochondria, ribosomes and other organelles *in vitro*, and this defect is overcome via treatment with carbonyl cyanide 3-chlorophenylhydrazone (CCCP), which is a mitochondrial uncoupler that produces ROS and causes membrane depolarization [12,13]. Ulk1 interacts with the Hsp90-Cdc37 complex to promote its stability and activation. In addition, this interaction is conducive to Ulk1-directed phosphorylation and the recruitment of Atg13 to damaged mitochondria. As a Hsp90 antagonist, 17-allylamino-17-demethoxygeldanamycin (17AAG) is the synthetic derivative of geldanamycin, that can inhibit ATP binding and hydrolysis, and block the formation of chaperone complexes. When differentiating erythroid cells are treated with 2.5 μM 17AAG, they display significantly decreased Ulk1 protein levels, but Ulk1 mRNA levels are not affected. Although this treatment does not affect reticulocyte maturation, it notably reduces reticulocytes harboring mitochondria containing autophagosomes. Hsp90-Cdc37, Ulk1 and Atg13 are all required for mitophagy during erythroid differentiation [14].

2.3. Autophagy-Related 4 (Atg4)

Autophagy is induced in polychromatic erythroblasts, and autophagosomes remain abundant until enucleation, which stimulates the expression of Atg4 family members (Atg4A and Atg4D) and Atg8. The quantitative electron microscope assay has shown that compared to wild-type, fewer autophagosomes are assembled in Atg4 cysteine mutant Atg4B (C74A)-expressing progenitor cells, suggesting that the roles of Atg4 family members (particularly Atg4B) are important for autophagosome fusion during the differentiation of human erythroblasts [15].

2.4. Autophagy-Related 7 (Atg7)

Atg7 plays a critical role in mitochondrial autophagy in the mammalian hematopoietic system and has a unique pro-apoptotic effect on lysosome dysfunction. A previous study has shown that Atg7 is essential for the self-renewal, proliferation and normal functioning of HSCs [16]. According to this study, Vav-Atg7$^{-/-}$ mice showed reductions in hematopoietic stem cells and progenitors of multiple lineages. Furthermore, Atg7-deficient Lin^{-}Sca1^{+}c-Kit^{+} (LSK) cells accumulate mitochondria and ROS, causing DNA damage, which suggests that mitophagy is important to the regulation of HSCs maintenance. In Atg7$^{-/-}$ erythroid cells, the mitochondria are targeted to form autophagosomes, but autophagosome elongation is impaired, and mitochondrion engulfment is inhibited [16,17]. It has been shown that half of Atg7$^{-/-}$ fetal liver cell-transplanted mice die, and the surviving mice display anemia, reticulocytosis, and lymphopenia [17]. Vav-Atg7$^{-/-}$ mice generated using Atg7 Flox/Flox and Vav-iCre mice have been reported to show severe anemia and shortened lifespans [18]. Additionally, the transferrin receptor is up-regulated, and mitochondrial loss is initiated in Ter119^{+}/CD71^{-} elevated cells in the bone marrow of Vav-Atg7$^{-/-}$ mice. The loss of Atg7-mediated mitophagy in Atg7$^{-/-}$ erythroblasts leads to the accumulation of damaged mitochondria with the increased formation of isolation membranes, resulting in cell death. In the absence of Atg7, mitochondrial proteins are selectively removed by mitophagy, but proteins associated with the endoplasmic reticulum and ribosomes are unaffected [18,19]. An additional study has shown that the number of mitochondria and mitochondrial ROS in developing red blood cells (RBCs) are increased in Vav-Atg7$^{-/-}$ mice, and the developing RBCs display phosphatidylserine at their surfaces and undergo caspase 3-mediated apoptosis [20]. A recent study showed that when Atg7 was deleted from erythroid progenitors of wild-type and mtDNA-mutator mice, the genetic disruption of autophagy did not cause anemia in wild-type mice but accelerated the mitochondrial respiration decline and induced macrocytic anemia in the mtDNA-mutator mice [21].

2.5. Bcl-2 Family: Bcl-2/Adenovirus E1B 19 kDa Interacting Protein 3-Like (Nix/Binp3L)

The Bcl-2 family is known to play a key role in apoptosis, and its function as a regulator of autophagy has also received increasing interest. Bcl-2 functions as an antiautophagy protein via interacting with the conserved autophagy protein Beclin 1 [43]. Nix (also named Bnip3L), which is a mitochondrial outer membrane protein, is the BH3-only member of the Bcl-2 family that inhibits the proliferation of tumor cell lines [44,45]. Nix activity is mediated through the minimal essential region (MER) in its cytoplasmic domain. The mutation of the central leucine residue of MER causes loss of Nix activity and deters rescue of mitochondrial clearance in reticulocytes [46]. Nix is also a selective receptor that combines with mammalian Atg8 homologs, including microtubule-associated protein light chain 3 (LC3/GABARAP) and ubiquitin-like modifiers, which are indispensable for the maturation of phagophores and autophagosomes [22]. Through its *N*-terminal LC3-interacting region, Nix can recruit GABARAP-L1 to depolarize impaired mitochondria. LC3, which is a mammalian homolog of Atg8, has unmodified (LC3-I) and lipid-modified (LC3-II) forms [23].

In erythroid cells, Nix is upregulated during reticulocyte maturation [47], which is essential for mitochondrial membrane potential dispersion and autophagosome formation. Nix promotes the conversion of LC3-I to LC3-II. In addition, the elimination of the Nix: LC3/GABARAP interaction delays mitochondrial clearance in erythrocytes [24,25]. A previous study has shown that the reticulocytes of $Nix^{-/-}$ mice exhibit markedly abnormal mitochondrial residues. $Nix^{-/-}$ mice display hemolytic anemia and erythroid hyperplasia and increased levels of caspase activation and phosphatidylserine due to the increased production of ROS [25,26]. Nix may function through the ancillary release of cytochrome c or the interaction with other mitochondrial effector molecules. In the absence of Nix, mitochondria are not incorporated into autophagosomes in a timely manner for clearance, leading to an erythroid maturation defect. The Nix-dependent clearance of mitochondria has also been detected in human K562 cells that have been induced to undergo erythroid lineage maturation [26]. Nix may be activated to signal into mitochondria to dissipate their mitochondrial transmembrane potential ($\Delta\Psi m$) during erythroid cells maturation. Nix interacts with other molecules in mitochondria leading to selective sequestration of mitochondria into autophagosomes [27].

2.6. Transcription Factors and KAP1

Transcription factors are regarded as additional essential elements that are in autophagy during erythropoiesis and include erythroid-specific genes, such as GATA1, which play a critical role in erythroid differentiation [28,29]. It has been found that GATA1 directly upregulates the transcription of genes encoding autophagy-related components, such as LC3B and its homologs. In murine erythroid cells, GATA1 activates autophagy-related genes, increasing their expression levels during human erythropoiesis [30]. The forkhead protein FoxO3 is required for the GATA1-mediated induction of LC3 and the formation of autophagosomes in erythroid cells [31,32]. Recently, McIver *et al.* found that GATA-1/FoxO3 could repress the expression of Exosc8, a pivotal component of the exosome complex. When downregulated in primary erythroid precursor cells, Exosc8 could induce erythroid cell maturation [33]. GATA1 establishes a dependent pathway to activate the formation of LC3 and autophagosomes for mitochondrial clearance during erythropoiesis.

KRAB-associated protein 1 (KAP1), which is also named tripartite motif containing 28 (TRIM28), transcription intermediary factor 1 β (TIF1β) or KRAB-interacting protein 1 (KRIP-1), is a transcriptional intermediary factor that acts as a scaffold in transcription complexes. The KRAB/KAP1-miRNA regulatory cascade controls mitophagy during human erythropoiesis [34]. KAP1-depleted erythroblasts exhibit erythrocyte maturation defects and accumulate mitochondria. A luciferase reporter assay performed using mouse erythroleukemia (MEL) cells has shown that miR-351 targets the Nix 3'-UTR. Overexpression of miR-351 inhibits erythroid differentiation and causes mitochondrial accumulation in MEL cells [34]. In human erythroleukemia (HEL) cells, knockdown of Kap1 also leads to the impairment of erythroid differentiation, increased mitochondria and the blockage of autophagy effectors, including Nix. Additionally, hsa-miR-125a-5p expression was increased in KAP1-depleted HEL cells. When hsa-miR-125a-5p is over-expressed, the downregulation of Nix and increased numbers of mitochondria are also observed [34]. Therefore, multi-factorial molecules interact with miRNAs to form a regulatory network during mitophagy for the control of erythropoiesis.

2.7. Other Modulators: FIP200, Ca2$^+$ and 15-Lipoxygenase

FIP200 (200-kDa focal adhesion kinase family-interacting protein) is known to play an essential role in mammalian autophagy and diverse cellular functions. Several studies have shown that FIP200 is an important part of the Ulk1-Atg13-FIP200 complex in autophagosome formation [35]. The deletion of FIP200 results in increased HSCs cycling, the loss of HSCs reconstituting capacities, aberrant myeloid expansion and the blocking of erythroid maturation. Furthermore, FIP200-null HSCs exhibit abnormal accumulation of mitochondria and have increased ROS levels. These studies suggest that FIP200 is a key regulator of fetal HSCs and plays a potential role in autophagy for the maintenance of fetal hematopoiesis [35,36].

Lipoxygenase is the key enzyme involved in unsaturated fatty acid metabolism, and it can translate arachidonic acid, linoleic acid and other fatty acids into their bioactive metabolites, affecting cell structure, metabolism and signal transduction. 15-lipoxygenase acts in response to oxidative damage and modulates the clearance of mitochondria during reticulocyte maturation [37]. A previous study has indicated that 15-lipoxygenase sediments with the mitochondrial fraction in rabbit reticulocytes [38]. Ca^{2+} promotes the binding of 15-lipoxygenase to reticulocyte mitochondria and stimulates the lipid peroxidation of mitochondrial lipids and free linoleic acid. Therefore, Ca^{2+} is important for regulating the 15-lipoxygenase-mediated degradation of mitochondria in reticulocytes [39].

Mammalian autophagy signaling pathways are complicated and the mTOR-dependent pathway is the most prominent. AMPK (AMP-activated protein kinase) and mTOR (mammalian target of rapamycin) regulate autophagy through the direct phosphorylation of Ulk1. In nutrient-deficient conditions, AMPK promotes autophagy through the activation of Ulk1, and conversely, mTOR activity prevents Ulk1 activation under normal situations [48]. The implicated roles of the intracellular autophagy pathway in erythroid cell differentiation and maturation are summarized in Figure 1.

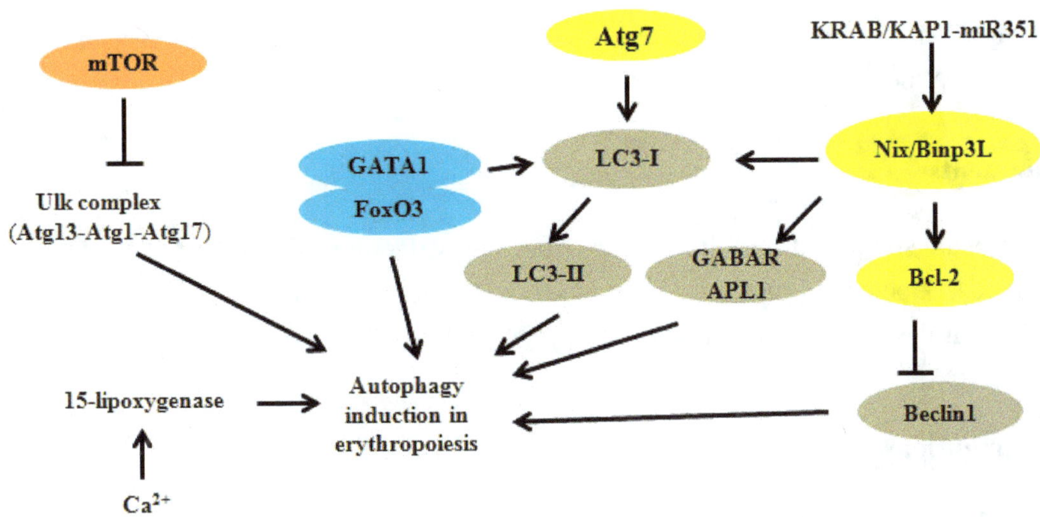

Figure 1. Autophagy-related factors are involved in the regulation of signal pathways in erythroid cells. The mTOR pathway is an important pathway that directly modulates the Ulk1 complex, and the inhibition of mTOR represses autophagy-related processes. Atg7 and Nix/Bnip3L are required for the removal of mitochondria, inducing the conversion of LC3-I to its lipid-modified form, LC3-II, to promote autophagy. miRNAs can regulate the expressions of key transcriptional components, and Ca^{2+} promotes the binding of 15-lipoxygenase to reticulocyte mitochondria.

3. Autophagy and β-Thalassemia

Several recent reports have indicated the key role of autophagy in red cell disorders, including β-thalassemia and myelodysplasia syndrome (MDS) [49–51]. When cultured $CD34^+$ erythroid progenitor cells from peripheral blood obtained from normal and β-thalassemia patients are induced to erythroid differentiation, autophagy is increased in the erythroblasts from the β-thalassemia patients compared with the normal erythroblasts, and this increase is mediated by the high levels of Ca^{2+} in the β-thalassemia erythroblasts. Normal erythroblasts show increased apoptosis following treatment with L-asparagine, which is an autophagy inhibitor, but this is not observed in erythroblasts from patients with β-thalassemia. Furthermore, reduced Ca^{2+} levels cause decreases in both autophagy and apoptosis. The high levels of autophagy may contribute to the increased apoptosis, leading to anemia and ineffective erythropoiesis in erythroblasts from β-thalassemia patients [51]. Notably, it has also been demonstrated that the occurrence of autophagy and early differentiation are linked in hESCs [52].

4. Perspectives

Autophagy is important in maintaining a cellular homeostatic environment. Autophagy has been implicated in many diseases, including various cancers, central nervous system (CNS)-related disorders, neurodegenerative disorders and heart disease in addition to aging. Autophagy studies are developing rapidly in the field of biology, especially in the erythroid research field. If autophagy is impaired during erythroid differentiation, erythroid maturation will be deficient. In this review, we summarize the relevant modulators of autophagy involved in the regulation of erythroid differentiation, which imply that autophagy is an important and complex process during erythroid differentiation.

Understanding the modulators of autophagy in normal and pathologic erythropoiesis may facilitate the prevention and treatment of red blood cell-related disorders.

Acknowledgments

This work was supported by the grants from New Century Excellent Talents in University (NCET-11-0518), Doctoral Fund of Ministry of Education of China (No. 20120162110054) and the National Natural Science Foundation of China (No. 81270576, 81470362 and 31101686).

Author Contributions

Jing Liu, Xiuli An and Jieying Zhang designed the paper; Jieying Zhang, Kunlu Wu, Xiaojuan Xiao, Jiling Liao, Qikang Hu, Huiyong Chen, Jing Liu and Xiuli An participated in collecting and analyzing the data; Jieying Zhang and Jing Liu wrote the paper; and Jing Liu, Xiuli An and Jieying Zhang revised the paper.

Abbreviation

17AAG, 17-allylamino-17-demethoxygeldanamycin; AMPK, AMP-activated protein kinase; Atg, autophagy-related; Bcl-2, B-cell lymphoma 2; BFU-E, burst-forming-unit erythroid; CCCP, carbonyl cyanide 3-chlorophenylhydrazone; Cdc37, cell division cycle 37; CFU-E, colony-forming-unit erythroid; CNS, central nervous system; FIP200, focal adhesion kinase family-interacting protein of 200-kDa; FoxO3, forkhead box O3; GABARAP, gamma aminobutyric acid A receptor-associated protein; GATA1, globin transcription factor; HEL, human erythroleukemia; HSCs, hematopoietic stem cells; Hsp90, 90 kDa heat shock protein; KRAB, Krueppel-associated box; KAP-1, KRAB-associated protein 1; KRIP-1, KRAB-interacting protein 1; LC3, light chain 3; LSK, $Lin^{-}Sca1^{+}c\text{-}Kit^{+}$; MCV, mean cell volume; MCH, mean corpuscular hemoglobin; MDS, myelodysplasia syndrome; MEL, mouse erythroleukemia; MER, minimal essential region; mTOR, mammalian target of rapamycin; Nix/Binp3L, Bcl-2/adenovirus E1B 19 kDa interacting protein 3-like; PE, phosphatidylethanolamine; RBCs, red blood cells; ROS, reactive oxygen species; RDW, relative distribution width; TRIM28, tripartite motif containing 28; TIF1β, transcription intermediary factor 1 beta; Ulk1, uncoordinated 51-like autophagy activating kinase 1.

References

1. Stephenson, J.R.; Axelrad, A.A.; McLeod, D.L.; Shreeve, M.M. Induction of colonies of hemoglobin-synthesizing cells by erythropoietin *in vitro*. *Proc. Natl. Acad. Sci. USA* **1971**, *68*, 1542–1546.
2. Géminard, C.; de Gassart, A.; Vidal, M. Reticulocyte maturation: Mitoptosis and exosome release. *Biocell* **2002**, *26*, 205–215.

3. Ashford, T.P.; Porter, K.R. Cytoplasmic components in hepatic cell lysosomes. *J. Cell Biol.* **1962**, *12*, 198–202.
4. Deter, R.L.; DeDuve, C. Influence of glucagon, an inducer of cellular autophagy on some physical properties of rat liver lysosomes. *J. Cell Biol.* **1967**, *33*, 437–449.
5. Klionsky, D.J.; Emr, S.D. Autophagy as a regulated pathway of cellular degradation. *Science* **2000**, *290*, 1717–1721.
6. Komatsu, M.; Ichimura, Y. Selective autophagy regulates various cellular functions. *Genes Cells* **2010**, *10*, 923–933.
7. Lemasters, J.J. Selective mitochondrial autophagy, or mitophagy, as a targeted defense against oxidative stress, mitochondrial dysfunction, and aging. *Rejuvenation Res.* **2005**, *8*, 3–5.
8. Scheffler, I.E. A century of mitochondrial research: Achievements and perspectives. *Mitochondrion* **2001**, *1*, 3–31.
9. Duchen, M.R. Mitochondria in health and disease: Perspectives on a new mitochondrial biology. *Mol. Asp. Med.* **2004**, *25*, 365–451.
10. Zhang, J.; Ney, P.A. Reticulocyte mitophagy: Monitoring mitochondrial clearance in a mammalian model. *Autophagy* **2010**, *6*, 405–408.
11. Fader, C.M.; Colombo, M.I. Multivesicular bodies and autophagy in erythrocyte maturation. *Autophagy* **2006**, *2*, 122–125.
12. Chan, E.Y.; Kir, S.; Tooze, S.A. siRNA screening of the kinome identifies ULK1 as a multidomain modulator of autophagy. *J. Biol. Chem.* **2007**, *282*, 25464–25474.
13. Kundu, M.; Lindsten, T.; Yang, C.Y.; Wu, J.M.; Zhao, F.P.; Zhang, J.; Selak, M.A.; Ney, P.A.; Thompson, C.B. Ulk1 plays a critical role in the autophagic clearance of mitochondria and ribosomes during reticulocyte maturation. *Blood* **2008**, *112*, 1493–1502.
14. Joo, J.H.; Dorsey, F.C.; Joshi, A.; Hennessy-Walters, K.M.; Rose, K.L.; McCastlain, K.; Zhang, J.; Iyengar, R.; Jung, C.H.; Suen, D.F.; *et al.* Hsp90-Cdc37 chaperone complex regulates Ulk1- and Atg13-mediated mitophagy. *Mol. Cell* **2011**, *43*, 572–585.
15. Betin, V.M.; Singleton, B.K.; Parsons, S.F.; Anstee, D.J.; Lane, J.D. Autophagy facilitates organelle clearance during differentiation of human erythroblasts: Evidence for a role for Atg4 paralogs during autophagosome maturation. *Autophagy* **2013**, *9*, 881–893.
16. Mortensen, M.; Soilleux, E.J.; Djordjevic, G.; Tripp, R.; Lutteropp, M.; Sadighi-Akha, E.; Stranks, A.J.; Glanville, J.; Knight, S.; Jacobsen, S.E.; *et al.* The autophagy protein Atg7 is essential for hematopoietic stem cell maintenance. *J. Exp. Med.* **2011**, *208*, 455–467.
17. Zhang, J.; Randall, M.S.; Loyd, M.R.; Dorsey, F.C.; Kundu, M.; Cleveland, J.L.; Ney, P.A. Mitochondrial clearance is regulated by Atg7-dependent and -independent mechanisms during reticulocyte maturation. *Blood* **2009**, *114*, 157–164.
18. Komatsu, M.; Waguri, S.; Ueno, T.; Iwata, J.; Murata, S.; Tanida, I.; Ezaki, J.; Mizushima, N.; Ohsumi, Y.; Uchiyama, Y.; *et al.* Impairment of starvation-induced and constitutive autophagy in Atg7-deficient mice. *J. Cell Biol.* **2005**, *169*, 425–434.
19. Mortensen, M.; Ferguson, D.J.; Edelmann, M.; Kessler, B.; Morten, K.J.; Komatsu, M.; Simon, A.K. Loss of autophagy in erythroid cells leads to defective removal of mitochondria and severe anemia *in vivo*. *Proc. Natl. Acad. Sci. USA* **2010**, *107*, 832–837.

20. Mortensen, M.; Simon, A.K. Nonredundant role of Atg7 in mitochondrial clearance during erythroid development. *Autophagy* **2010**, *6*, 423–425.
21. Li-Harms, X.; Milasta, S.; Lynch, J.; Wright, C.; Joshi, A.; Iyengar, R.; Neale, G.; Wang, X.; Wang, Y.D.; Prolla, T.A.; *et al.* Mito-protective autophagy is impaired in erythroid cells of aged mtDNA-mutator mice. *Blood* **2015**, *125*, 162–174.
22. Novak, I.; Kirkin, V.; McEwan, D.G.; Zhang, J.; Wild, P.; Rozenknop, A.; Rogov, V.; Löhr, F.; Popovic, D.; Occhipinti, A.; *et al.* Nix is a selective autophagy receptor for mitochondrial clearance. *EMBO Rep.* **2010**, *11*, 45–51.
23. Schwarten, M.; Mohrluder, J.; Ma, P.; Stoldt, M.; Thielmann, Y.; Stangler, T.; Hersch, N.; Hoffmann, B.; Merkel, R.; Willbold, D.; *et al.* Nix directly binds to GABARAP: A possible crosstalk between apoptosisand autophagy. *Autophagy* **2009**, *5*, 690–698.
24. Schweers, R.L.; Zhang, J.; Randall, M.S.; Loyd, M.R.; Li, W.; Dorsey, F.C.; Kundu, M.; Opferman, J.T.; Cleveland, J.L.; Miller, J.L.; *et al.* NIX is required for programmed mitochondrial clearance during reticulocyte maturation. *Proc. Natl. Acad. Sci. USA* **2007**, *104*, 19500–19505.
25. Zhang, J.; Ney, P.A. Nix induces mitochondrial autophagy in reticulocytes. *Autophagy* **2008**, *4*, 354–356.
26. Sandoval, H.; Thiagarajan, P.; Dasgupta, S.K.; Schumacher, A.; Prchal, J.T.; Chen, M.; Wang, J. Essential role for Nix in autophagic maturation of erythroid cells. *Nature* **2008**, *454*, 232–235.
27. Chen, M.; Sandoval, H.; Wang, J. Selective mitochondrial autophagy during erythroid maturation. *Autophagy* **2008**, *4*, 926–928.
28. Welch, J.J.; Watts, J.A.; Vakoc, C.R.; Yao, Y.; Wang, H.; Hardison, R.C.; Blobel, G.A.; Chodosh, L.A.; Weiss, M.J. Gobal regulation of erythroid gene expression by transcription factor GATA-1. *Blood* **2004**, *104*, 3136–3147.
29. Layon, M.E.; Layon, C.J.; West, R.J.; Lowrey, C.H. Expression of GATA-1 in a non-hematopoietic cell line induces β-globin locus control region chromatin structure remode ling and an erythroid pattern of gene expression. *J. Mol. Biol.* **2007**, *366*, 737–744.
30. Kang, Y.A.; Sanalkumar, R.; O'Geen, H.; Linnemann, A.K.; Chang, C.J.; Bouhassira, E.E.; Farnham, P.J.; Keles, S.; Bresnick, E.H. Autophagy driven by a master regulator of hematopoiesis. *Mol. Cell. Biol.* **2012**, *32*, 226–239.
31. Spitali, P.; Grumati, P.; Hiller, M.; Chrisam, M.; Aartsma-Rus, A.; Bonaldo, P. Autophagy is impaired in the tibialis anterior of dystrophin null mice. *PLoS Curr.* **2013**, *5*, 1–9.
32. Bakker, W.J.; van Dijk, T.B.; Parren-van Amelsvoort, M.; Kolbus, A.; Yamamoto, K.; Steinlein, P.; Verhaak, R.G.; Mak, T.W.; Beug, H.; Löwenberg, B.; *et al.* Differential regulation of FoxO3a target genes in erythropoiesis. *Mol. Cell. Biol.* **2007**, *27*, 3839–3854.
33. McIver, S.C.; Kang, Y.A.; DeVilbiss, A.W.; O'Driscoll, C.A.; Ouellette, J.N.; Pope, N.J.; Camprecios, G.; Chang, C.J.; Yang, D.; Bouhassira, E.E.; *et al.* The exosome complex establishes a barricade to erythroid maturation. *Blood* **2014**, *124*, 2285–2297.
34. Barde, I.; Rauwel, B.; Marin-Florez, R.M.; Corsinotti, A.; Laurenti, E.; Verp, S.; Offner, S.; Marquis, J.; Kapopoulou, A.; Vanicek, J.; *et al.* A KRAB/KAP1-miRNA cascade regulates erythropoiesis through stage-specific control of mitophagy. *Science* **2013**, *340*, 350–353.

35. Jung, C.H.; Jun, C.B.; Ro, S.H.; Kim, Y.M.; Otto, N.M.; Cao, J; Kundu, M.; Kim, D.H. ULK-Atg13-FIP200 complexes mediate mTOR signaling to the autophagy machinery. *Mol. Biol. Cell* **2009**, *20*, 1992–2003.
36. Liu, F.; Lee, J.Y.; Wei, H.; Tanabe, O.; Engel, J.D.; Morrison, S.J.; Guan, J.L. FIP200 is required for the cell-autonomous maintenance of fetal hematopoietic stem cells. *Blood* **2010**, *116*, 4806–4814.
37. Kuhn, H.; Belkner, J.; Wiesner, R. Subcellular distribution of lipoxygenase products in rabbit reticulocyte membranes. *Eur. J. Biochem.* **1990**, *191*, 221–227.
38. Vijayvergiya, C.; DeAngelis, D.; Walther, M.; Kühn, H.; Duvoisin, R.M.; Smith, D.H.; Wiedmann, M. High-level expression of rabbit 15-lipoxygenase induces collapse of the mitochondrial pH gradient in cell culture. *Biochemistry* **2004**, *43*, 15296–15302.
39. Watson, A.; Doherty, F.J. Calcium promotes membrane association of reticulocyte 15-lipoxygenase. *Biochem. J.* **1994**, *298*, 377–383.
40. Xie, Z.; Klionsky, D.J. Autophagosome formation: Core machinery and adaptations. *Nat. Cell Biol.* **2007**, *9*, 1102–1109.
41. Geng, J.; Klionsky, D.J. The Atg8 and Atg12 ubiquitin-like conjugation systems in macroautophagy: Protein modifications: Beyond the usual suspects' review series. *EMBO Rep.* **2008**, *9*, 859–864.
42. Walls, K.C.; Ghosh, A.P.; Franklin, A.V.; Klocke, B.J.; Ballestas, M.; Shacka, J.J.; Zhang, J.; Roth, K.A. Lysosome dysfunction triggers Atg7-dependent neural apoptosis. *J. Biol. Chem.* **2010**, *285*, 10497–10507.
43. Pattingre, S.; Tassa, A.; Qu, X.; Garuti, R.; Liang, X.H.; Mizushima, N.; Packer, M.; Schneider, M.D.; Levine, B. Bcl-2 antiapoptotic proteins inhibit Beclin 1-dependent autophagy. *Cell* **2005**, *122*, 927–939.
44. Boyd, J.M.; Malstrom, S.; Subramanian, T.; Venkatesh, L.K.; Schaeper, U.; Elangovan, B.; D'Sa-Eipper, C.; Chinnadurai, G. Adenovirus E1B 19 kDa and Bcl-2 proteins interact with a common set of cellular proteins. *Cell* **1994**, *79*, 341–351.
45. Matsushima, M.; Fujiwara, T.; Takahashi, E.; Minaguchi, T.; Eguchi, Y.; Tsujimoto, Y.; Suzumori, K.; Nakamura, Y. Isolation, mapping, and functional analysis of a novel human cDNA (BNIP3L) encoding a protein homologous to human NIP3. *Genes Chromosomes Cancer* **1998**, *21*, 230–235.
46. Zhang, J.; Loyd, M.R.; Randall, M.S.; Waddell, M.B.; Kriwacki, R.W.; Ney, P.A. A short linear motif in BNIP3L (Nix) mediates mitochondrial clearance in reticulocytes. *Autophagy* **2012**, *8*, 1325–1332.
47. Aerbajinai, W.; Giattina, M.; Lee, Y.T.; Raffeld, M.; Miller, J.L. The proapoptotic factor Nix is coexpressed with Bcl-xL during terminal erythroid differentiation. *Blood* **2003**, *102*, 712–717.
48. Kim, J.; Kundu, M.; Viollet, B.; Guan, K.L. AMPK and mTOR regulate autophagy through direct phosphorylation of Ulk1. *Nat. Cell Biol.* **2011**, *13*, 132–141.
49. Watson, A.S.; Mortensen, M.; Simon, A.K. Autophagy in the pathogenesis of myelodysplastic syndromeand acute myeloid leukemia. *Cell Cycle* **2011**, *10*, 1719–1725.

50. Fang, J.; Rhyasen, G.; Bolanos, L.; Rasch, C.; Varney, M.; Wunderlich, M.; Goyama, S.; Jansen, G.; Cloos, J.; Rigolino, C.; *et al.* Cytotoxic effects of bortezomib in myelodysplastic syndrome/acute myeloidleukemia depend on autophagy-mediated lysosomal degradation of TRAF6 and repression of PSMA1. *Blood* **2012**, *120*, 858–867.
51. Lithanatudom, P.; Wannatung, T.; Leecharoenkiat, A.; Svasti, S.; Fucharoen, S.; Smith, D.R. Enhanced activation of autophagy in β-thalassemia/Hb E erythroblasts during erythropoiesis. *Ann. Hematol.* **2011**, *90*, 747–758.
52. Tra, T.; Gong, L.; Kao, L.P.; Li, X.L.; Grandela, C.; Devenish, R.J.; Wolvetang, E.; Prescott, M. Autophagy in human embryonic stem cells. *PLoS One* **2011**, *6*, e27485.

11

Mitochondrial Transcription Factor A and Mitochondrial Genome as Molecular Targets for Cisplatin-Based Cancer Chemotherapy

Kimitoshi Kohno [1,*], Ke-Yong Wang [2], Mayu Takahashi [3], Tomoko Kurita [4], Yoichiro Yoshida [5], Masakazu Hirakawa [6], Yoshikazu Harada [7], Akihiro Kuma [8], Hiroto Izumi [9] and Shinji Matsumoto [1]

1 Asahi Matsumoto Hospital, Kokuramimami-ku Tsuda, Kitakyushu-shi 800-0242, Japan; E-Mail: sm.kokura.1949@docomo.ne.jp

2 Shared-Use Research Center, University of Occupational and Environmental Health School of Medicine, Yahatanishi-ku, Kitakyushu-shi 807-8555, Japan; E-Mail: kywang@med.uoeh-u.ac.jp

3 Department of Neurosurgery, University of Occupational and Environmental Health School of Medicine, Yahatanishi-ku, Kitakyushu-shi 807-8555, Japan; E-Mail: maytak@med.uoeh-u.ac.jp

4 Department of Gynecology, University of Occupational and Environmental Health School of Medicine, Yahatanishi-ku, Kitakyushu-shi 807-8555, Japan; E-Mail: t-kurita@med.uoeh-u.ac.jp

5 Department of Gastroenterological Surgery, School of Medicine, Fukuoka University, Fukuoka 814-0180, Japan; E-Mail: yy4160@yahoo.co.jp

6 Department of Radiology, Beppu Hospital, Kyushu University, Beppu 874-0838, Japan; E-Mail: hirakawa@beppu.kyushu-u.ac.jp

7 Department of Pathology and Cell Biology, University of Occupational and Environmental Health School of Medicine, Yahatanishi-ku, Kitakyushu-shi 807-8555, Japan; E-Mail: y8harada@gmail.com

8 Second Department of Internal Medicine, University of Occupational and Environmental Health School of Medicine, Yahatanishi-ku, Kitakyushu-shi 807-8555, Japan; E-Mail: akihiro_k@me.com

9 Department of Occupational Pneumology, Institute of Industrial Ecological Science, University of Occupational and Environmental Health, Yahatanishi-ku, Kitakyushu-shi 807-8555, Japan; E-Mail: h-izumi@med.uoeh-u.ac.jp

* Author to whom correspondence should be addressed; E-Mail: k-kohno@med.uoeh-u.ac.jp

Academic Editors: Lars Olson, Jaime M. Ross and Giuseppe Coppotelli

Abstract: Mitochondria are important cellular organelles that function as control centers of the energy supply for highly proliferative cancer cells and regulate apoptosis after cancer chemotherapy. Cisplatin is one of the most important chemotherapeutic agents and a key drug in therapeutic regimens for a broad range of solid tumors. Cisplatin may directly interact with mitochondria, which can induce apoptosis. The direct interactions between cisplatin and mitochondria may account for our understanding of the clinical activity of cisplatin and development of resistance. However, the basis for the roles of mitochondria under treatment with chemotherapy is poorly understood. In this review, we present novel aspects regarding the unique characteristics of the mitochondrial genome in relation to the use of platinum-based chemotherapy and describe our recent work demonstrating the importance of the mitochondrial transcription factor A (mtTFA) expression in cancer cells.

Keywords: mtTFA; mtDNA; cisplatin; cancer chemotherapy; prognosis

1. Introduction

Mitochondria produce the energy required to maintain cellular functions as well as command the process of apoptosis in order to inhibit these functions [1,2]. Cancer cells are immortalized as a result of various genetic and epigenetic changes in the genome. One of the major causes of cancer is the acquisition of the driving force of cellular proliferation induced by a mutation and/or the methylation of oncogenes and tumor suppressor genes. Another factor is resistance to apoptosis, which is also acquired due to genetic or epigenetic changes in apoptosis-related genes. These processes are closely related to the onset of drug resistance and the phenotypic characteristics of cancer stem cells. However, it is unknown whether the direct cytotoxic effects or apoptosis-inducing ability of anti-cancer agents are effective as cancer chemotherapy. Cancer cells are resistant to conventional chemotherapy and radiotherapy regimens, and various molecular mechanisms underlying the onset of resistance to therapy have been proposed [3,4]. For example, cellular survival signal transduction systems protect mitochondrial integrity against drug-induced stress, and apoptosis-related molecules are orchestrated around mitochondria to induce and execute apoptosis in cancer cells. These molecules also play important roles in the development of drug resistance in cancer cells. It is therefore intriguing to explore how the crosstalk of apoptosis-related molecules regulates various mitochondrial functions leading to drug resistance. This review describes some of the putative mechanisms of action of cisplatin at the site of the mitochondrial genome and in terms of the mtTFA expression, which may play an important role in the cellular functions of cancer cells and the prognosis of cancer patients.

2. Cisplatin and DNA

Cisplatin, cisplatinum or *cis*-diamminedichloroplatinum (II) is a well-known chemotherapeutic drug [5,6]. Among various anti-cancer agents, cisplatin is one of the most effective and widely used anti-cancer agents for the treatment of several types of solid tumors [7]. Special attention has been paid to its molecular mechanisms of action [8]. The cytotoxic effects of cisplatin are thought to be mediated primarily by the generation of nuclear DNA adducts, which, if not repaired, cause cell death as a consequence of the inhibition of DNA replication and transcription. However, the ability of cisplatin to induce nuclear DNA (nDNA) damage per se is not sufficient to explain its high degree of effectiveness.

3. Calcium Ion, Endoplasmic Reticulum (ER) Stress, ROS/Redox System and Apoptosis

Bcl-2 family proteins control mitochondrial outer membrane permeability and regulate apoptosis [9,10]. Among Bcl-2 family proteins, anti-apoptotic proteins, Bcl-2 and Bcl-xL localize at the mitochondria outer membrane and thereby inhibit cytochrome c release. On the other hand, apoptosis inducing pro-apoptotic proteins, Bad, Bid and Bax, localize from cytoplasm to mitochondria following various apoptosis signals and promote cytochrome c release [11]. Calcium ion is well-known to play an important role in the mitochondrial control of apoptosis [12].

It has been reported that cisplatin induces calpain-mediated Bid cleavage in association with increased intracellular calcium ion levels and calpain-cleaved Bid induces cytochrome c release from mitochondria [13]. The endoplasmic reticulum (ER) is a specialized organelle for the folding and trafficking of proteins and a cellular target of cisplatin [14]. ER stress signaling is involved in cisplatin-induced apoptosis, as cisplatin has been shown to induce the calpain-dependent activation of ER-specific caspase leading to apoptosis [15]. ER stress is also linked to the production of reactive oxygen species (ROS). ROS/Redox signaling is prominently associated with the progression of human malignancies [16]. Furthermore, ROS/Redox signaling is a critical stress response pathway associated with cancer chemotherapy. Oxidative damage is observed *in vivo* following exposure to cisplatin in several tissues, suggesting the role of oxidative stress in the pathogenesis of cisplatin-induced dose-limiting toxicities [17,18]. However, the mechanisms underlying the cisplatin-induced generation of ROS and their contribution to cisplatin cytotoxicity in normal and cancer cells remain poorly understood.

ROS generation occurs independently of the amount of cisplatin-induced nDNA damage and takes place in mitochondria as a consequence of the impairment of protein synthesis. The contribution of cisplatin-induced mitochondrial dysfunction in determining the cytotoxic effects of this drug varies among cells and depends on the mitochondrial redox status and integrity of mitochondrial DNA (mtDNA). Cisplatin-induced cellular pathways and responses are summarized in Figure 1.

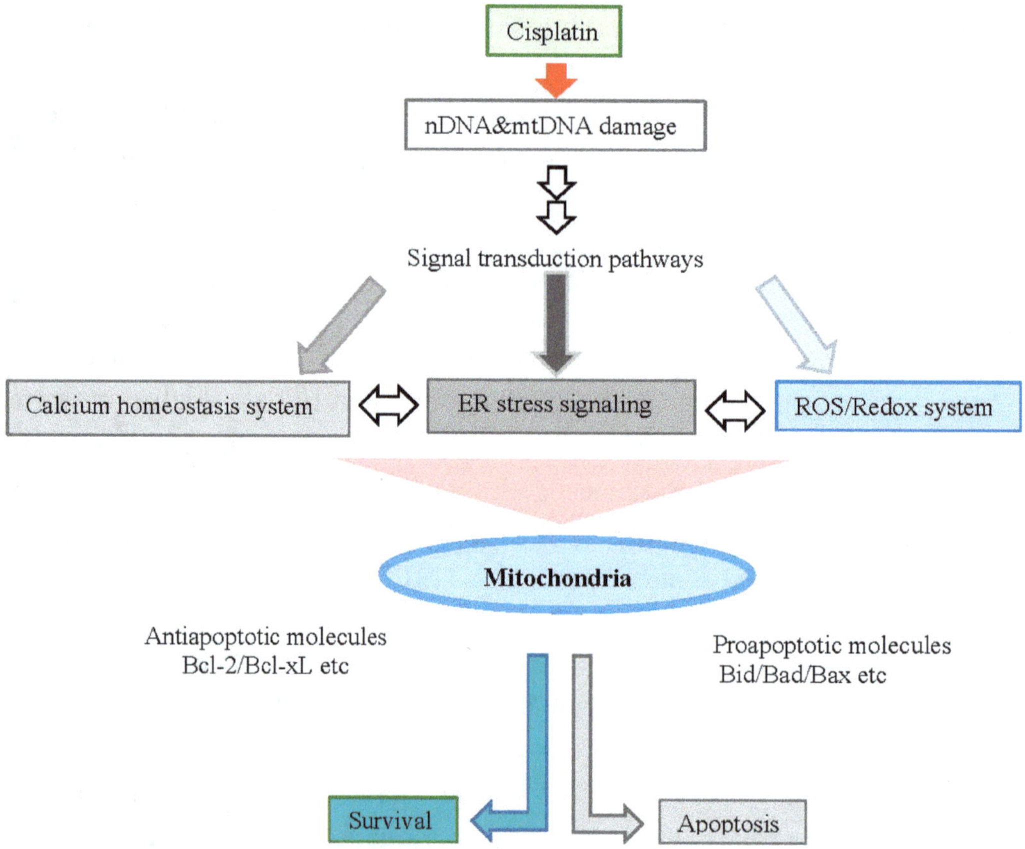

Figure 1. Cellular pathways and apoptosis induced by cisplatin. Cisplatin activates signal transduction pathways and may directly interact with mitochondria, which can induce apoptosis.

4. mtTFA and Cisplatin Resistance

The development of resistance to cisplatin is a major obstacle in terms of the clinical outcomes of cancer patients [19–21]. Several mechanisms are thought to be involved in the onset of cisplatin resistance, including decreased intracellular drug accumulation, increased levels of cellular thiols, an increased nucleotide excision-repair activity and a decreased mismatch-repair activity [22,23]. Two transporters, copper-transporting P-type ATPase 7B (ATP7B) and ATP-binding cassette, sub-family C, member 2 (ABCC2), may be involved in cisplatin efflux and resistance [22,23]. Genome-wide analyses have been shown to be a powerful method for understanding drug resistance. A recent report showed that the Son of sevenless/the mitogen-activated protein kinase/extracellular signal-regulated kinase (SOS/MAPK/ERK) pathway is activated in cisplatin-resistant cells [24]. This pathway mediates the degradation of the proapoptotic molecule Bim resulting in the inhibition of mitochondria-dependent apoptotic pathways [25].

One antioxidative factor, the level of cellular glutathione, has been shown to be involved in the development of cisplatin resistance. Interestingly, upregulation of the ABC transporter and cellular glutathione is a characteristic of cancer stem cells (CSCs) [26–28]. Therefore, CSCs show multidrug resistance and radioresistance. Recently, it has been demonstrated that the expression of CD44, especially variant isoforms (CD44v) among major CSC markers, contributes to ROS defenses via upregulation of the synthesis of reduced glutathione [29]. These observations suggest that CSCs exhibit natural drug-resistant phenotypes and that drug-induced resistant cells display acquired resistant phenotypes. These data also indicate that CSCs mitochondria might differ from those of non-CSCs. However, little is known about the mitochondrial features related to energy production and the ROS/Redox system of CSCs. Therefore, defining these features will be critical for developing mitochondria-targeted therapeutics [30].

In general, the molecules responsible for these phenotypes are upregulated in cisplatin-resistant cells, which indicates that the transcription factors activated in response to cisplatin may play crucial roles in the development of drug resistance [19,20]. Interestingly, we have found that mitochondrial transcription factor A (mtTFA) is upregulated in cisplatin-resistant cells [31]. To date, several nuclear transcription factors involved in the development of resistance against platinum-containing agents have been identified, including Y-box binding protein-1 (YB-1) [32], nuclear factor I/B [33], activating transcription factor 4 (ATF4) [34], zinc-finger factor 143 (ZNF143) [35], and Clock [36]. Several of these transcription factors are regulated by E-box binding transcription factors, which regulate the epithelial-mesenchymal transition [37]. Furthermore, both YB-1 and ZNF143 lack the high-mobility group (HMG) domain and are capable of binding preferentially to cisplatin-modified DNA in addition to HMG domain proteins, such as mtTFA. It has therefore been proposed that various mitochondrial functions may be regulated by the circadian clock system [38], which controls cancer cell proliferation and angio/stromagenesis via WNT signaling [39,40].

Mitochondrial dysfunction is associated not only with cancer progression, but also with chemoresistance [41], and a relationship of mtDNA mutations with cisplatin-induced apoptosis and/or cisplatin resistance has been reported. These mtDNA mutations endow cancer cells with chemoresistance [42]. It has also recently been reported that nuclear co-activators, including peroxisome proliferator-activated receptor gamma co-activator-1 (PGC-1), are upregulated and compensate for respiratory chain defects due to mtDNA mutations in cisplatin-resistant cells [43]. Interestingly, a frameshift mutation of the nuclear mtTFA gene has been reported in colorectal cancer cells with microsatellite instability [44]. These cancer cells express truncated mtTFA and show more resistance to cisplatin-induced apoptosis. These data also show that mitochondrial dysfunction due to genetic changes in both mtDNA and nDNA is closely associated with chemoresistance.

5. Mitochondrial DNA as a Target for Cisplatin

Mitochondria are closely related to carcinogenesis, cancer progression and chemoresistance. The genes encoded by mtDNA are tightly packed together with minimal noncoding regions. Although the number of mitochondria and mtDNA in cancer cells has not been extensively studied, a reduction in the mtDNA copy number has been reported in various human cancers [45]. Furthermore, it is well known that mtDNA is more susceptible than nDNA to damage from reactive oxygen species and chemicals, including anti-cancer agents, due to either a limited capacity for DNA repair or the presence of nucleosome-free structures [46,47].

Cisplatin is a major DNA-targeting agent and the most potent key drug for treating solid tumors among anti-cancer agents. It has been shown that the mitochondrial DNA adduct levels are higher than the nDNA adduct levels and that both a higher degree of initial binding and lack of removal of cisplatin-DNA adducts appear to contribute to the preferential formation of cisplatin-mtDNA adducts. The effects of cisplatin arise from its ability to damage DNA, with the major adducts formed being intrastrand d(GpG) crosslinks. Therefore, the target DNA sequence of cisplatin is G-stretch. Table 1 shows the number of G-stretch sequences in the mtDNA of various species in comparison with that observed in nDNA. As shown in Table 1, the number of G-stretch sequences is higher in primates than in other species. Furthermore, the number of G-stretch sequences in mtDNA is significantly higher than that noted in nDNA. These findings indicate that the number of G-stretch sequences may be related to the higher initial binding capacity of cisplatin to mtDNA.

Table 1. Number of cisplatin-targeted DNA sequences in mitochondrial DNA.

Species	Total Number of mtDNA		GG	GGG	GGGG	GGGGG
Human	16,565	L chain	426	73	15	4
		H chain	1772	624	224	69
		total	2198	697	239	73
Gorilla	16,364	L chain	425	71	16	5
		H chain	1712	596	216	72
		total	2137	667	232	77
Rat	16,300	L chain	397	63	14	4
		H chain	1299	377	99	32
		total	1696	440	113	36
Mouse	16,300	L chain	397	58	11	3
		H chain	1104	288	72	19
		total	1501	346	83	22
Xenopus	17,553	L chain	445	72	12	1
		H chain	1091	259	60	8
		total	1536	331	72	9
Drosophila	16,019	L chain	599	230	16	1
		H chain	770	358	78	30
		total	1369	588	94	31
Expectation		Double strand	2071	518	129	32
		Single strand	1035	259	65	16
Nuclear DNA		Double strand	1919 ± 319	501 ± 77	118 ± 25	35 ± 14

Expectation indicates the calculated number of each G-stretch sequence in the same number of nucleotide sequences of human mitochondria. Nuclear DNA indicates the average number of G-stretch sequences in human *YB-1*, *Sp-1* and *ZNF143* genes by way of example.

It has been demonstrated that the formation of DNA adducts is increased under acidic conditions [48]. In general, perturbation of the intracellular pH of highly proliferative cancer cells is observed [49]. If the pH around mtDNA is higher than that noted around nDNA, this finding may explain the higher initial binding capacity of cisplatin to mtDNA.

6. Cellular Functions of mtTFA

mtTFA is a 25-kDa protein encoded by a nuclear gene and imported to the mitochondria, where it is required for both the transcription and maintenance of mitochondrial DNA. mtTFA preferentially recognizes cisplatin-damaged DNA as well as oxidized DNA. Increased apoptosis is observed in mtTFA knockout animals, suggesting that mtTFA is involved in the process of apoptosis. However, the roles of mtTFA have not been extensively studied in cancer cells [50]. We recently reported the nuclear localization of mtTFA [51]. In addition, the proportion of nuclear-localized mtTFA varies among different cancer cells, and DNA microarray and chromatin immunoprecipitation assays have shown that mtTFA regulates the transcription of nuclear genes. Furthermore, the overexpression of mtTFA enhances the growth of cancer cell lines, whereas the downregulation of mtTFA inhibits the growth of these cells by regulating mtTFA target genes, such as baculoviral IAP repeat-containing 5 (BIRC5; also known as survivin) [52] and BCL2L [53]. Moreover, knockdown of the mtTFA expression induces p21-dependent G1 cell cycle arrest. These results imply that mtTFA functions in both nuclei and mitochondria to promote cell growth.

mtTFA is upregulated by treatment with cisplatin as well as the tumor suppressor p53. p53 is a multifunctional tumor suppressor protein that interacts with a variety of proteins with both positive and negative effects. Furthermore, mutation of the p53 gene is often observed in a wide variety of tumors. Mutated p53 gene products are stabilized and accumulate in the cytoplasm of cancer cells. A fraction of p53 proteins localize in the mitochondria at the onset of p53-dependent apoptosis, although not during p53-independent apoptosis [54]. Using immunochemical coprecipitation, we observed the binding of mtTFA with p53 [55]. The interaction between mtTFA and p53 requires the high mobility group-box B1 or high mobility group-box B2 of mtTFA and amino acids 363–376 of p53. In addition, the binding of mtTFA to cisplatin-modified DNA is significantly enhanced by p53, whereas binding to oxidized DNA is inhibited. Meanwhile, mtTFA preferentially recognizes cisplatin-damaged DNA, as well as oxidized DNA, and the conformational alteration induced by 8-oxo-dG in DNA differs from that induced by cisplatin. Therefore, the different effects of p53 on the binding of mtTFA may depend on structural differences in damaged DNA. Further analyses of the mechanisms by which mtTFA recognizes damaged DNA would be desirable for understanding the differential modulation of mtTFA binding by p53.

The interaction between mtTFA and p53 may be involved in transcriptional regulation, and, similar to transcription factors, both proteins have a DNA-binding domain. Our findings suggest that the interaction of p53 with mtTFA may play an important role in apoptosis. In addition, apoptosis is increased in cells lacking mtTFA [56]. These results suggest a model in which p53 inhibits the ability of mtTFA to execute cell death signaling. Hence, it is possible that manipulating the mtTFA function may be used as a mitochondria-targeted cancer treatment to enhance apoptosis.

It has previously been demonstrated that the expression levels of both mtTFA and mitochondrial antioxidant protein thioredoxin2 (TRX2) are upregulated in cisplatin-resistant cell lines [31]. In addition, TRX2 directly interacts with mtTFA and enhances its damaged DNA binding activity. These results suggest that TRX2 functions as an antioxidant as well as supporting the mtTFA function.

ZNF143 has been shown to be a cisplatin-inducible gene that regulates the mitochondrial ribosomal protein S11 [57]. Interestingly, there is one ZNF143 binding region in the promoter region of the mtTFA gene. The involvement of ZNF143 in cell growth and the protection of cells from oxidative damage, as well as the effects of cisplatin treatment, has recently been reported [35,58,59]. Furthermore, a strong ZNF143 expression was previously found to show a significant correlation with pathologically moderate to poor differentiation and highly invasive characteristics in 183 paraffin-embedded tumor samples from patients with lung adenocarcinoma [60]. Based on these reports, we next discuss the clinical implications of mtTFA. A schematic summary of the mutual relationship of transcription factors, including mtTFA, is shown in Figure 2 and the major factors described in this review are listed with detailed information in Table 2.

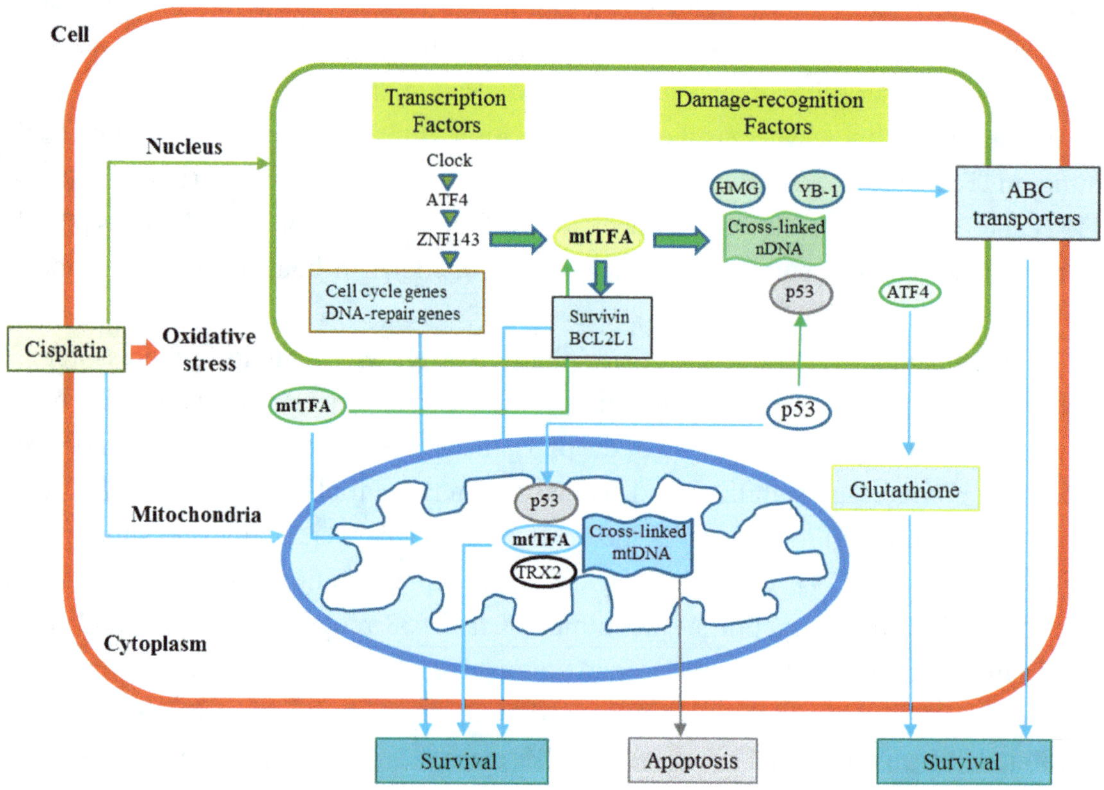

Figure 2. Cisplatin-induced signaling, cisplatin resistance and transcription system in cancer cells discussed in this review. mtTFA functions in both nuclei and mitochondria to not only interact with cisplatin-modified DNA, but also regulate the nuclear and mitochondrial gene expression.

Table 2. Selected transcription factors associated with drug resistance and target genes. "+" means that factors can recognize cisplatin-crosslinks or can render cells resistant.

Factors	Selected Target Genes	Selected Interacting Factors	Damage Recognition	Drug Resistance	Other Functions	References
mtTFA	*Survivin BcLL2*	p53, TRX2	+	+	Cell growth, anti-apoptosis	[54,55]
ZNF143	*mtTFA*	p73	+	+	Cell cycle, DNA repair	[35,57]
ATF4	*ZNF143*			+	Glutathione biosynthesis	[34,36]
Clock	*ATF4, Tip60*			+	Circadian rhythm, DNA repair	[36,38]
YB-1	*MDR1*	p53, PCNA, Topo1	+	+	Endothelial cell growth	[61–63]

7. Clinical Implications of the mtTFA Expression in Tumors

Mitochondrial transcription factor A (mtTFA) is necessary for both the transcription and maintenance of mitochondrial DNA (mtDNA). However, mtTFA is also localized in the nucleus and regulates nuclear genes. To date, the expression of mtTFA has not been thoroughly elucidated in the clinical setting.

Studies of clinical specimens have recently investigated the relationships between clinicopathological factors, the prognosis and the immunohistochemical expression of mtTFA, as shown in Table 3.

Table 3. Associations of the mTFA expression and malignant characteristics in human cancers. B cell lymphoma 2 like 1 (BCL2L1), Folinic acid plus 5-fluorouracil plus oxaliplatin (FOLFOX).

Tumor	No. of Cases	Malignant Characteristics	Reference
Serous ovarian	60	Poor prognosis, up-regulation of BCL2L1 expression	[64]
Pancreatic ductal adenocarcinoma	70	Poor prognosis, up-regulation of survivin expression	[65]
Colorectal	105	Poor prognosis	[66]
Metastatic colorectal	59	Poor clinical outcome with FOLFOX treatment	[67]
Endometrial	245	Invasion and metastasis, p53 mutation, poor prognosis	[68]

An immunohistochemical analysis of the mtTFA expression in 60 tissue samples of serous ovarian cancer showed 56.7% of the serous ovarian cancer patients to be positive for mtTFA, whereas 43.3% were negative [64]. A significant correlation was also reported between the nuclear mtTFA expression and the BCL2L1 expression in seven ovarian cancer cell lines as well as specimens of clinical ovarian cancer. Furthermore, a univariate survival analysis showed that the overall five-year survival rate is significantly worse for patients with mtTFA-positive cancer *versus* mtTFA-negative cancer.

The correlations between the mtTFA expression and the survivin index as well as a poor prognosis were recently assessed using 70 paraffin-embedded tumor samples from patients with surgically-resected pancreatic adenocarcinoma [65]. The results suggested that mtTFA is a prognostic factor for a poor outcome of human cancer and may function as an antiapoptotic factor, regulating target genes, such as BCL2L1 and survivin. In another study, clinical specimens from 105 colorectal patients were immunohistochemically stained using a polyclonal anti-mtTFA antibody [66]. Consequently, a total of 47 (44.8%) of the 105 patients with colorectal cancer were determined to have a positive mtTFA expression, and a positive expression of mtTFA was found to significantly correlate with both lymph node and distant metastasis in addition to an advanced TNM stage. Furthermore, the survival of the patients with a positive mtTFA expression was significantly worse than that of the patients with a negative mtTFA expression. Therefore, a positive mtTFA expression appears to be a useful marker of tumor progression and a poor prognosis in patients with colorectal cancer.

Whether the expression of mtTFA predicts the clinical outcomes of patients with metastatic colorectal cancer treated with modified 5-fluorouracil, leucovorin and oxaliplatin 6 (mFOLFOX6) was recently evaluated [67]. In that study, 59 patients with metastatic lesions of colorectal cancer treated with mFOLFOX6 were analyzed. As a result, a strong expression of mtTFA was detected in eight of 33 cases of a complete response/partial response (24.2%) and 18 of 26 cases of SD/PD (69.2%), indicating that the mtTFA expression significantly correlates with the response to chemotherapy ($p < 0.01$). These results suggest that immunohistochemical studies of mtTFA may be useful for predicting the clinical outcomes of metastatic colorectal cancer patients treated with FOLFOX.

The relationships between the immunohistochemical expression of mtTFA and various clinicopathological variables in 245 cases of endometrioid adenocarcinoma were also recently evaluated [68]. In that report, the mtTFA expression in the endometrioid adenocarcinomas was shown to be significantly associated with the surgical stage, myometrial invasion, lymphovascular space invasion, cervical invasion and lymph node metastasis. In addition, a correlation analysis between the mtTFA and p53 expression levels using the Pearson test showed a significant correlation, and a univariate survival analysis showed that the 10-year overall survival rate of the patients with mtTFA-positive endometrioid adenocarcinoma was significantly worse than that of the patients with mtTFA-negative endometrioid adenocarcinoma. Therefore, a positive mtTFA expression may be a useful marker of tumor progression and a poor prognosis in patients with endometrioid adenocarcinoma.

8. Conclusions and Perspectives

Recent studies have demonstrated the important role of mitochondria in cancer biology. This review focused on the mitochondrial genome and transcription factor A. Since mtTFA proteins have been shown to be highly expressed in cancer and drug-resistant cells compared to normal cells, and the mtTFA expression is upregulated by signals of oxidative and DNA damage stress, this protein may potentially serve as a promising target in cancer chemotherapy. Promising strategies include inhibition of the expression and/or function of mtTFA specifically in cancer cells. Small interference RNA can inhibit specific gene expression levels. mtTFA functions with other transcription factors such as mtTFB1 and B2. Inhibiting the interaction with mtTFBs is expected to be a good strategy using peptides, and both RNA and peptide drugs are good candidates using cancer-specific drug delivery systems [69,70]. Furthermore, the mtTFA expression and mitochondrial genome possess attractive characteristics for platinum-based chemotherapy, and the mtTFA expression reflects a poor prognosis in patients with solid cancers. These findings indicate that further research on mitochondria may provide novel and unique therapeutic interventions for overcoming cancer.

Acknowledgments

This work was partly supported by Grants-in-Aids for Scientific Research from the Ministry of Education, Culture, Sports, Science and Technology of Japan (No. 24501323).

Author Contributions

All authors participated in developing the ideas presented in this manuscript, researching the literature and writing parts of the text. Kimitoshi Kohno and Ke-Yong Wang performed final editing of both text and images.

References

1. Wallace, D.C. Mitochondria and cancer. *Nat. Rev. Cancer* **2012**, *12*, 685–698.
2. Guha, M.; Avadhani, N.G. Mitochondrial retrograde signaling at the crossroads of tumor bioenergetics, genetics and epigenetics. *Mitochondrion* **2013**, *13*, 577–591.

3. Igney, F.H.; Krammer, P.H. Death and anti-death: Tumour resistance to apoptosis. *Nat. Rev. Cancer* **2002**, *2*, 277–288.
4. Holohan, C.; van Schaeybroeck, S.; Longley, D.B.; Johnston, P.G. Cancer drug resistance: An evolving paradigm. *Nat. Rev. Cancer* **2013**, *13*, 714–726.
5. Cohen, S.M.; Lippard, S.J. Cisplatin: From DNA damage to cancer chemotherapy. *Prog. Nucl. Acid Res. Mol. Biol.* **2001**, *67*, 93–130.
6. Wang, D.; Lippard, S.J. Cellular processing of platinum anticancer drugs. *Nat. Rev. Drug Discov.* **2005**, *4*, 307–320.
7. Kelland, L. The resurgence of platinum-based cancer chemotherapy. *Nat. Rev. Cancer* **2007**, *7*, 573–584.
8. Zamble, D.B.; Lippard, S.J. Cisplatin and DNA repair in cancer chemotherapy. *Trends Biochem. Sci.* **1995**, *20*, 435–439.
9. Correia, C.; Lee, S.H.; Meng, X.W.; Vincelette, N.D.; Knorr, K.L.; Ding, H.; Nowakowski, G.S.; Dai, H.; Kaufmann, S.H. Emerging understanding of Bcl-2 biology: Implications for neoplastic progression and treatment. *Biochim. Biophys. Acta* **2015**, *1853*, 1658–1671.
10. Frenzel, A.; Grespi, F.; Chmelewskij, W.; Villunger, A. Bcl2 family proteins in carcinogenesis and the treatment of cancer. *Apoptosis* **2009**, *14*, 584–596.
11. Lopez, J.; Tait, S.W. Mitochondrial apoptosis: Killing cancer using the enemy within. *Br. J. Cancer* **2015**, *112*, 957–962.
12. Orrenius, S.; Gogvadze, V.; Zhivotovsky, B. Calcium and mitochondria in the regulation of cell death. *Biochem. Biophys. Res. Commun.* **2015**, *460*, 72–81.
13. Mandic, A.; Viktorsson, K.; Strandberg, L.; Heiden, T.; Hansson, J.; Linder, S.; Shoshan, M.C. Calpain-mediated Bid cleavage and calpain-independent Bak modulation: Two separate pathways in cisplatin-induced apoptosis. *Mol. Cell. Biol.* **2002**, *22*, 3003–3013.
14. Linder, S.; Shoshan, M.C. Lysosomes and endoplasmic reticulum: Targets for improved, selective anticancer therapy. *Drug Resist. Updat.* **2005**, *8*, 199–204.
15. Xu, Y.; Wang, C.; Li, Z. A new strategy of promoting cisplatin chemotherapeutic efficiency by targeting endoplasmic reticulum stress. *Mol. Clin. Oncol.* **2014**, *2*, 3–7.
16. Simran, S.S.; Paul, T.S. Mitochondrial ROS in cancer: Initiators, amplifiers or an Achilles' heel? *Nat. Rev. Cancer* **2014**, *14*, 709–721.
17. Chirino, Y.I.; Pedraza, C.J. Role of oxidative and nitrosative stress in cisplatin-induced nephrotoxicity. *Exp. Toxicol. Pathol.* **2009**, *61*, 223–242.
18. Rybak, L.P.; Whitworth, C.A.; Mukherjea D.; Ramkumar, V. Mechanisms of cisplatin-induced ototoxicity and prevention. *Hear. Res.* **2007**, *226*, 157–167.
19. Torigoe, T.; Izumi, H.; Ishiguchi, H.; Yoshida, Y.; Tanabe, M.; Yoshida, T.; Igarashi, T.; Niina, I.; Wakasugi, T.; Imaizumi, T.; *et al.* Cisplatin resistance and transcription factors. *Curr. Med. Chem. Anticancer Agents* **2005**, *5*, 15–27.
20. Kohno, K.; Uchiumi, T.; Niina, I.; Wakasugi, T.; Igarashi, T.; Momii, Y.; Yoshida, T.; Matsuo, K.; Miyamoto, N.; Izumi, H. Transcription factors and drug resistance. *Eur. J. Cancer* **2005**, *41*, 2577–2586.

21. Galluzzi, L.; Vitale, I.; Michels, J.; Brenner, C.; Szabadkai, G.; Harel-Bellan, A.; Castedo, M.; Kroemer, G. Systems biology of cisplatin resistance: Past, present and future. *Cell Death Dis.* **2014**, *5*, e1257.
22. Shen, D.W.; Pouliot, L.M.; Hall, M.D.; Gottesman, M.M. Cisplatin resistance: A cellular self-defense mechanism resulting from multiple epigenetic and genetic changes. *Pharmacol. Rev.* **2012**, *64*, 706–721.
23. Galluzzi, L.; Senovilla, L.; Vitale, I.; Michels, J.; Martins, I.; Kepp, O.; Castedo, M.; Kroemer, G. Molecular mechanisms of cisplatin resistance. *Oncogene* **2012**, *31*, 1869–1883.
24. Kong, L.R.; Chua, K.N.; Sim, W.J.; Ng, H.C.; Bi, C.; Ho, J.; Nga M.E.; Pang, Y.H.; Ong, W.R.; Soo, R.A.; *et al.* MEK Inhibition overcomes cisplatin resistance conferred by SOS/MAPK pathway activation in squamous cell carcinoma. *Mol. Cancer Ther.* **2015**, *14*, 1750–1760.
25. Crawford, N.; Chacko, A.D.; Savage, K.I.; McCoy, F.; Redmond, K.; Longley, D.B.; Fennell, D.A. Platinum resistant cancer cells conserve sensitivity to BH3 domains and obatoclax induced mitochondrial apoptosis. *Apotosis* **2011**, *16*, 311–320.
26. Nguyen, L.V.; Vanner, R.; Dirks, P.; Eaves, C.J. Cancer stem cells: An evolving concept. *Nat. Rev. Cancer* **2012**, *12*, 133–143.
27. Visvader, J.E.; Lindeman, G.J. Cancer stem cells in solid tumours: Accumulating evidence and unresolved questions. *Nat. Rev. Cancer* **2008**, *8*, 755–768.
28. Dean, M.; Fojo, T.; Bates, S. Tumour stem cells and drug resistance. *Nat. Rev. Cancer* **2005**, *5*, 275–284.
29. Nagano, O.; Okazaki, S.; Saya, H. Redox regulation in stem-like cancer cells by CD44 variant isoforms. *Oncogene* **2013**, *32*, 5191–5198.
30. Song, I.S.; Jeong, J.Y.; Jeong, S.H.; Kim, H.K.; Ko, K.S.; Rhee, B.D.; Kim, N.; Han, J. Mitochondria as therapeutic targets for cancer stem cells. *World J. Stem Cells* **2015**, *7*, 418–427.
31. Kidani, A.; Izumi, H.; Yoshida, Y.; Kashiwagi, E.; Ohmori, H.; Tanaka, T.; Kuwano, M.; Kohno, K. Thioredoxin2 enhances the damaged DNA binding activity of mtTFA through direct interaction. *Int. J. Oncol.* **2009**, *35*, 1435–1440.
32. Kohno, K.; Izumi, H.; Uchiumi, T.; Ashizuka, M.; Kuwano, M. The pleiotropic functions of the Y-box-binding protein, YB-1. *Bioessays* **2003**, *25*, 691–698.
33. Kashiwagi, E.; Izumi, H.; Yasuniwa, Y.; Baba, R.; Doi, Y.; Kidani, A.; Arao, T.; Nishio, K.; Naito, S.; Kohno, K. Enhanced expression of nuclear factor I/B in oxaliplatin-resistant human cancer cell lines. *Cancer Sci.* **2011**, *102*, 382–386.
34. Tanabe, M.; Izumi, H.; Ise, T.; Higuchi, S.; Yamori, T.; Yasumoto, K.; Kohno, K. Activating transcription factor 4 increases the cisplatin resistance of human cancer cell lines. *Cancer Res.* **2003**, *63*, 8592–8595.
35. Wakasugi, T.; Izumi, H.; Uchiumi, T.; Suzuki, H.; Arao, T.; Nishio, K.; Kohno, K. ZNF143 interacts with p73 and is involved in cisplatin resistance through the transcriptional regulation of DNA repair genes. *Oncogene* **2007**, *26*, 5194–5203.
36. Igarashi, T.; Izumi, H.; Uchiumi, T.; Nishio, K.; Arao, T.; Tanabe, M.; Uramoto, H.; Sugio, K.; Yasumoto, K.; Sasaguri, Y.; *et al.* Clock and ATF4 transcription system regulates drug resistance in human cancer cell lines. *Oncogene* **2007**, *26*, 4749–4760.

37. Lamouille, S.; Xu, J.; Derynck, R. Molecular mechanisms of epithelial-mesenchymal transition. *Nat. Rev. Mol. Cell Biol.* **2014**, *15*, 178–196.
38. Bailey, S.M.; Udoh, U.S.; Young, M.E. Circadian regulation of metabolism. *J. Endocrinol.* **2014**, *222*, R75–R96.
39. Yasuniwa,Y.; Izumi, H.; Wang, K.Y.; Shimajiri, S.; Sasaguri, Y.; Kawai, K.; Kasai, H.; Shimada, T.; Miyake, K.; Kashiwagi, E.; *et al.* Circadian disruption accelerates tumor growth and angio/stromagenesis through a Wnt signaling pathway. *PLoS ONE* **2010**, *5*, e15330.
40. Izumi, H.; Wang, K.Y.; Morimoto, Y.; Sasaguri, Y.; Kohno, K. Circadian disruption and cancer risk: A new concept of stromal niche (review). *Int. J. Oncol.* **2014**, *44*, 364–370.
41. Guaragnella, N.; Giannattasio, S.; Moro, L. Mitochondrial dysfunction in cancer chemoresistance. *Biochem. Pharmacol.* **2014**, *92*, 62–72.
42. Mizutani, S.; Miyato, Y.; Shidara, Y.; Asoh, S.; Tokunaga, A.; Tajiri, T.; Ohta, S. Mutations in the mitochondrial genome confer resistance of cancer cells to anticancer drugs. *Cancer Sci*. **2009**, *100*, 1680–1687.
43. Yao, Z.; Jones, A.W.; Fassone, E.; Sweeney, M.G.; Lebiedzinska, M.; Suski, J.M.; Wieckowski, M.R.; Tajeddine, N.; Hargreaves, I.P.; Yasukawa, T.; *et al.* PGC-1β mediates adaptive chemoresistance associated with mitochondrial DNA mutations. *Oncogene* **2013**, *32*, 2592–2600.
44. Guo, J.; Zheng, L.; Liu, W.; Wang, X.; Wang, Z.; Wang, Z.; French, A.J.; Kang, D.; Chen, L.; Thibodeau, S.N.; *et al.* Frequent truncating mutation of TFAM induces mitochondrial DNA depletion and apoptotic resistance in microsatellite-unstable colorectal cancer. *Cancer Res*. **2011**, *71*, 2978–2987.
45. Campbell, C.T.; Kolesar, J.E.; Kaufman, B.A. Mitochondrial transcription factor A regulates mitochondrial transcription initiation, DNA packaging, and genome copy number. *Biochim. Biophys. Acta* **2012**, *1819*, 921–929.
46. Olivero, O.A.; Semino, C.; Kassim, A.; Lopez-Larraza, D.M.; Poirier, M.C. Preferential binding of cisplatin to mitochondrial DNA of Chinese hamster ovary cells. *Mutat. Res.* **1995**, *346*, 221–230.
47. See comment in PubMed Commons below Olivero, O.A.; Chang, P.K.; Lopez-Larraza, D.M.; Semino-Mora, M.C.; Poirier, M.C. Preferential formation and decreased removal of cisplatin-DNA adducts in Chinese hamster ovary cell mitochondrial DNA as compared to nuclear DNA. *Mutat. Res.* **1997**, *391*, 79–86.
48. Murakami, T.; Shibuya, I.; Ise, T.; Chen, Z.S.; Akiyama, S.; Nakagawa, M.; Izumi, H.; Nakamura, T.; Matsuo, K.; Yamada, Y.; *et al.* Elevated expression of vacuolar proton pump genes and cellular pH in cisplatin resistance. *Int. J. Cancer* **2001**, *93*, 869–874.
49. Reshkin, S.J.; Greco, M.R.; Cardone, R.A. Role of pHi, and proton transporters in oncogene-driven neoplastic transformation. *Philos. Trans. R. Soc. Lond. B* **2014**. *369*, 20130100.
50. Yoshida, Y.; Hoshino, S.; Izumi, H.; Kohno, K.; Yamashita, Y. New roles of mitochondrial transcription factor A in cancer. *J. Phys. Chem. Biophys.* **2011**, *1*, 101.
51. Han, B.; Izumi, H.; Yasuniwa, Y.; Akiyama, M.; Yamaguchi, T.; Fujimoto, N.; Matsumoto, T.; Wu, B.; Tanimoto, A.; Sasaguri, Y.; *et al.* Human mitochondrial transcription factor A functions in both nuclei and mitochondria and regulates cancer cell growth. *Biochem. Biophys. Res. Commun.* **2011**, *408*, 45–51.

52. Altieri, D.C. Survivin, cancer networks and pathway-directed drug discovery. *Nat. Rev. Cancer* **2008**, *8*, 61–70.
53. Cory, S.; Adams, J.M. The Bcl2 family: Regulators of the cellular life-or-death switch. *Nat. Rev. Cancer* **2002**, *2*, 647–656.
54. Vaseva, A.V.; Moll, U.M. The mitochondrial p53 pathway. *Biochim. Biophys. Acta* **2009**, *1787*, 414–420.
55. Yoshida, Y.; Izumi, H.; Torigoe, T.; Ishiguchi, H.; Itoh, H.; Kang, D.; Kohno, K. P53 physically interacts with mitochondrial transcription factor A and differentially regulates binding to damaged DNA. *Cancer Res.* **2003**, *63*, 3729–3734.
56. Wang, J.; Silva, J.P.; Gustafsson, C.M.; Rustin, P.; Larsson, N.G. Increased *in vivo* apoptosis in cells lacking mitochondrial DNA gene expression. *Proc. Natl Acad. Sci. USA* **2001**, *98*, 4038–4043.
57. Ishiguchi, H.; Izumi, H.; Torigoe, T.; Yoshida, Y.; Kubota, H.; Tsuji, S.; Kohno, K. ZNF143 activates gene expression in response to DNA damage and binds to cisplatin-modified DNA. *Int. J. Cancer* **2004**, *111*, 900–909.
58. Izumi, H.; Yasuniwa, Y.; Akiyama, M.; Yamaguchi, T.; Kuma, A.; Kitamura, N.; Kohno, K. Forced expression of ZNF143 restrains cancer cell growth. *Cancers* **2011**, *3*, 3909–3920.
59. Izumi, H.; Wakasugi, T.; Shimajiri, S.; Tanimoto, A.; Sasaguri, Y.; Kashiwagi, E.; Yasuniwa, Y.; Akiyama, M.; Han, B.; Wu, Y.; *et al.* Role of ZNF143 in tumor growth through transcriptional regulation of DNA replication and cell-cycle-associated genes. *Cancer Sci.* **2010**, *101*, 2538–2545.
60. Kawatsu, Y.; Kitada, S.; Uramoto, H.; Li, Z.; Takeda, T.; Kimura, T.; Horie, S.; Tanaka, F.; Sasaguri, Y.; Izumi, H.; *et al.* The combination of strong expression of ZNF143 and high MIB-1 labelling index independently predicts shorter disease-specific survival in lung adenocarcinoma. *Br. J. Cancer.* **2014**, *110*, 2583–2592.
61. Takahashi, M.; Shimajiri, S.; Izumi, H.; Hirano, G.; Kashiwagi, E.; Yasuniwa, Y.; Wu, Y.; Han, B.; Akiyama, M.; Nishizawa, S.; *et al.* Y-box binding protein-1 is a novel molecular target for tumor vessels. *Cancer Sci.* **2010**, *101*, 1367–1373.
62. Wu, Y.; Wang, K.Y.; Li, Z.; Liu, Y.P.; Izumi, H.; Uramoto, H.; Nakayama, Y.; Ito, K.I.; Kohno, K. Y-box binding protein 1 enhances DNA topoisomerase 1 activity and sensitivity to camptothecin via direct interaction. *J. Exp. Clin. Cancer Res.* **2014**, *33*, 112–117.
63. Wu, Y.; Wang, K.Y.; Li, Z.; Liu, Y.P.; Izumi, H.; Yamada, S.; Uramoto, H.; Nakayama, Y.; Ito, K.I.; Kohno, K. Y-box binding protein 1 expression in gastric cancer subtypes and association with cancer neovasculature. *Clin. Transl. Oncol.* **2015**, *17*, 152–159.
64. Kurita, T.; Izumi, H.; Kagami, S.; Kawagoe, T.; Toki, N.; Matsuura, Y.; Hachisuga, T.; Kohno, K. Mitochondrial transcription factor A regulates *BCL2L1* gene expression and is a prognostic factor in serous ovarian cancer. *Cancer Sci.* **2012**, *103*, 239–444.
65. Kimura, T.; Kitada, S.; Uramoto, H.; Li, Z.; Kawatsu, Y.; Takeda, T.; Horie, S.; Nabeshima, A.; Noguchi, H.; Sasaguri, Y.; *et al.* The combination of strong immunohistochemical mtTFA expression and a high survivin index predicts a shorter disease-specific survival in pancreatic ductal adenocarcinoma. *Histol. Histopathol.* **2015**, *30*, 193–204.
66. Nakayama, Y.; Yamauchi, M.; Minagawa, N.; Torigoe, T.; Izumi, H.; Kohno, K.; Yamaguchi, K. Clinical significance of mitochondrial transcription factor A expression in patients with colorectal cancer. *Oncol. Rep.* **2012**, *27*, 1325–1330.

67. Yoshida, Y.; Hasegawa, J.; Nezu, R.; Kim, Y.K.; Hirota, M.; Kawano, K.; Izumi, H.; Kohno, K. Clinical usefulness of mitochondrial transcription factor A expression as a predictive marker in colorectal cancer patients treated with FOLFOX. *Cancer Sci.* **2011**, *102*, 578–582.
68. Toki, N.; Kagami, S.; Kurita, T.; Kawagoe, T.; Matsuura, Y.; Hachisuga, T.; Matsuyama, A.; Hashimoto, H.; Izumi, H.; Kohno, K. Expression of mitochondrial transcription factor A in endometrial carcinomas: Clinicopathologic correlations and prognostic significance. *Virchows Arch.* **2010**, *456*, 387–393.
69. Burnett, J.C.; Rossi, J.J. RNA-based therapeutics: Current progress and future prospects. *Chem. Biol.* **2012**, *19*, 60–71.
70. Pirogova, E.; Istivan, T.; Gan, E.; Cosic, I. Advances in methods for therapeutic peptide discovery, design and development. *Curr. Pharm. Biotechnol.* **2011**, *12*, 1117–1127.

Permissions

All chapters in this book were first published by MDPI; hereby published with permission under the Creative Commons Attribution License or equivalent. Every chapter published in this book has been scrutinized by our experts. Their significance has been extensively debated. The topics covered herein carry significant findings which will fuel the growth of the discipline. They may even be implemented as practical applications or may be referred to as a beginning point for another development.

The contributors of this book come from diverse backgrounds, making this book a truly international effort. This book will bring forth new frontiers with its revolutionizing research information and detailed analysis of the nascent developments around the world.

We would like to thank all the contributing authors for lending their expertise to make the book truly unique. They have played a crucial role in the development of this book. Without their invaluable contributions this book wouldn't have been possible. They have made vital efforts to compile up to date information on the varied aspects of this subject to make this book a valuable addition to the collection of many professionals and students.

This book was conceptualized with the vision of imparting up-to-date information and advanced data in this field. To ensure the same, a matchless editorial board was set up. Every individual on the board went through rigorous rounds of assessment to prove their worth. After which they invested a large part of their time researching and compiling the most relevant data for our readers.

The editorial board has been involved in producing this book since its inception. They have spent rigorous hours researching and exploring the diverse topics which have resulted in the successful publishing of this book. They have passed on their knowledge of decades through this book. To expedite this challenging task, the publisher supported the team at every step. A small team of assistant editors was also appointed to further simplify the editing procedure and attain best results for the readers.

Apart from the editorial board, the designing team has also invested a significant amount of their time in understanding the subject and creating the most relevant covers. They scrutinized every image to scout for the most suitable representation of the subject and create an appropriate cover for the book.

The publishing team has been an ardent support to the editorial, designing and production team. Their endless efforts to recruit the best for this project, has resulted in the accomplishment of this book. They are a veteran in the field of academics and their pool of knowledge is as vast as their experience in printing. Their expertise and guidance has proved useful at every step. Their uncompromising quality standards have made this book an exceptional effort. Their encouragement from time to time has been an inspiration for everyone.

The publisher and the editorial board hope that this book will prove to be a valuable piece of knowledge for researchers, students, practitioners and scholars across the globe.

List of Contributors

Gjumrakch Aliev
School of Health Science and Healthcare Administration, University of Atlanta, 6685 Peachtree Industrial Blvd., Atlanta, Georgia, 30360, USA
Department of Nutrition and Biochemistry, Faculty of Sciences, Javeriana University, Bogotà D.C., Colombia
Stress Relief and Memory Training Center, Brooklyn, New York, NY 11235, USA

Ludis Morales
Department of Nutrition and Biochemistry, Faculty of Sciences, Javeriana University, Bogotà D.C., Colombia

Valentin Bragin
Stress Relief and Memory Training Center, Brooklyn, New York, NY 11235, USA

Hector H. Palacios
Department of Biology, College of Sciences, University of Texas at San Antonio, One UTSA Circle, San Antonio, TX 78249-1664, USA

Eldar Gasimov
Department of Cytology, Histology and Embryology, Azerbaijan Medical University, 25 Street Bakhikhanov, Baku AZ10 25, Azerbaijan

Mark E. Obrenovich
Department of Pathology, School of Medicine, Case Western Reserve University, WRB 5301, Cleveland, Ohio, 44106, USA

Jerzy Leszek
Department of Psychiatry, Wroclaw Medical University, 25 St. Kraszewskiego, Wroclaw, 50-229, Poland

Arturo Solís Herrera
Dirección de Investigación y desarrollo, Centro de Estudios de la Fotosíntesis Humana, S.C. López Velarde 108 y 109, Centro, Aguascalientes, Aguascalientes, 20000, México

Dmitry Gokhman
Department of Mathematics, College of Sciences, University of Texas at San Antonio, One UTSA Circle, San Antonio, TX 78249, USA

Giovanni Pagano, Annarita Aiello Talamanca and Giuseppe Castello
Istituto Nazionale Tumori Fondazione G. Pascale—Cancer Research Center at Mercogliano (CROM)—IRCCS, Naples I-80131, Italy

Mario D. Cordero
Research Laboratory, Dental School, Universidad de Sevilla, Sevilla 41009, Spain

Marco d'Ischia
Department of Chemical Sciences, University of Naples "Federico II", Naples I-80126, Italy

Maria Nicola Gadaleta
National Research Council, Institute of Biomembranes and Bioenergetics, Bari I-70126, Italy

Federico V. Pallardó
CIBERER (Centro de Investigación Biomédica en Red de Enfermedades Raras), University of Valencia—INCLIVA, Valencia 46010, Spain

Sandra Petrović
"Vinca" Institute of Nuclear Sciences, University of Belgrade, Belgrade 11001, Serbia

Luca Tiano
Biochemistry Unit, Department of Clinical and Dental Sciences, Polytechnical University of Marche, Ancona I-60131, Italy

Adriana Zatterale
Genetics Unit, Azienda Sanitaria Locale (ASL) Napoli 1 Centro, Naples I-80136, Italy

Raffaella Crescenzo, Francesca Bianco, Arianna Mazzoli, Antonia Giacco, Giovanna Liverini and Susanna Iossa
Department of Biology, University of Naples "Federico II", Napoli I-80126, Italy

Benjamin Baur, Christoph Solbach, Gordon Winter, Hans-Jürgen Machulla and Sven N. Reske
Clinic for Nuclear Medicine, University of Ulm, Albert-Einstein-Allee, 89081 Ulm, Germany

Elena Andreolli
Clinic for Nuclear Medicine, University of Ulm, Albert-Einstein-Allee, 89081 Ulm, Germany
Department of Health Sciences, University of Milano-Bicocca, Milan 20900, Italy

Naghia Ahmed, Dario Ronchi and Giacomo Pietro Comi
Neurology Unit, IRCCS Foundation Ca' Granda Ospedale Maggiore Policlinico, Dino Ferrari Centre, Department of Pathophysiology and Transplantation, Università degli Studi di Milano, via Francesco Sforza 35, Milan 20122, Italy

Ruey-Sheng Wang, Yi-Chen Wu and Chii-Reuy Tzeng
Graduate Institute of Clinical Medicine & Department of Obstetrics and Gynecology, College of Medicine, Taipei Medical University, Taipei 110, Taiwan

Heng-Yu Chang
Department of Biochemistry and Molecular Cell Biology, College of Medicine, Taipei Medical University, Taipei 110, Taiwan

Shu-Huei Kao
School of Medical Laboratory Science and Biotechnology, College of Medical Science and Technology, Taipei Medical University, Taipei 110, Taiwan

Cheng-Heng Kao
Center of General Education, Chang Gung University, Taoyuan 333, Taiwan

Shuyuan Yeh and Chawnshang Chang
George H. Whipple Lab for Cancer Research, Departments of Pathology, Urology and Radiation Oncology, University of Rochester Medical Center, Rochester, NY 14642, USA

Yu Luo, Alan Hoffer and Barry Hoffer
Department of Neurological Surgery, Case Western Reserve University, Cleveland, OH 44106, USA

Xin Qi
Department of Physiology, Case Western Reserve University, Cleveland, OH 44106, USA

Thierry Arnould and Patricia Renard
Laboratory of Biochemistry and Cell Biology (URBC), Namur Research Institute for Life Sciences (NARILIS), University of Namur, Rue de Bruxelles 61, 5000 Namur, Belgium

Sébastien Michel
Laboratory of Biochemistry and Cell Biology (URBC), Namur Research Institute for Life Sciences (NARILIS), University of Namur, Rue de Bruxelles 61, 5000 Namur, Belgium
Department of Physiology, University of Lausanne, Rue du Bugnon 7, CH-1005 Lausanne, Switzerland

Christoph Bode and Heiko Bugger
Department of Cardiology and Angiology, Heart Center Freiburg University, Hugstetter Str. 55, 79106 Freiburg, Germany

Silvia Alvarez
Institute of Biochemistry and Molecular Medicine, School of Pharmacy and Biochemistry, University of Buenos Aires-National Scientific and Technical Research Council (UBA-CONICET), Junín 956, C1113AAD Buenos Aires, Argentina

María Cecilia Cimolai
Department of Cardiology and Angiology, Heart Center Freiburg University, Hugstetter Str. 55, 79106 Freiburg, Germany
Institute of Biochemistry and Molecular Medicine, School of Pharmacy and Biochemistry, University of Buenos Aires-National Scientific and Technical Research Council (UBA-CONICET), Junín 956, C1113AAD Buenos Aires, Argentina

Kunlu Wu, Xiaojuan Xiao, Jiling Liao, Qikang Hu, Huiyong Chen and Jing Liu
State Key Laboratory of Medical Genetics & School of Life Sciences, Central South University, Changsha 410078, China

Jieying Zhang
State Key Laboratory of Medical Genetics & School of Life Sciences, Central South University, Changsha 410078, China
Red Cell Physiology Laboratory, New York Blood Center, New York, NY 10065, USA

Xiuli An
Laboratory of Membrane Biology, New York Blood Center, New York, NY 10065, USA
College of Life Science, Zhengzhou University, Zhengzhou 450001, China

Kimitoshi Kohno and Shinji Matsumoto
Asahi Matsumoto Hospital, Kokuramimami-ku Tsuda, Kitakyushu-shi 800-0242, Japan

Ke-Yong Wang
Shared-Use Research Center, University of Occupational and Environmental Health School of Medicine, Yahatanishi-ku, Kitakyushu-shi 807-8555, Japan

Mayu Takahashi
Department of Neurosurgery, University of Occupational and Environmental Health School of Medicine, Yahatanishi-ku, Kitakyushu-shi 807-8555, Japan

Tomoko Kurita
Department of Gynecology, University of Occupational and Environmental Health School of Medicine, Yahatanishi-ku, Kitakyushu-shi 807-8555, Japan

Yoichiro Yoshida
Department of Gastroenterological Surgery, School of Medicine, Fukuoka University, Fukuoka 814-0180, Japan

Masakazu Hirakawa
Department of Radiology, Beppu Hospital, Kyushu University, Beppu 874-0838, Japan

Yoshikazu Harada
Department of Pathology and Cell Biology, University of Occupational and Environmental Health School of Medicine, Yahatanishi-ku, Kitakyushu-shi 807-8555, Japan

Akihiro Kuma
Second Department of Internal Medicine, University of Occupational and Environmental Health School of Medicine, Yahatanishi-ku, Kitakyushu-shi 807-8555, Japan

Hiroto Izumi
Department of Occupational Pneumology, Institute of Industrial Ecological Science, University of Occupational and Environmental Health, Yahatanishi-ku, Kitakyushu-shi 807-8555, Japan

Index

P

R

S

T

U

www.ingramcontent.com/pod-product-compliance
Lightning Source LLC
LaVergne TN
LVHW081031210926
848485LV00004B/997